Emerging Technologies
for Food Quality
and Food Safety Evaluation

Contemporary Food Engineering

Series Editor

Professor Da-Wen Sun, Director

Food Refrigeration & Computerized Food Technology
National University of Ireland, Dublin
(University College Dublin)
Dublin, Ireland
http://www.ucd.ie/sun/

Emerging Technologies for Food Quality and Food Safety Evaluation, *edited by Yong-Jin Cho and Sukwon Kang* (2011)

Food Process Engineering Operations, *edited by George D. Saravacos and Zacharias B. Maroulis* (2011)

Biosensors in Food Processing, Safety, and Quality Control, *edited by Mehmet Mutlu* (2011)

Physicochemical Aspects of Food Engineering and Processing, *edited by Sakamon Devahastin* (2010)

Infrared Heating for Food and Agricultural Processing, *edited by Zhongli Pan and Griffiths Gregory Atungulu* (2010)

Mathematical Modeling of Food Processing, *edited by Mohammed M. Farid* (2009)

Engineering Aspects of Milk and Dairy Products, *edited by Jane Sélia dos Reis Coimbra and José A. Teixeira* (2009)

Innovation in Food Engineering: New Techniques and Products, *edited by Maria Laura Passos and Claudio P. Ribeiro* (2009)

Processing Effects on Safety and Quality of Foods, *edited by Enrique Ortega-Rivas* (2009)

Engineering Aspects of Thermal Food Processing, *edited by Ricardo Simpson* (2009)

Ultraviolet Light in Food Technology: Principles and Applications, *Tatiana N. Koutchma, Larry J. Forney, and Carmen I. Moraru* (2009)

Advances in Deep-Fat Frying of Foods, *edited by Serpil Sahin and Servet Gülüm Sumnu* (2009)

Extracting Bioactive Compounds for Food Products: Theory and Applications, *edited by M. Angela A. Meireles* (2009)

Advances in Food Dehydration, *edited by Cristina Ratti* (2009)

Optimization in Food Engineering, *edited by Ferruh Erdoğdu* (2009)

Optical Monitoring of Fresh and Processed Agricultural Crops, *edited by Manuela Zude* (2009)

Food Engineering Aspects of Baking Sweet Goods, *edited by Servet Gülüm Sumnu and Serpil Sahin* (2008)

Computational Fluid Dynamics in Food Processing, *edited by Da-Wen Sun* (2007)

Contemporary Food
Engineering Series
Da-Wen Sun, Series Editor

Emerging Technologies for Food Quality and Food Safety Evaluation

Edited by
Yong-Jin Cho

CRC Press
Taylor & Francis Group
Boca Raton   London   New York

CRC Press is an imprint of the
Taylor & Francis Group, an **informa** business

CRC Press
Taylor & Francis Group
6000 Broken Sound Parkway NW, Suite 300
Boca Raton, FL 33487-2742

First issued in paperback 2016

© 2011 by Taylor and Francis Group, LLC
CRC Press is an imprint of Taylor & Francis Group, an Informa business

No claim to original U.S. Government works

ISBN 13: 978-1-138-19913-2 (pbk)
ISBN 13: 978-1-4398-1524-3 (hbk)

Visit the Taylor & Francis Web site at
http://www.taylorandfrancis.com

and the CRC Press Web site at
http://www.crcpress.com

Contents

List of Figures

List of Tables

Series Preface

Food engineering is the multidisciplinary field of applied physical sciences combined with the knowledge of product properties. Food engineers provide the technological knowledge transfer essential to the cost-effective production and commercialization of food products and services. In particular, food engineers develop and design processes and equipment in order to convert raw agricultural materials and ingredients into safe, convenient, and nutritious consumer food products. However, food engineering topics are continuously undergoing changes to meet diverse consumer demands, and the subject is being rapidly developed to reflect market needs.

In the development of food engineering, one of the many challenges is to employ modern tools and knowledge, such as computational materials science and nanotechnology, to develop new products and processes. Simultaneously, improving quality, safety, and security remains a critical issue in the study of food engineering. New packaging materials and techniques are being developed to provide more protection to foods, and novel preservation technologies are emerging to enhance food security and defense. Additionally, process control and automation regularly appear among the top priorities identified in food engineering. Advanced monitoring and control systems are developed to facilitate automation and flexible food manufacturing. Furthermore, energy savings and minimization of environmental problems continue to be important issues in food engineering, and significant progress is being made in waste management, efficient utilization of energy, and reduction of effluents and emissions in food production.

The Contemporary Food Engineering book series, which consists of edited books, attempts to address some of the recent developments in food engineering. Advances in classical unit operations in engineering related to food manufacturing are covered as well as such topics as progress in the transport and storage of liquid and solid foods; heating, chilling, and freezing of foods; mass transfer in foods; chemical and biochemical aspects of food engineering and the use of kinetic analysis; dehydration, thermal processing, nonthermal processing, extrusion, liquid food concentration, membrane processes, and applications of membranes in food processing; shelf-life, electronic indicators in inventory management, and sustainable technologies in food processing; and packaging, cleaning, and sanitation. These books are aimed at professional food scientists, academics researching food engineering problems, and graduate-level students.

The editors of these books are leading engineers and scientists from all parts of the world. All of them were asked to present their books in such a manner as to address the market needs and pinpoint the cutting-edge technologies in food engineering. Furthermore, all contributions are written by internationally renowned experts who have both academic and professional credentials. All authors have attempted to

xxiii

provide critical, comprehensive, and readily accessible information on the art and science of a relevant topic in each chapter, with reference lists for further information. Therefore, each book can serve as an essential reference source to students and researchers in universities and research institutions.

Da-Wen Sun
Series Editor

Series Editor

Born in southern China, Professor Da-Wen Sun is a world authority in food engineering research and education; he is a member of the Royal Irish Academy, which is the highest academic honor that can be attained by scholars and scientists working in Ireland. His main research activities include cooling, drying, and refrigeration processes and systems; quality and safety of food products; bioprocess simulation and optimization; and computer vision technology. Especially, his innovative studies on vacuum cooling of cooked meats, pizza quality inspection by computer vision, and edible films for shelf-life extension of fruits and vegetables have been widely reported in national and international media. The results of his work have been published in over 500 papers including 200 peer reviewed journal papers. According to Thomson Scientific's *Essential Science Indicator*[SM] updated as of July 1, 2010 based on data derived over a period of 10 years and 4 months (January 1, 2000 to April 30, 2010) from the ISI Web of Science, a total of 2,554 scientists are among the top 1% of the most frequently cited scientists in the category of Agricultural Sciences, and professor Sun tops the list with his ranking of 31.

Sun received his B.Sc. honors (first class), his M.Sc. in mechanical engineering, and his Ph.D. in chemical engineering in China before working in various universities in Europe. He became the first Chinese national to be permanently employed in an Irish university when he was appointed as college lecturer at the National University of Ireland, Dublin (University College Dublin) in 1995, and was then continuously promoted in the shortest possible time to senior lecturer, associate professor, and full professor. He is currently the professor of food and biosystems engineering and the director of the Food Refrigeration and Computerized Food Technology Research Group at University College Dublin.

Sun has contributed significantly to the field of food engineering as a leading educator in this field. He has trained many Ph.D. students who have made their own contributions to the industry and academia. He has also regularly given lectures on advances in food engineering in international academic institutions and delivered keynote speeches at international conferences. As a recognized authority in food engineering, he has been conferred adjunct/visiting/consulting professorships from 10 top universities in China including Zhejiang University, Shanghai Jiaotong University, Harbin Institute of Technology, China Agricultural University, South China University of Technology, and Jiangnan University. In recognition of his significant contribution to food engineering worldwide and for his outstanding leadership in this field, the International Commission of Agricultural and Biosystems Engineering (CIGR) awarded him the CIGR Merit Award in 2000 and again in 2006. The Institution of Mechanical Engineers (IMechE) based in the United Kingdom named him Food Engineer of the Year 2004. In 2008, he was awarded the CIGR Recognition Award in honor of his distinguished achievements in the top 1% of agricultural engineering scientists in the world. In 2007, he was presented with the AFST (I) Fellow Award by the Association of Food Scientists and Technologists (India), and in 2010 he was presented with the CIGR Fellow Award. The title of Fellow is the highest honor in CIGR, and is conferred upon individuals who have made sustained, outstanding contributions worldwide.

Sun is a Fellow of the Institution of Agricultural Engineers and a Fellow of Engineers Ireland (the Institution of Engineers of Ireland). He has received numerous awards for teaching and research excellence, including the President's Research Fellowship and the President's Research Award of University College Dublin on two occasions. He is the CIGR Incoming President in 2011–2012, President in 2013–2014, and Past President in 2015–2016, and a member of the CIGR Presidium and CIGR Executive Board. He is also the editor-in-chief of *Food and Bioprocess Technology—An International Journal* (Springer); the former editor of *Journal of Food Engineering* (Elsevier); and an editorial board member for *Journal of Food Engineering* (Elsevier), *Journal of Food Process Engineering* (Blackwell), *Sensing and Instrumentation for Food Quality and Safety* (Springer), and *Czech Journal of Food Sciences*. He is a chartered engineer.

Preface

The evaluation of food quality and safety is one of the key tasks in the manufacturing process of foods and beverages as well as in their R&D. Even though food quality may depend on its hedonic aspect, its quantitative evaluation becomes more important in the food industry and academia. Fortunately, it is possible to definitely evaluate food quality, owing to the advent of some sophisticated systems, including nondestructive testing techniques and emerging technology. The aim of this book is to review and introduce the state-of-the-art technology and systems for quantitative evaluation of food quality, particularly those suitable for definite measurement of it. Also, this book will cover detecting systems for food safety because good foods relate directly to safe foods.

This book has eleven chapters, including an introductory chapter to describe concept and scope of food quality parameters such as color, texture, shape and size, chemical compositions, flavor, safety, and so on. Chapters 2 to 11 deal with individual systems for quality parameters, i.e., instruments for textural and rheological properties, artificial intelligent systems for sensory evaluation, computer vision, near infrared (NIR), nuclear magnetic resonance (NMR) and magnetic resonance imaging (MRI), sonic and ultrasonic systems, multispectral and hyperspectral imaging techniques, electronic noses, biosensors, and nanotechnology in food quality and safety evaluation systems. Each chapter describes the principles related to each system and applications to foods.

Especially, biosensors in Chapter 10 and nanotechnology in Chapter 11 are recognized as new tools for laboratory analysis and *on-site* industrial use. Conventionally speaking, food biosensors are typically used for the biochemical evaluation of food, such as quality analysis and safety monitoring, rather than for the determination of physical properties. Recently, nanotechnology is being applied in enhancement of sensitivity and detection limit in measurement systems for food quality and safety.

It is hoped that this book will encourage those already working in the food industry and scientists in universities and research institutes, and help clarity the needs of the food industry and stimulate new ideas, either for importing technologies from other disciplines or for the development of new approaches to food quality evaluation.

Yong-Jin Cho

The Editors

Yong-Jin Cho, born in Korea, received his B.S., M.S., and Ph.D. degrees from Seoul National University, Korea. During his graduate courses in Seoul National University, he studied diffusion in biomaterials and bioinstrumentation. Also, he researched computer vision systems and the physical-physiological properties of biomaterials as a visiting scholar at Clemson University, USA.

Since joining the Korea Food Research Institute in 1990, he has studied food systems, process engineering, and bioinstrumentation including biosensor and food nanotechnologies. Currently he is serving as the head of the Bio-Nanotechnology Research Center, Korea Food Research Institute, and is an adjunct professor of the University of Science and Technology in Korea.

Dr. Cho has held several positions including the Chair of Division of Bioprocess Engineering as Korean Society for Agricultural Machinery, the Chair of the Research Group for Bio-molecules Measurement, and the Secretary General of the Federation of Korea Food Related Societies. Presently, he is the Secretary General of the Korea Society for Food Engineering, and the trustee of the Korea Nanotechnology Research Society.

He is the author of 111 peer-reviewed papers published in international and domestic journals and holds 34 patents. Dr. Cho has presented at numerous invited-lectures and he is the coauthor of three books on instrumentation and physical properties.

Sukwon Kang received his B.A. and M.S. degrees in agricultural engineering from the Seoul National University and his Ph.D. degree in biological and environmental engineering from Cornell University.

Dr. Kang was a research associate of the Agricultural Research Service at the U.S. Department of Agriculture and a researcher at Rutgers, the State University of New Jersey. He is currently a researcher at the National Academy of Agricultural Science at the Rural Development Administration since 2002.

He has conducted research and given presentations on advanced in food engineering. His current areas of research include optical engineering, such as near-infrared spectroscopy, hyperspectral and multispectral imaging techniques and computer vision, all with an emphasis on solving practical problems in food engineering.

The Contributors

Osvaldo H. Campanella
Department of Agricultural and
    Biological Engineering and Whistler
    Carbohydrate Research Center
Purdue University
West Lafayette, Indiana

Diane E. Chan
Environmental Microbial and Food
    Safety Laboratory
Beltsville Agricultural Research Center
Agricultural Research Service
USDA
Beltsville, Maryland

Kuanglin Chao
Environmental Microbial and Food
    Safety Laboratory
Beltsville Agricultural Research Center
Agricultural Research Service
USDA
Beltsville, Maryland

Byoung-Kwan Cho
Department of Biosystems Machinery
    Engineering
Chungnam National University
Daejeon, Korea

Yong-Jin Cho
Bio-Nanotechnology Research Center
Korea Food Research Institute
Seongnam, Korea

Stephen R. Delwiche
Food Quality Laboratory
Beltsville Agricultural Research Center
Agricultural Research Service
USDA
Beltsville, Maryland

Sandra P. Garcia
Agilent Technologies
Walnut Creek, California

Patrick Jackman
FRCFT
University College Dublin
National University of Ireland
Agriculture and Food Science Centre
Belfield, Dublin, Ireland

Sukwon Kang
National Academy of Agricultural
    Science
Rural Development Administration
Suwon, Korea

Jae-Ho Kim
Department of Molecular Science and
    Technology
Ajou University
Suwon, Korea

Ki-Bok Kim
Center for Safety Measurement
Korea Research Institute of Standards
    and Science
Daejeon, Korea

Moon S. Kim
Environmental Microbial and Food
    Safety Laboratory
Beltsville Agricultural Research Center
Agricultural Research Service
USDA
Beltsville, Maryland

Namsoo Kim
Bio-Nanotechnology Research Center
Korea Food Research Institute
Seongnam, Korea

Seongmin Kim
Department of Bioindustrial Machinery
 Engineering
College of Agriculture and Life
 Sciences
Chonbuk National University
Jeonju, Korea

Sanghoon Ko
Department of Food Science and
 Technology
Sejong University
Seoul, Korea

Seung Ju Lee
Department of Food Science and
 Biotechnology
Dongguk University
Seoul, Korea

Alan M. Lefcourt
Environmental Microbial and Food
 Safety Laboratory
Beltsville Agricultural Research Center
Agricultural Research Service
USDA
Beltsville, Maryland

Michael J. McCarthy
Department of Food Science and
 Technology
University of California
Davis, California

Rebecca R. Milczarek
USDA Agricultural Research
 Service
Western Regional Research Center
Processed Foods Research Unit
Albany, California

Bongsoo Noh
Division of Food Science
Seoul Women's University
Seoul, Korea

Bosoon Park
U.S. Department of Agriculture
Agricultural Research Service
Russell Research Center
Athens, Georgia

Da-Wen Sun
FRCFT
University College Dublin
National University of Ireland
Agriculture and Food Science
 Centre
Belfield, Dublin, Ireland

Chun-Chieh Yang
Environmental Microbial and Food
 Safety Laboratory
Beltsville Agricultural Research Center
Agricultural Research Service
USDA
Beltsville, Maryland

1 Introduction

Yong-Jin Cho

CONTENTS

1.1 THE ROLE OF FOOD QUALITY AND SAFETY EVALUATION

Evaluation of food quality and safety is essential to fresh foods such as meat, fruits and vegetables, as well as processed foods. Particularly, quantitative evaluation of fresh food is becoming more important. Even though certain features of food are attractive to consumers, which is often a subjective attribute, quantitative information on food quality and safety will increase the consumers' acceptance and belief in the reliability of food products.

Currently, one of the most important concerns to food customers is related to the reliability of food. A customer's reliability on foods is formed via various routes. For instance, it is formed when customers consume a food product over a long-term period. In addition, when trustworthy companies release new products, they can acquire reliability through an indirect route that is not dependent on the food product itself, but rather the reputation of the company that released the product. Furthermore, through advertisement, which is provided through several types of media, including broadcasting, newspapers, or Internet, customers may develop reliability on products. However, reliability on food products can also be acquired through numeric values that reliably measure the products' quality and safety. From this aspect, a quantitative assessment of the quality and safety of food products can be stated to be an essential workup.

On the other hand, several worldwide countries have implemented product liability (PL). The most important factor to PL is that accurate and objective data about the quality of products be provided. In the case of food products, when PL is implemented through a quantitative assessment of its quality, measurements that are reliable to both the manufacturers and customers are provided.

Accordingly, a quantitative assessment of the quality and safety of food products is considered highly important from the perspective of both the manufacturers and customers. In this book, new emerging technology for the quantitative assessment of the quality and safety of food products, which has recently been of increasing interest, will be introduced.

1.2 MEASURING PARAMETERS RELATED TO ENERGY SOURCES

For the quantitative assessment of the quality of food products, an understanding of the energy sources needed to assess certain quality factors is mandatory. The physicochemical properties of foods are too diverse to be explained by only one or two factors. Foods mainly have a solid, a liquid, and a semisolid phase, and their constitutions and structure are very complex. Therefore, there are differences in the available technologies used to assess the quality of food that depend on the quality factors one is interested in examining. Accordingly, the appropriate technology should be selectively introduced for the quantitative assessment of the corresponding quality factors.

Table 1.1 lists the range of technologies, according to the energy source, that are currently available to measure different quality characteristics. Technologies that are commonly used for quantitative assessment of the quality of food products are based on mechanical energy, electromagnetic energy, and electrical energy. Technologies based on mechanical energy include vibration, impact, ultrasonic, and creep/stress relaxation technologies. Using these methods, the firmness, elasticity, and viscoelasticity of food products can be effectively measured (Han et al., 1997;

TABLE 1.1
Measuring Parameters Regarding Energy Sources for the Evaluation of Food Quality

Energy Source	Method	Measuring Parameter
Mechanical energy	Vibration	Firmness, viscoelasticity
	Impact	Firmness, internal cavity
	Ultrasonic	Elasticity, internal texture
	Creep/stress relaxation	Viscoelasticity
Electromagnetic energy	X-ray	Internal texture
	UV spectroscopy	Chemical composition
	Visible spectroscopy	Chemical composition, color
	Near-infrared (NIR) spectroscopy	Chemical composition, internal texture
	Visible imaging technology	Color, shape, dimension, surface texture
	NIR imaging technology	Temperature, surface texture
	Laser scattering	Particle size, texture
	Nuclear magnetic resonance (NMR)/ magnetic resonance imaging (MRI)	Chemical composition
Electrical energy	Impedance	Moisture, internal texture
	Potentiometric	Freshness
Others	Gas sensor	Volatile component
	Biosensor	Biomolecule, cell

Kim et al., 1998; Kress-Rogers and Brimelow, 2001; and Kim et al., 2006). In the case of electromagnetic energy, a diverse range of different technologies have been developed and applied. These technologies have been used to measure the chemical composition, internal texture, surface texture, color, shape, dimension, temperature, and particle size. The impedance and potentiometric methods are electrical-energy-based technologies that have also been developed and applied. These technologies are used to measure the moisture content, internal texture, and freshness. In addition, gas sensor and biosensor technologies have recently been developed and are used to detect and measure volatile components, biomolecules, and cells.

1.3 MEASUREMENT FOR BIOLOGICAL AND CHEMICAL RISK FACTORS

For the effective evaluation of food safety, methods for quantitatively measuring the biological and chemical risk factors are needed. On a laboratory level, many methods that can quantitatively measure biological detrimental factors and chemical factors through cell culture and fine instrumental analysis have already been introduced. Nevertheless, these methods require several hours to days, and the specimens must be subjected to pretreatment steps. There are gradually increasing demands for technologies that can quantitatively measure factors on location. With the recent technical advancements of technologies, such as biotechnology, nanotechnology, or fusion technology, various types of methods that can promptly measure detrimental factors in food products have been developed. Table 1.2 summarizes the latest status of these technical developments.

TABLE 1.2
Technologies for the Detection of Biological and Chemical Risk Factors and their Strengths and Weaknesses

Method or Technology	Strength	Weakness
Impedance/potentiometric	Accurate and simple	Homogeneous sample necessary
Enzyme-linked immunosorbent assay (ELISA)	Highly sensitive kit	Narrow detection span
Polymerase chain reaction (PCR)	High sensitivity and linearity	Error due to nontarget DNA amplification
Fluorescence	High sensitivity	Photo bleaching
Colorimetry	High sensitivity	Complicated usage
Surface plasmon resonance (SPR)	High sensitivity without label	Sophisticated device necessary
Microcantilever	Low detection limit	Low repeatability in liquid sample
Quartz crystal microbalance (QCM)	Miniaturization	Low sensitivity
Surface acoustic wave (SAW)	Miniaturization	Low sensitivity and repeatability

1.4 FUTURE TECHNOLOGY

Foods are characterized by the presence of a complex matrix. It is therefore not easy to quantitatively measure the factors associated with the quality and safety of products. Nevertheless, future technologies for the assessment of the quality and safety of food products will require additional essential elements. These include low detection limit, high sensitivity, high specificity, miniaturization for portable use, simple sample preparation, and networking for sensor communication.

Requirements for a low detection limit and a high measuring sensitivity are mandatory for obtaining the reliability of customers. A high degree of specificity is also required because specific constituents or substances should selectively be detected in such conditions as a complex food matrix. In addition, miniaturization and simple sample preparation are essential elements because specific types of technologies and equipment should be designed so that they can be used at the location where the food products are produced or consumed. On the other hand, our society is ubiquitous, and there is a high possibility that these goals can be achieved in the near future. A network where the information that has been obtained through sensors can be transmitted for mutual communication would be an essential factor in the development of future technologies.

Thanks to the remarkable growth of biotechnology and nanotechnology, sensor technology for the assessment of the quality and safety of food products has already been developed to a great extent. Nevertheless, to fulfill the current demands with future technologies, a lot of tasks still need to be completed to resolve the issues facing currently available technologies. The latest research trends in this area have mainly focused on the development of fusion technology. In addition to the development of such technologies as biotechnology (BT), nanotechnology (NT), and information technology (IT), there is a higher level of interest in such fusion technologies as BT-NT, BT-IT, and BT-NT-IT. If this is accomplished, it is expected that effective fusion technology for the assessment of the quality and safety of food products would be universally used by the food industry (Cho et al., 2008).

REFERENCES

Cho, Y. J., C. J. Kim, N. Kim, C. T. Kim, and B. Park. 2008. Some cases in applications of nanotechnology to food and agricultural systems. *Biochip Journal* 2(3):183–85.

Han, Y. J., Y. J. Cho, W. S. Rial, and W. E. Lambert. 1997. Nondestructive techniques for quality inspection of fruits and vegetables. *Journal of Food Science and Nutrition* 2(3):269–79.

Kim, M. S., D. H. Keum, K. B. Kim, M. H. Kim, S. H. Noh, Y. J. Cho, and H. K. Cho. 2006. *Physical and engineering properties of biological materials* (in Korean). Seoul: Moonundang.

Kim, M. S., C. S. Kim, S. H. Noh, B. Park, Y. H. Bae, S. I. Cho, Y. J. Cho, H. K. Cho, and H. Hwang. 1998. *Non-destructive measurement of physical properties of biological resources: Principles and applications* (in Korean). Seoul: Moonundang.

Kress-Rogers, E., and C. J. B. Brimelow (eds.). 2001. *Instrumentation and sensors for the food industry*. Boca Raton, FL: CRC Press.

2 Instrumental Techniques for Measurement of Textural and Rheological Properties of Foods

Osvaldo H. Campanella

CONTENTS

2.1   INTRODUCTION

2.1.1   RHEOLOGY DEFINITION

Rheology is the science that studies the deformation and flow of materials (Steffe, 1996) due to the action of forces. It has a significant importance in the area of food science and engineering because it is relevant to aspects related to the processing of foods as well as their quality evaluation, notably their texture. Another aspect of singular importance in the study of the rheology of foods is closely related to the complex nature and the wide spectrum of rheological properties of existing foods. This has given place to a large number of experimental techniques utilized for food rheological characterization.

Whether or not it is necessary to measure either the flow or the deformation of the food material upon the action of forces will depend on its mechanical characteristics. That will be discussed in the following sections, but it is the relationship between deformation or flow and the applied forces that is used to determine the rheological properties of the food material.

Other important aspects during rheological testing of foods are the definitions of rheological parameters associated with flow, deformation, and forces. Rheological properties are intrinsic properties of the material at given deformations/flow and force conditions; thus, in order to obtain properties that are only dependent on the material, it is necessary to eliminate the effects of sample size and geometry on the measured rheological properties. Rheological information that is not influenced by the sample size and geometry, and the type of test used are known as fundamental rheological data and fundamental tests, respectively.

Often, it is not possible, or practical, to determine accurately the geometry or size of a food sample, and it becomes difficult to define fundamental rheological properties. Tests in which geometric factors or sample size become important and affect the measured properties are known as empirical rheological tests. These tests yield rheological properties that are empirical in nature, which depend on the sample size, shape, and morphology, as well as testing conditions, such as deformation rates; thus, they cannot be considered fundamental properties.

From a process design standpoint it is more convenient to have information of fundamental rheological properties of the foods; however, empirical rheological properties have singular significance in applications related to process control, quality assurance, and quality control, as well as textural evaluation of foods. That is, for applications where the main objective is to assess whether the mechanical and textural properties of the food do not vary over time.

Another goal in the food rheology area is trying to relate instrumentally measured food texture with sensory characteristics of the food.

Methods to evaluate fundamental and empirical rheological properties of foods are described in several books dedicated to the study of the rheology of foods (Steffe, 1996; Rao, 1999; Bourne, 2002). These methods will be briefly described in Section 2.2.

The rheological properties of foods are closely associated with their composition and structure and vary significantly among them. Table 2.1 describes rheological behavior of typical foods and how they relate to their composition and structure, as well as the relevant rheological parameters that are needed to characterize those foods. From a rheological standpoint, solutions of small molecules or monomers represent the simplest food materials, which can be characterized by a single value of viscosity that is a function of temperature, type of molecule, and concentration. That viscosity is independent of the flow or rate at which the deformation is performed. Liquid foods with those characteristics are known as Newtonian liquids; many oils, milk at low or moderate solid concentrations, solutions of simple sugars, and fruit and vegetable juices without pulp all exhibit Newtonian behavior. The viscosity of those Newtonian fluids decreases with increases in temperatures following an Arrhenius relationship such as the one shown below:

$$\mu = Ae^{B/T}$$

(2.1)

TABLE 2.1
Typical Examples of Observed Rheological Phenomena and their Relation to the Structure and Composition of the Food

Molecular Characteristics	Rheological Behavior	Examples	Measured Rheological Properties
Solutions of monomers, low molecular weight components	Liquid—Newtonian	Milk, low-concentration sugar solutions	Viscosity
Solutions of high molecular weight polymers, low-concentration solid-liquid dispersions	Liquid—Non-Newtonian shear thinning fluids, viscoelastic liquids	Tomato sauce, ketchup, fruit puree, dough	Apparent viscosity, elastic and viscous moduli
Solutions of very high molecular weight polymers, high-concentration solid-liquid dispersions	Viscoelastic solids	Gelatin gel, pudding deserts	Elastic and viscous moduli
Macrostructural materials, colloidal suspensions	Shear/time-dependent, structure-dependent materials, structure recovery and breaking	Tomato paste, mayonnaise, foams, yogurt	Viscosity changes with time and shear
Solids with negligible porosity	Viscoelastic solids	Biscuits, carrots, snacks	Elastic modulus

A and *B* are empirical constants that depend on the food, and *T* is the absolute temperature.

Food components of large molecular sizes or polymeric materials forming solutions may exhibit liquid properties, but their behaviors are more complex. These liquids, known as non-Newtonian, have viscosities that are not constant. Often these viscosities are a function of the flow or deformation rate.

Other food products, such as dough, exhibit properties that resemble those of soft materials. Depending on the timescale, these materials can exhibit either solid or liquid behavior—or both behaviors simultaneously. These materials, known as viscoelastic, will be described in Section 2.2.3. Viscoelastic behavior is generally observed in foods whose components are large polymeric molecules.

Foods formed by extremely large macromolecules, like colloidal dispersions, which in addition can also interact among them, are characterized by structures that are dependent on the deformation intensity or the magnitude of the stress applied. The structure of these foods can be strong or weak and depends on the food composition and type of macromolecules. Suspensions containing particles may exhibit deformation/flow-dependent macrostructures. The rheological properties of these foods are difficult to evaluate because their structure depends on the intensity and time during which the deformation or flow is applied, or the intensity of the applied force. Examples of the type of foods are mayonnaise, tomato paste, foams, and yogurt.

2.1.2 RHEOLOGICAL CALCULATIONS

The rheological properties of many foods range between the properties of true liquids and true solids, for which very specific tests for their characterization have been developed. Most foods, however, have properties that resemble those of soft materials, and often they are referred to as semiliquid or semisolid materials. More specifically, those materials are called viscoelastic materials, which, depending on time, can exhibit either elastic (solid) or viscous (liquid) behavior, or both.

Properties of liquids and solids are based on two models described by equations known as the Newton law of viscosity and the Hooke law applied to liquids and solids, respectively. Those equations are relationships among applied forces and the resulting deformation and flows. But to characterize the intrinsic properties of a food material, forces and deformations or flows need to be made independent of the dimensions and size of the sample.

In rheological terms the conversion of forces and deformations, which are affected by the sample characteristics, into quantities that are independent of these sample characteristics is well established. For forces the conversion is simply done by calculating the force per unit of area in which the force is acting, and that quantity is called stress. It is important to note that a force can be applied tangentially to the area or normally, which gives place to the definition of shear and normal stresses (pressure is a typical example of normal stress). In the International System of Units (SI) stress is expressed in units of Newtons (N) per square (m^2), which is known as Pascal (Pa). Similarly, deformation and deformation rate (flow) need to be converted to sample size or geometry-independent quantities. The standard procedure is to express these deformations or deformation rates per unit of a characteristic length of the sample. This relative deformation or deformation rate (flow) is a rheological parameter known as strain when it is associated to the deformation, and strain rate when it is associated to the deformation rate or flow. Like with the definition of *stress*, there are many types of deformations, and among them the more known and used for rheological characterization are the shear strain, or the shear strain rate, and the normal strain, or normal strain rate. Given the fact that strain is a relationship between two lengths, it has no units, whereas strain rate gives the change of strain per unit of time, so it has units of reciprocal seconds.

The properties of the ideal liquid and solid models described by the Newton law of viscosity and the Hooke law can be determined using shear tests, which are schematically illustrated in Figure 2.1. As indicated in the figure, for a shear test the force acts tangentially to the area used to calculate the stress. For the case of liquids, the shear test can be simply described by the flow originated in a fluid confined between two plates, one of which is moving with a constant velocity, V, through the application of a force, F (Figure 2.1a). By assuming perfect nonslip conditions at the liquid-plate interface, it can be demonstrated that the velocity profile, $v_x(y)$, is linear and changes from a value that is equal to the plate velocity at the interface between the moving plate and the liquid to a value of zero at the interface between the fixed plate and the liquid. This simple velocity profile facilitates the calculation of the liquid strain rate, more known in these tests as shear rate, $\dot{\gamma}$, whereas shear stress, τ, is calculated by dividing the pushing force, F, by the area in where this force is acting, A (Figure 2.1a). These rheological quantities are used to write the well-known Newton law of viscosity (Equation 2.4), whereas the shear rate and shear stress are calculated by Equations (2.2) and (2.3), respectively.

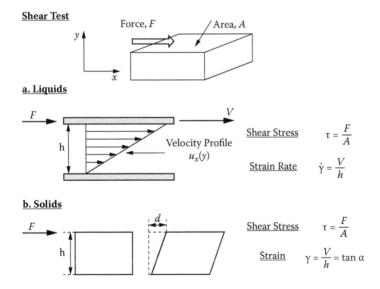

FIGURE 2.1 Schematic description of a shear test for (a) liquids and (b) solids.

$$\dot{\gamma} = \frac{V}{h} \tag{2.2}$$

$$\tau = \frac{F}{A} \tag{2.3}$$

$$\tau = \mu\dot{\gamma} \tag{2.4}$$

μ is the viscosity of the liquid and has units of Pa.s in the SI system of units.

A solid food does not flow when it is deformed by the action of a force, so that a strain instead of a strain rate should be used to quantify the deformation of the sample. As illustrated in Figure 2.1b, the food is displaced by a distance d due to the action of the force and the strain for this test is defined as the deformation per unit of a characteristic length conveniently chosen as the thickness of the sample, and is calculated by the following equation:

$$\gamma = \frac{d}{h} \tag{2.5}$$

The shear stress calculation is similar to that used for the liquid material (Equation 2.3). The Hooke Law for a solid food can now be expressed by the following equation:

$$\tau = G\gamma \tag{2.6}$$

where G is the shear modulus of the material.

It is important to note that for rheological testing of liquid foods the shear test is more commonly used, but there exist other rheological tests, such as those based on extensional flows, that may have significance for the characterization of soft viscoelastic foods and relation to some processing flows and deformations (Campanella and Peleg, 2002). In the development of fundamental rheological methods for foods, significant efforts have been focused on the design of methods able to apply only one type of deformation or flow. However, deformations or flows occurring during food processing are far from being of only one kind, and thus it is important to know the response of food materials to different types of deformations and flows. Several books devoted to the rheology of polymeric materials describe methods to characterize liquid, solid, and viscoelastic materials (Ferry, 1980; Macosko, 1994) under different kinds of deformations and flows.

2.1.3   USES OF RHEOLOGICAL PROPERTIES IN FOOD SCIENCE AND ENGINEERING

Knowledge of the rheological properties of foods has significant importance in many aspects related to the processing and development of new foods, as well as use as an analytical tool. Figure 2.2 depicts some of the uses and applications of rheological

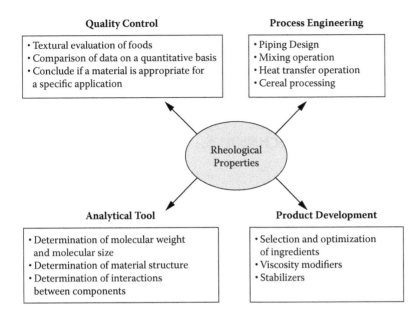

FIGURE 2.2 Schematic diagram indicating the potential use of rheological properties in the food science and engineering areas.

data. In processing design, extreme care must be given to the rheological properties of liquid foods, in particular when they are non-Newtonian, because these properties have a large influence in the behavior of the flow, which differs notably from the flows produced by the simpler Newtonian liquids. That applies to many units operations involving not only the mechanics of fluids, but also other transport phenomena, such as heat transfer, where the viscosity of the liquid has an important role. Mixing operations are also closely related to the rheology of liquid foods, and a good characterization of the rheological properties of the material at different flow conditions is necessary. During processing of many cereals by operations like extrusion, flaking, and pelletizing, the rheology of the processed material may influence not only the process but also the quality of the final product.

Product development makes good use of rheological data, especially when new ingredients such as viscosity modifiers and stabilizers have to be selected and tested to develop new texturally sound foods.

In regards to analytical methods to characterize food polymers, rheological measurements of diluted solutions of food macromolecules may help to characterize the makeup of many food components, thus providing information on molecular weight and size. For more concentrated solutions or suspension, rheological properties may provide information on the food structure as well as potential interactions among their components.

Textural properties of foods are one of the most important aspects that define the quality of a food product. Thus, rheological methods and rheological characterization of foods play a very important role in this area. Ultimately, texture of foods is what a consumer senses, and it should be studied using sound sensory evaluation techniques. However, possible correlations between sensory evaluations and rheological instrumental assessment of texture have some economical and practical benefits, discussed in Section 2.5, making the development of new instrumental rheological techniques to assess food texture an ongoing research area. Rheological techniques can also be used for quality assessment of intermediate products and raw materials.

2.2 INSTRUMENTAL TECHNIQUES

2.2.1 Background

As discussed, rheological properties of most foods range between the two well-defined behaviors of liquids and solids. However, many foods have rheological properties that resemble more those of soft matter and often are defined from a rheological standpoint as semiliquid or semisolid materials. The presence of elastic effects in the rheological response of these soft food materials is an important factor that can influence the selection and use of a specific rheological instrumental technique. Whereas in semisolid materials the presence of elasticity is unavoidable and instrumental techniques used should take into account this phenomenon, semiliquid or liquid foods may exhibit negligible or no elasticity at all, and techniques could overlook these effects. Thus, it is convenient to classify instrumental techniques to characterize semiliquid and semisolid foods in two different groups: (1) nonelastic liquid foods and (2) elastic semiliquid or semisolid foods. The first group encompasses mostly beverages, such as fruit purees, oils, and tomato products, to name a few, in which elasticity is negligible or has only a minor effect on the food quality or during the processing of these foods. The second group includes semiliquid and semisolid foods where the presence of elasticity may affect their quality perception or their processing. For conciseness, these foods will be called soft viscoelastic foods without differentiation if they are either semiliquids or semisolids.

2.2.2 Nonelastic Liquid Food

The relationship between shear stress and shear rate, which defines the rheological behavior of a nonelastic liquid food, is usually plotted in a diagram of shear stress versus shear rate, which is called the flow curve of that liquid food. The form of that flow curve enables the classification of nonelastic liquid foods in two main groups: Newtonian (lineal flow curve) and non-Newtonian (nonlineal flow curve) liquid foods. Non-Newtonian liquid foods include three different kinds of materials: time independent, time dependent, and plastic fluids (Figure 2.3).

Whatever the behavior of the liquid food is, the important information to be obtained from rheological testing of these liquids is their flow curve, which provides their rheological signature.

2.2.2.1 Time-Independent Non-Newtonian Liquid and Plastic Foods

Figure 2.4 schematically illustrates the possible flow curves that can be obtained in terms of shear stress versus shear rate, or viscosity versus shear rate or stress

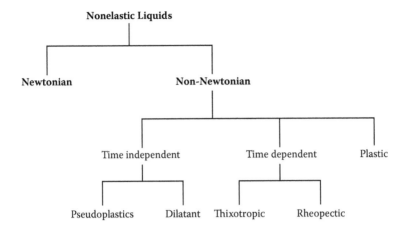

FIGURE 2.3 Schematic indicating different rheological behavior and classification, from a rheological standpoint, of nonelastic liquid foods.

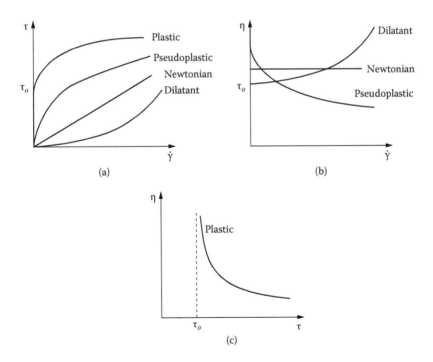

FIGURE 2.4 Schematic representation of typical flow curves for liquid foods in terms of (a) shear stress versus shear rate and (b) viscosity versus shear rate. (c) Typical viscosity versus applied viscosity for a plastic material exhibiting a yield stress.

(for plastic fluids). The concept of viscosity as a property purely dependent on the material, which has a sense for Newtonian fluids, needs to be redefined for non-Newtonian fluids. As illustrated in Figure 2.4 for non-Newtonian liquids, viscosity is a function of the shear rate; thus, it would be more appropriate to define an apparent viscosity η, which from Equation 2.4 can be defined as

$$\eta = \frac{\tau}{\dot{\gamma}}$$

(2.7)

Several rheological models have been used to describe the nonlineal flow curves of non-Newtonian liquids, but one of the better known and used is the power law model, which is described by

$$\tau = k\,\dot{\gamma}^n$$

(2.8)

and the apparent viscosity, η, can be calculated as

$$\eta = k\,\dot{\gamma}^{n-1}$$

(2.9)

k and n are known as the consistency index and the flow index, respectively. In SI units, k has units of Pa.s^n, whereas n has no units. It should be noted that the Newtonian liquid is a special case of the power law model when $n = 1$ and $k = \mu$. That the units of the consistency index k depend on the flow index n has been the object of continuous criticism to the power model. However, its use has been widespread in the area of rheology of food and nonfoods because of practical convenience. Typical flow curves in terms of apparent viscosity as a function of shear rate are depicted in Figure 2.4b.

Figure 2.4a and 2.4c show typical flow curves of a plastic material. As noted in Figure 2.4a, if the material does not reach a minimum stress, known as yield stress, τ_0, it will not flow and the shear rate will be zero. For those conditions apparent viscosity is not defined. Thus, Figure 2.4c clearly shows that values of viscosity are obtainable only for stresses larger than the yield stress τ_0. For plastic materials, once the stress overcomes the yield value and the material starts to flow, the flow curve could be concave downward, as shown in Figure 2.4a, concave upward, or linear. The behavior illustrated in Figure 2.4a is the more commonly observed in foods known as Herschel-Bulkley fluids. The linear case is rare within foods, but common in some nonfood materials such as asphalt, and is known as Bingham plastics. A rheological model that describes the behavior of plastic foods is given below:

$$\tau = \tau_0 + k\,\dot{\gamma}^n$$

(2.10)

Other models have been used to describe the behavior of non-Newtonian fluids, and they are discussed in several books focused on the rheology of polymers (Ferry, 1980;

Bird et al., 1987; Macosko, 1994; Morrison, 2001) and foods (Steffe, 1996; Rao, 1999; Bourne, 2002). The shear rate, depending on the viscosity of shear thickening and shear thinning (more common in foods), properly described by the power law model, can be observed at moderate shear rates, which are comparable with the shear rates applied during the processing of foods. In that sense, the power law model has been of great use to describe the rheology of many liquid foods and their relation to processing. For a wider range of shear rates, however, the rheology typical power law behavior in that a of liquids deviates from the typical power law behavior in that a double-logarithmic plot of apparent viscosity versus shear rate is represented by a straight line. This behavior, which is schematically illustrated in Figure 2.5, can be explained by the characteristics of these liquid foods. In general, they are dispersions of particles of different sizes in an aqueous media. The shear thinning behavior (pseudoplastic liquids) is explained by the alignment of these particles in the shear flow. When the particles are aligned due to the high shear rates, the resistance to the flow becomes lower, which results in lower viscosities. At extremely low shear rates the shear forces are not high enough to overcome the natural randomness of the particles created by kinetics events such as Brownian motion or thermal effects. Thus, at low shear rates the liquid food exhibits Newtonian behavior. In the other extreme, at extremely high shear rates, aligning of the particles has reached its maximum level so a Newtonian behavior is observed again. As illustrated, power law behavior is only observed at intermediate shear rates.

2.2.2.2 Time-Dependent Non-Newtonian Liquids

As described in Table 2.1, liquids that exhibit interaction among their forming macromolecules may have a shear depending structure. These structures are fragile and may break upon application of shear, showing an important decrease in viscosity. Testing of these liquids can be done in different ways, but a typical test is one in which shear rates are increased (up curves) to a maximum value and after decreased (down curves) at a particular rate. Figure 2.6a shows a schematic of a typical result for the case in that breakage of the liquid structure decreases the apparent viscosity;

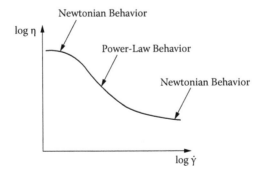

FIGURE 2.5 Schematic flow curve, in a double-logarithmic plot of viscosity versus shear rate, of a nonelastic non-Newtonian liquid over a wide shear range. The flow curve exhibits power law behavior at moderate shear rates and Newtonian behavior at extremely low and high shear rates.

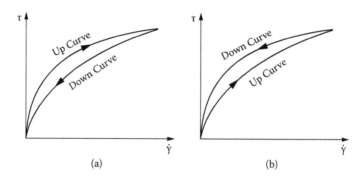

FIGURE 2.6 Flow curves of time-dependent non-Newtonian liquids obtained by increasing and decreasing the shear rates at a given rate. (a) Thixotropic liquids. (b) Rheopectic liquids.

the behavior is called *thixotropy* and the liquids *thixotropics*. Figure 2.6b illustrates the opposite behavior (not very common within foods) known as *rheopexy*, and the liquids that exhibit it *rheopectics*. When these behavior results are observed, it is said that the liquids exhibit hysteresis. The area of the hysteresis curve was used to study different types of starches, in particular those with the presence of granule bound starch synthase (GBSS), and its effect in the shear breakdown of gelatinized pastes of these starches (Han et al., 2002). During shearing thixotropic liquids two events may occur. First, there is a breakage of the linkage between the liquid components, and the rate of structural breakdown at a given shear rate depends on the number of linkages present, which by the effect of shear decreases with time. The breakage of those linkages has been mathematically described following a second-order reaction-like equation. The second event is the simultaneous rebuilding of the linkages, and the liquid structure increases with time. Once equilibrium between breakage and rebuilding is established at a given shear rate, the viscosity of the liquid remains constant. The value of this viscosity depends on the applied shear. Since the breakage and rebuilding of the liquid structure during a shear test occur simultaneously, the rate at which the shear rate is increased and decreased is very important, because if it is too small, it may give the structure of the material time to recover, and for those cases hysteresis can disappear. Conversely, if the rate or decrease of shear rate is fast, the material will not have time to recover its structure and hysteresis will be noticeable. The capacity of recovery of a thixotropic liquid food can be determined by the test schematically illustrated in Figure 2.7. Rheological studies focused on the functionality of whey proteins making use of this rheological method (Tang et al., 1993). The thixotropic behavior of liquid foods has many implications in food processing. Thixotropic behavior, for instance, may cause problems at the start of the manufacturing operations. Due to the high initial apparent viscosity, pumps and other liquid handling equipment (e.g., mixers), must be greatly oversized to handle the extra pressure load at start-up. Once these liquids have been worked (sheared) for some time, their viscosity falls and this extra power rating is no longer necessary. As a practical rule of thumb, thixotropic liquids are often run continuously to maintain the liquids at manageable viscosity. In extreme

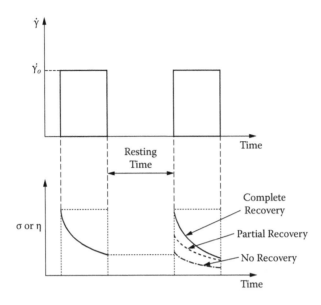

FIGURE 2.7 Typical test applied to thixotropic liquids to study recovery of their structure with time.

cases, heating is required at pumps/mixers to reduce the initial viscosity to manageable levels.

Often the processing sequence adopted depends on the rate of recovery of the liquid structure. In some industrial processes the structure takes a long time to repair—hours or even days—so a typical process would consist first in shearing the liquid violently to break up the structure, then pumping, atomizing, or processing in equipment suitable for low-viscosity liquids. Then, the product must be placed and the structure allowed to recover. Models to describe thixotropy, a description of the phenomenon, differences with viscoelasticity (described in the following section), suitable testing for thixotropic materials, and the role in the processing of food and nonfood materials are discussed by Barnes (1997).

2.2.3 SOFT VISCOELASTIC FOODS

These materials may exhibit either viscous (liquid) or elastic (solid) behavior and, under certain conditions, both behaviors simultaneously. It is not easy to distinguish viscoelastic from plastic food materials, which also exhibit either liquid or solid behavior, although in this case depending on the stress applied rather than time. Viscoelastic foods can be purely liquids at any applied stress, but they can exhibit elastic/solid behavior when the time involved in the deformation is very short. Viscoelastic behavior is commonly represented by simple mechanical analogs or combinations of them, some of which are schematically illustrated in Figure 2.8. These analogs are based on the simple models representing pure liquid and solid materials, accurately described by Equations 2.4 and 2.6. As

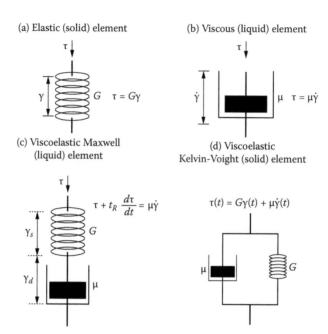

FIGURE 2.8 Mechanical analogs to study viscoelasticity of foods along with their rheological models. (a) Elastic element. (b) Viscous (liquid) element. (c) Viscoelastic Maxwell (liquid) element. (d) Viscoelastic Kelvin-Voight (solid) element.

the behavior of viscoelastic materials depends on time, rheological methods used to characterize them may need to involve that variable. Thus, the common approach is to apply either strains or stresses that vary with time, which gives way to a number of rheological tests that are briefly described below. Rheological equations that describe to the mechanical analogs depicted in Figure 2.8 can be derived from Equations 2.4 and 2.6. For the Maxwell model (Figure 2.8c), it is assumed that the stress acting in each of the single elements forming it is the same, whereas the strains are additives. The assumption yields the following rheological constitutive equation:

$$\lambda \frac{d\tau(t)}{dt} + \tau(t) = \mu\dot\gamma(t) \tag{2.11}$$

λ is known as the relaxation time of the Maxwell model and is defined as $\lambda = \mu/G$.

For the Kelvin-Voight model (Figure 2.8d) it can be assumed that the strain applied to each element of the model is the same, whereas the stresses are additives. The assumption yields the following equation:

$$\tau(t) = G\gamma(t) + \mu\dot\gamma(t) \tag{2.12}$$

A more detailed description of these methods and other mechanical analog models is given in the general rheology literature (Ferry, 1980; Morrison, 2001) and other books more specific to food rheology (Steffe, 1996; Bourne, 2002).

The other aspect to be taken into account when describing rheological methods for characterizing soft viscoelastic materials is the type of deformation/strain or stress applied to the sample. If the strains are small enough to not disturb the structure of the sample, they are called small deformation tests and it is assumed that the mechanical response of the material is not affected by the magnitude of either the strain or the stress applied. It is also said that the mechanical behavior of the material is within the lineal viscoelastic response. Although the approach facilitates the analysis of the data, often it is not closely related to the behavior of the material under real processing conditions, or conditions associated to the textural perception of the food during consumption, which in general involve large strains or stresses that disturb the structure of the samples. Despite that, tests performed within the lineal viscoelastic behavior range provide relevant information on the structure, composition, effect of processing, temperature, and other variables of the sample, and are often used for the characterization of many viscoelastic foods.

2.2.3.1 Small-Amplitude Oscillatory Tests

With the advent of better hardware and software these tests have had extensive use over the last fifteen years. The tests more extensively used are those based on shear deformations. Basically, either a strain or a stress that sinusoidally varies with time is imposed. Modern instruments have the ability to provide both options in what is called either stress control or strain control modes. Instruments of moderate costs provide a true control stress mode, whereas control strain is performed by using the instrument control software. True control strain instruments are more expensive and are also suitable for other viscoelastic tests, such as the strain relaxation test (Section 2.2.3.2), whereas stress control instruments are adequate to perform creep test experiments (Section 2.2.3.3). Small-amplitude oscillatory tests, however, have found large applications to characterize the viscoelastic properties of foods. Regardless of the mode utilized, the small-amplitude oscillatory test requires the input of the amplitude of the stress (τ_0) in the stress control mode or the strain (γ_0) in the strain mode control, the frequency (ω) or range of frequencies of the oscillation to be applied, and a group of parameters that are associated to the analysis and interpretation of the data that are specific to the instrument manufacturer. In addition, the software allows the user to enter a number of rheological parameters that can be obtained from the test. Analysis of data is simplified by the use of the mechanical analogs, such as those illustrated in Figure 2.8 and their corresponding rheological models. For instance, if a strain control mode is used, the input of strain to the test will be $\gamma(t) = \gamma_0 \sin \omega t$, where γ_0 and ω are the parameters fixed by the user. In particular, γ_0 has to be selected within the lineal viscoelastic range of the sample (see strain/stress sweep test) to get reliable data. If the testing is performed on a true liquid material, the corresponding stress response to the test can be calculated from Equation 2.4 as

$$\tau = \mu \dot{\gamma}(t) = \mu \frac{d\gamma(t)}{dt} = \mu \frac{d}{dt}(\gamma_0 \sin \omega t) = \mu \gamma_0 \omega \cos \omega t = \tau_0 \cos \omega t \qquad (2.13)$$

whereas for a true solid material, by using Equation 2.6, the stress response is

$$\tau = G\gamma(t) = G\gamma_0 \sin\omega t = \tau_0 \sin\omega t \qquad (2.14)$$

The important conclusion from the results of the test hypothetically applied to true liquids and true solids is that for true liquids, the stress response varies following a *cosine* relationship that is 90° out of phase with the applied strain, whereas for true solids both strain and stress vary sinusoidally, so they are in phase. In other words, the phase angle, δ, is 0°. For viscoelastic materials the phase angle between the applied strain and stress response, or vice versa, is $0 < \delta < 90°$ and the corresponding stress response is

$$\tau = \tau_0 \sin(\omega t + \delta) \qquad (2.15)$$

The measurement of either the stress (in control strain mode) or the strain (in control stress mode) and the phase angle δ are used to calculate the viscoelastic moduli of the sample, more commonly known as storage modulus G' and loss modulus G'', calculated as shown below:

$$G' = \frac{\tau_0}{\gamma_0} \cos\delta \qquad (2.16)$$

$$G'' = \frac{\tau_0}{\gamma_0} \sin\delta \qquad (2.17)$$

Other viscoelastic parameters can be derived from these moduli and are listed elsewhere in the literature (Steffe, 1996; Bourne, 2002). It is important to clarify that these derived parameters do not provide further information than the one given by Equations 2.14 and 2.15; however, they often provide ways to represent the data more clearly. There is ample literature reporting viscoelastic parameters obtained with this test and possible relationships with food texture (Dolz et al., 2006; Stern et al., 2007; Holm et al., 2009).

A number of tests using small-amplitude oscillatory tests are performed to characterize viscoelastic foods. They can be classified as (1) strain sweep test, (2) frequency sweep test, (3) temperature sweep test and (4) time sweep test:

1. *Strain sweep test.* This test is commonly used to identify the linear viscoelastic range of the food material. With the instrument working on a strain mode, the test consists of changing the strain applied to the sample at a constant frequency (generally at 1 Hertz) and monitoring the storage and the loss moduli, or other related parameters, as a function of the applied strain (Figure 2.9a). As discussed, within the linear viscoelastic range the measured viscoelastic properties are independent of the applied

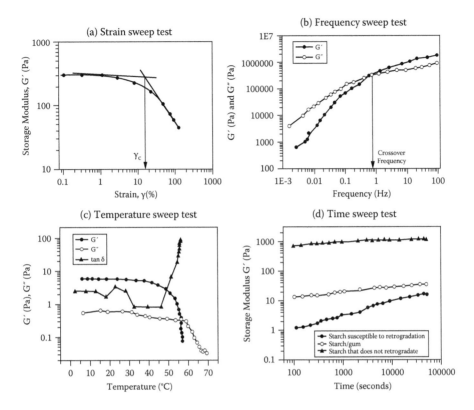

FIGURE 2.9 Typical results of small-amplitude oscillatory tests using shear flows. (a) Strain sweep test. (b) Frequency sweep test. (c) Temperature sweep test. (d) Time sweep test.

strain, so values of the storage and loss moduli should be constants. Although the purpose of the strain sweep test is mainly to identify the linear viscoelastic region of the sample, the test has also been used to identify textural properties of gel-like foods. For instance, the strain at which the viscoelastic storage and loss moduli change from constant values to strain-dependent values, defined as a critical strain, has been used to characterize the texture of mayonnaise prepared with different concentrations of alginate and alginates of different molecular weights (Mancini et al., 2002). A similar approach was used to characterize concentrated xanthan gum suspensions (Song et al., 2006) and distinguish the type of gels that can be formed at different xanthan gum concentrations and xanthan gums produced with different strains. A stress sweep test is performed in the same way but varying the stress instead of the strain.

2. *Frequency sweep test.* Frequency sweep tests are intended to characterize the viscoelastic properties of foods, in particular to identify their elastic or viscous characteristics. Food gels are often evaluated by comparing the storage and the loss moduli of the samples; gels are considered those materials in which the storage modulus is significantly higher than the loss

modulus. The effect of frequency on these moduli is also important to study the characteristics of gel-like food materials. Pure elastic foods exhibit no effect of frequency, whereas more liquid materials are significantly affected by the oscillation frequency. The crossover frequency, defined as the frequency at which the storage modulus is equal to the loss modulus, has been defined as a gelling condition for many materials. Wang and Zhang (2009), among other tests, studied the effect of concentration of a new gum in the crossover frequency.

The test has also found great use in studying the interaction of different macromolecules such as starch and gums. For instance, Kaur et al. (2008) studied the interaction of normal and waxy starches with cassia gum. The effect of resting of starch samples to investigate retrogradation kinetics of different starches was also studied by using this test (Matalanis et al., 2009). A typical result of a frequency sweep test is illustrated in Figure 2.9b.

3. *Temperature sweep test.* The small strains applied during oscillatory tests in the controlled strain mode enable the study of rheological changes produced by gelation or other moderately slow reactions because the sample is minimally disturbed while the reaction is occurring, and thus it has a negligible effect on the measurements. The effect of temperature on physical gels like those produced by gums, or a mixture of gums in which the elastic nature of these materials is created by the entanglement of their very large molecules and formation of junction zones, can be studied by the temperature sweep test. Figure 2.9c illustrates the rheological behavior of a ι-carragenaan gel (unpublished data) during a temperature sweep test. The significant decrease in G' and significant increase in tanδ at a certain temperature indicate the possibility of melting of junction zones, which breaks the elastic nature of these gels. Changes in the loss modulus are not as significant, and therefore less used to identify these molecular changes. Other studies involving the interaction of food macromolecules, such as starch and proteins, have been conducted using temperature sweep tests (Yang and Zhu, 2006; Wang et al., 2009).

4. *Time sweep test.* Time sweep tests are very useful to study structural changes in foods at isothermal and nonisothermal conditions. For isothermal conditions, a temperature at which the structural changes want to be studied is set and the sample is deformed in an oscillatory manner using either low strains or low-stress amplitudes. The use of low amplitudes is essential to ensure that the changes observed in the sample are due to changes in structure produced by the physicochemical transformations (e.g., starch retrogradation, polymer aging, enzymatic reactions) and not the applied deformation. The oscillation frequency is fixed (often to 1 Hertz), as well as the amplitude of strain or the stress applied specifically to values that are within the linear viscoelastic region. Cho and Yoo (2010) determined the viscoelastic properties of sweet potato at temperatures in the presence of several sugars. The effect of aging time at 4°C was studied using this approach, as well as viscoelastic properties of these systems at

room temperature. Other studies using this test have been utilized to assess the influence of structural modifications (e.g., retrogradation) and the addition of food ingredients on the viscoelastic and sensorial properties of many foods (Biliaderis and Tonogai, 1991; Prokopowich and Biliaderis, 1995; van Soest et al., 1996; Yuan and Thompson, 1998; Arocas et al., 2009a; Dolores Alvarez et al., 2009; Mariotti et al., 2009; Matalanis et al., 2009). Studies concerning structural changes such as retrogradation, aging, gelatinization, denaturation, and aggregation, which are produced by the presence of polymeric components (e.g., starch and proteins) in the food, can be studied using two different approaches. In one of the approaches kinetic studies are performed under controlled conditions of temperature and relative humidity in chambers where samples are extracted at different times and tested rheologically. The second approach involves performing the studies of the effects of storage time or reactions on the food structure directly in the rheological instrument. The former has some advantages because it allows the study under well-controlled conditions, with the drawback that many samples are needed (one sample is used for each testing time). In the second approach, and although temperatures can be accurately controlled in modern rheological instruments, control of relative humidity is more difficult and samples can be dehydrated over the period of study, thus introducing some inconsistencies in the results.

A typical time sweep rheological test under nonisothermal conditions is the determination of starch gelatinization kinetics using the Rapid Visco Analyzer (RVA). However, an RVA test imposes large shear rates and cannot be classified within this group. The introduction of pasting cells in modern rheometers (e.g., TA Instruments, Paar Physica) is providing the opportunity to perform tests where alternate application of steady shear and oscillation under controlled varying temperatures enables studies of effects of shear and temperatures on the viscoelasticity of foods (Arocas et al., 2009b).

2.2.3.2 Step Strain Relaxation Test

This test was widely used in the past when only universal testing machines were available to measure viscoelasticity of foods. However, with the advent of modern rheometers its use has been significantly reduced. Ideally, the test is based on the application of a sudden strain to the sample and maintained for a given time, whereas measuring the relaxation of the stress. It is known that a purely elastic material will be able to sustain the strain with a constant stress as long as the strain is maintained, as indicated by Equation 2.6, and no relaxation of the stress is observed. Conversely, a purely liquid material will not be able to sustain any strain because it is only responsive to a strain rate (Equation 2.4). Viscoelastic materials will exhibit a relaxation of the stress while the strain is being applied (Figure 2.10a). Although ideal strain relaxation tests are indicated in Figure 2.10a as a full line, in practical terms the application of a sudden strain may create some measurement artifacts mainly associated with the inertia of the measurement system. Practical application of the test should be as fast as possible but without compromising the results. More real tests are indicated

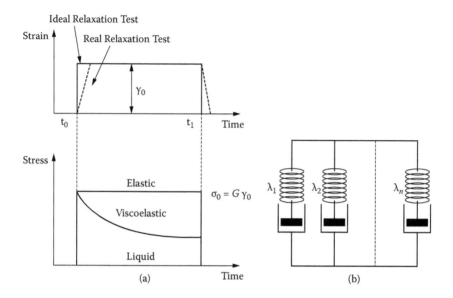

FIGURE 2.10 (a) Schematic diagram illustrating a step strain relaxation test. (b) Generalized Maxwell mechanical analog used to describe step strain relaxation data.

as dashed lines in Figure 2.10a. Mechanical analogs are used to describe the stress relaxation of viscoelastic materials. For a Maxell fluid the stress response to a step strain can be mathematically obtained by setting $\dot{\gamma} = 0$ and $\tau = \tau_0$ at $t = t_0$ and integrating Equation 2.11 to yield:

$$\tau(t) = \tau_0 e^{-\left(\frac{t-t_0}{\lambda}\right)} \tag{2.18}$$

Although many real semiliquid food materials exhibit stress relaxation, their behaviors do not follow the trend predicted by the Maxwell model (Equation 2.18). Instead, models like the one depicted in Figure 2.10b, known as a generalized Maxwell model, are used to describe stress relaxation data. The rheological equation that describes a generalized Maxwell model is

$$\tau(t) = \tau_1 e^{-\left(\frac{t-t_0}{\lambda_1}\right)} + \tau_2 e^{-\left(\frac{t-t_0}{\lambda_2}\right)} + \ldots\ldots + \tau_n e^{-\left(\frac{t-t_0}{\lambda_n}\right)} \tag{2.19}$$

τ_i and λ_i, with $i = 1, n$, are free parameters of the model that can be obtained by curve fitting using standard nonlinear regression techniques; λ_i, as discussed, are known as relaxation times of the material. The number of Maxwell elements can be varied, but in general three or four Maxwell elements are enough to appropriately fit stress relaxation experimental data of many semiliquid foods. For more semisolid-like foods the stress relaxes to an asymptotic nonzero value, which would correspond to a material with a component having a relaxation time being infinitely large. For those cases Equation 2.19 should be changed to

$$\tau(t) = \tau_1 e^{-\left(\frac{t-t_0}{\lambda_1}\right)} + \tau_2 e^{-\left(\frac{t-t_0}{\lambda_2}\right)} + \ldots\ldots + \tau_{residual} \tag{2.20}$$

For semiliquid-like viscoelastic materials, $\tau_{residual} \rightarrow 0$.

Generalized Maxwell models have been used to describe the stress relaxation of many semiliquid and semisolid foods, such as gels, bread, cheese, and meat (Del Nobile et al., 2007). One advantage of this test is that the strain applied can be small enough to be within the linear viscoelastic region, but large strains can also be applied. For the latter case, the stress relaxation response is known as nonlinear relaxation data because it will depend on time and the strain applied (Quintana et al., 2002). Although the approach has been of some use in food rheology, the selection of an appropriate number of elements in the generalized Maxwell creates certain uncertainty and does not give a real physical meaning of the parameters. This type of problem is defined as an ill-defined problem in rheology (Honerkamp, 1989). Peleg (1980) proposed a normalization of the stress relaxation data that reduces the number of parameters to two and has been widely used to analyze stress relaxation data (Peleg and Normand, 1983; Kampf and Nussinovitch, 1997; Joshi et al., 2004; Markowski et al., 2006; Andres et al., 2008).

2.2.3.3   Creep Test

This test is commonly used to characterize the viscoelastic properties of semisolid foods. It consists in the application of a sudden stress to the sample while monitoring the strain (Figure 2.11a). The figure illustrates typical responses of purely liquid and elastic materials as well as those of viscoelastic materials. The test can be slightly modified to show the recovery of viscoelastic materials (Figure 2.11b) in a test known as a recoil test. As illustrated in the figure, purely elastic materials exhibit full strain recovery when the stress is taken off, which indicates the perfect rheological memory of these materials, whereas purely liquid materials do not have strain recovery, an indication of no elastic memory at all for these materials. For viscoelastic materials partial recovery can be observed, which is characteristic of the fading memory of these materials. The concept of memory is well discussed in books of polymer rheology (Bird et al., 1987; Morrison, 2001). It can be noted that except for elastic materials, all the other materials exhibit a permanent strain or deformation that depends on the duration of the applied stress and on the material itself. That is a characteristic of the fading memory of these materials. The simplest model used to describe creep behavior of semisolid foods is the Kelvin-Voight model (Figure 2.8). However, that simple model is insufficient to describe real creep data. Other models, such as the four-element Burger model, have found application to describe the creep response of many food materials, such as cheese and rice gels (Quemada, 1999; Diez-Sales et al., 2007; Xu et al., 2008; Olivares et al., 2009), and Burger type models to describe the creep behavior of processed cheese (Pereira et al., 2001). A number of models having a significant number of lower rheological parameters and based on normalization of the creep data were proposed to facilitate the analysis (Peleg, 1980; Purkayastha et al., 1984, 1985). Like with the relaxation test, the stress applied could be small enough so the material response is within the linear viscoelastic behavior. However, large stresses can be applied, but the material responds

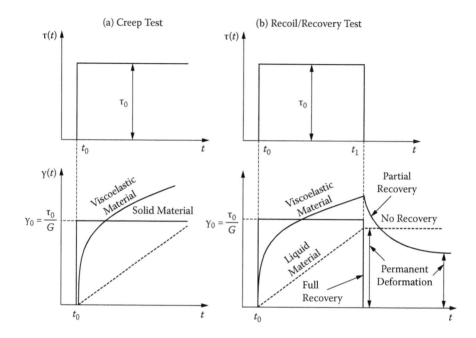

FIGURE 2.11 (a) Schematic diagram illustrating a step stress creep test. (b) Schematic diagram illustrating a recoil/recovery test.

nonlinearly to the imposed stress, yielding a nonlinear creep behavior. Nonlinear creep behavior is more difficult to analyze because the strain response is a function of the applied stress in addition to time and the material itself. For these large stress tests and their resulting nonlinear creep response, the normalization of creep data proposed by Peleg (1980) is preferable.

2.2.3.4 Start-Up Flow

Although this test is based on a steady shear flow, the fact that the start of the flow is done in a very short time can put in evidence viscoelastic phenomena if the tested material exhibits some degree of elasticity. For this test, a steady and constant shear rate is applied suddenly, whereas the stress evolution is recorded (Figure 2.12). For viscoelastic materials, a stress overshoot, with a magnitude and time where it occurs closely related to the imposed shear rate, is observed. For high shear rates the stress overshoot is high and occurs at short times, whereas at very small shear rates the presence of overshoot is not observed. This behavior can be analyzed in terms of the Deborah number, a dimensionless parameter that provides an indication of the extent of elasticity of the sample. The Deborah number is calculated as the ratio between the sample typical relaxation time, λ, which is a material property already defined in Section 2.2.3.2, and an experimental time, t_{exp}:

$$N_{De} = \frac{\lambda}{t_{exp}} \tag{2.21}$$

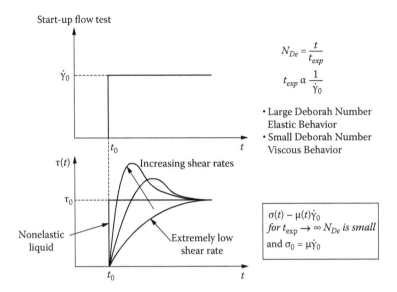

FIGURE 2.12 Schematic diagram illustrating the start-up flow test.

The magnitude of the Deborah number gives an indication of the influence of elasticity on the test. Tests performed under conditions of large Deborah numbers show viscoelastic behavior (e.g., stress overshoot), whereas tests characterized by low Deborah numbers do not show it. It is important to note that the relaxation time, λ, is a material property and viscoelastic foods can have more than one relaxation time, but in the definition of the Deborah number a characteristic relaxation time is considered. Thus, for a given material the magnitude of the Deborah number can be modified if the experimental time is changed. The characteristic relaxation time of nonelastic liquids is negligible, so viscoelastic effects like the one described in this test are not evident (see Figure 2.12 for the behavior of a nonelastic liquid). The larger the characteristic relaxation time, the more evident becomes the overshoot phenomenon. In this test the experimental time can be calculated as $t_{exp} = 1/\dot{\gamma}_0$, so large shear rates imply shorter experimental times, and thus the stress overshoot is more notorious, as schematically illustrated in Figure 2.12.

The start-up flow resembles the flow originated in the mouth cavity when a semi-liquid or semisolid food is consumed, so there have been several attempts to relate results of this test with those related to sensory evaluation of foods (Kokini and Dickie, 1981, 1982; Dickie and Kokini, 1983; Bistany and Kokini, 1983a, 1983b; Kokini and Cussler, 1983).

The test has been used to characterize the properties of semiliquid foods such as mayonnaise (Campanella and Peleg, 1987a), and to study the effect of freezing on the viscoelastic properties of starch pastes gelatinized in the presence of lipids and other ingredients, such as xanthan gum (Navarro et al., 1995). Other applications of the test were to characterize the viscoelastic properties of dextrin systems (Byars et al., 2003) and potential relationships between steady-state shear tests and dynamic viscoelastic measurements (Kokini et al., 1984). A rheological model that combines steady shear data with viscoelastic time-dependent effects present during the start-up of the test is a typical nonlinear model, which is introduced through the use of a damping function (Rolon-Garrido and Wagner, 2009), widely used in the rheology of polymeric plastics.

2.3 AVAILABLE METHODS TO CHARACTERIZE NONELASTIC LIQUID AND SOFT VISCOELASTIC FOODS

2.3.1 INTRODUCTION

Although the term *rheology* encompasses all aspects of deformation and flow of matter, with liquids it is generally accepted that the viscosity of Newtonian liquids and the apparent viscosity at selected shear rates for non-Newtonian non-elastic liquids are of primary importance to characterize these types of food. As described in Section 2.2.2, shear flow is almost solely used to characterize liquid foods, and the significance of the shear viscosity on the rheology of liquids is supported by the large number of commercial instruments available for its mea-surement. One of the most common instruments used by the food industry is the Brookfield™ viscometer, which, like most other instruments that employ a rota-tional shear, imposes a shear rate on the material and records the resulting torque. That torque eventually, and through the knowledge of the rotating geometry, can be transformed in shear stress. More advanced instruments have been introduced in the market during the last twenty years. They are generically called rheometers or mechanical spectrometers because they are able to measure other rheological properties in addition to shear viscosity, in particular viscoelastic properties using methods such as those described in Section 2.2.3. Figure 2.13 shows a classifi-cation of the different commercial instruments used to measure the rheological properties of nonelastic liquids and soft viscoelastic foods using either shear flow or strains. For testing foods, rotational rheometers are more commonly used than capillary viscometers because they can also measure viscoelastic and transient properties like those described in Section 2.2.3, whereas the latter is able to only measure steady shear properties. Extensional properties of foods (described in

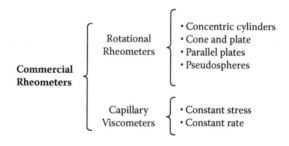

FIGURE 2.13 Classification of commercial viscometers to determine the shear-based rheological properties of foods.

Section 2.4.1) can be determined using capillary viscometry, but uses have been limited. Rotational rheometers have three important advantages over capillary viscometry. First, a sample can be sheared for as long as desired, so changes over time can be followed (Section 2.2.2—time-dependent liquids). Second, with an appropriate design (e.g., cone and plate and pseudosphere geometries, explained below), a nearly uniform shear rate can exist through the sample. Conversely, in a capillary tube the shear rate changes from zero at the center of the tube to a maximum value at the tube wall. Third, as discussed, rotational rheometers can also measure other rheological properties in addition to viscosity. Nevertheless, capillary viscometers do have their advantages; they are generally simpler than rotational theometers; can cover a wider range of shear rates; and because of the design, can perform tests at very high temperatures with fewer problems of sample dehydration. Tests at conditions similar to those found in some food processes, notably extrusion, can be obtained using capillary viscometry (Macosko, 1994; Pai et al., 2009).

The description of the rotational geometries is focused on steady flow calculations where one of the elements of the geometry is rotated at different angular velocities, whereas the torque is measured. Viscoelastic properties are also measured with these rotating geometries, but the elements are moved in an oscillatory manner as described in Section 2.2.3.1. Complete details of calculations, geometries, and assumptions can be found elsewhere in the literature (Ferry, 1980; Macosko, 1994; Steffe, 1996; Bourne, 2002), but for completeness are briefly described below.

2.3.2 CONCENTRIC CYLINDERS GEOMETRY

This geometry is commonly used for liquids of low and medium viscosity. The material is confined between concentric cylinders, one of which rotates (often the inner cylinder) with an angular velocity, Ω, whereas the other cylinder remains stationary. The torque, M, necessary to rotate the moving inner cylinder is measured. This geometry can be used to measure rheological properties of Newtonian and non-Newtonian liquids, but data are worked differently. Details of how to calculate the shear rate and shear stress from the corresponding angular speed and torque are described in many food rheology specialized books (Steffe, 1996; Bourne, 2002). It must be noted that concurrent with new developments in rheological instrumentation has been the incorporation of software that is freeing rheologists from manual tabulation and calculations to get fundamental rheological parameters. However, it is worthwhile to know how data are worked out by the software, and the references mentioned above should be consulted. Because of the difference in radii, the shear stress and the shear rate are not constant throughout the sample, and it is convenient for the calculations to use the shear rate and the shear stress evaluated at the inner rotating cylinder. Since for Newtonian liquids the viscosity does not depend on the shear rate, the nonuniform shear field does not affect the measurements. However, it could cause appreciable error when a non-Newtonian liquid is tested. Another problem associated with this geometry is when liquids with yield stress (plastic materials) are tested. Depending on the shear applied, regions in the gap where the shear stress is smaller than the yield stress can exist, so the liquid moves as a core without having a velocity gradient. For this case, calculations based on a velocity gradient through the gap between the cylinders are invalidated. There are corrections for these cases that are properly described by Steffe (1996). A practical solution to this problem is the use of small gaps between cylinders or high shear rates. A summary of the equations used to calculate shear rates and shear stresses in the concentric cylinders geometry for Newtonian and power law liquids (Equations 2.8 and 2.9) is given in Table 2.2. Details of the geometry and dimensions are also given in the table. It should be noted that the calculation of the shear rate for power law liquids includes the flow index n, which is one of the parameters that must be calculated from the test. That, unfortunately, forces the calculation of the shear rate to be performed in an iterative manner. The other negative aspect of the approach is that the rheological model that describes the liquid may not obey a power law relationship, which would make this approach invalid. When the rheological model describing the rheology of the liquid is unknown, a different approach can be followed. The approach can be also used for power law liquids and Newtonian liquids, and it is based on the following equations:

- From torque (M) measurements the shear stress at the inner cylinder (bob) is calculated as

$$\tau_b = \frac{M}{2\pi R_b^2 h} \tag{2.22}$$

- From measurements of the angular velocity Ω, an abbreviated shear rate at the inner cylinder (bob) is calculated as

$$\dot{\gamma}_b = \frac{\Omega}{\ln \varepsilon}\left(1 + \ln \varepsilon \cdot \frac{d(\log \Omega)}{d(\log \tau_b)}\right) = \frac{\Omega}{\ln \varepsilon}(1 + \ln \varepsilon \cdot m) \tag{2.23}$$

- A more complete equation can be used to estimate the shear rate at the inner cylinder (bob):

$$\dot{\gamma}_b = \frac{\Omega}{\ln \varepsilon}\left(1 + m \cdot \ln \varepsilon + q \cdot \frac{(\ln \varepsilon)^2}{3\Omega}\right) \tag{2.24}$$

where $\eta = (R_c/R_b)$ and $(d(\log \Omega)/d(\log \tau_b)) = m$ can be calculated as the slope of a plot of $\log \Omega$ versus $\log \tau_b$, whereas q is determined as the slope of a $\log m$ versus $\log \tau_b$. For most food and biological materials a plot of $\log \Omega$ versus $\log \tau_b$ gives a straight line, which indicates a power law behavior with a slope $n = 1/m$.

The evaluation of the rheological parameters for the liquid using this approach is as follows:

1. Calculate τ_b using Equation 2.22.
2. Plot $\log \Omega$ versus $\log \tau_b$ and calculate m at different values of τ_b.

TABLE 2.2
Geometries Developed to Measure the Properties of Nonelastic Liquids

Geometry	Shear Rate (1/s)		Shear Stress (Pa)	Schematic
	Newtonian	Power Law	Newtonian and Power Law	
Concentric cylinders	$\dot{\gamma}_b = 2\Omega \dfrac{R_c^2}{R_c^2 - R_b^2}$	$\dot{\gamma}_b = 2\Omega \dfrac{R_c^{2/n}}{R_c^{2/n} - R_b^{2/n}}$	$\tau_b = \dfrac{M}{2\pi R_b^2}$	Ω, M h R_b R_c

Geometry	Shear Rate (1/s)	Shear Stress (Pa)	Schematic
Cone and plate	Newtonian and non-Newtonian liquid $$\dot{\gamma} = \frac{\Omega}{\alpha}$$	Newtonian and non-Newtonian liquid $$\tau = \frac{3M}{2\pi R^3}$$	M Ω R α
Parallel plates	Newtonian and non-Newtonian liquid $$\dot{\gamma}_R = \frac{R\Omega}{h}$$	Newtonian and non-Newtonian liquid $$\tau_R = \frac{3M}{2\pi R^3}\left(1 + \frac{1}{3}\frac{d(\log M)}{d(\log \dot{\gamma}_R)}\right)$$	M Ω R h α
Pseudospheres	Newtonian and non-Newtonian liquid $$\dot{\gamma} = \lambda \frac{\Omega}{h}$$	Newtonian and non-Newtonian liquid $$\tau_R = \frac{M}{\pi R(R_o^2 - R_i^2)}$$	Calvo and Campanella (1990)

3. Calculate $m.\ln\varepsilon$.
   - If $m.\ln\varepsilon < 0.2$, use Equation 2.23.
   - If $m.\ln\varepsilon > 0.2$, use Equation 2.24.
4. Draw flow curve τ_b versus $\dot{\gamma}_b$ and fit a suitable rheological model.

2.3.3 Cone and Plate Geometry

Calculations of shear rate and shear stress are independent of the material rheology, and as they are uniform, it is not necessary to define a region in which to calculate them. That characteristic is advantageous for testing plastic foods having large yield stress. Table 2.2 shows equations to calculate shear rate and shear stress for the cone and plate geometry as well the relevant geometric dimensions utilized in the calculations. The range of shear rates that can be applied with this geometry is limited by the maximum angular speed that can be achieved before spinning the sample out of the geometry. Another limitation is that the geometry is not appropriate for suspensions of large-size particles, in particular if these particles flow to the region of the cone apex.

2.3.4 Parallel Plates Geometry

Equations to calculate shear rate and shear stress along the characteristic geometric constants useful for the calculations are given in Table 2.2. The shear field in this geometry is not uniform, having the shear rate at maximum value at the plate edge and a value of zero at the center of the geometry. In consequence, the shear rate is evaluated at the plate edge. To calculate the shear stress at the plate edge, the torque M and the geometric factors are used in addition to the slope of a log M versus log $\dot{\gamma}_R$ (Table 2.2). Large ranges of shear rates can be obtained by reducing the gap between the plates h at the expense of a nonuniform shear field through the sample.

2.3.5 Pseudosphere Geometry

This geometry exploits advantages of both the cone and plate geometry (uniform shear field) and the parallel plates geometry (adjustable shear rate range by changing the gap). The geometry was tested with Newtonian and non-Newtonian liquids. It can also be used with low-viscosity liquids (Newtonian liquids) by only adjusting the gap h. Very reproducible results were obtained when compared to those obtained with a cone and plate geometry. Simple equations allow the calculation of the shear rate and the shear stress (Table 2.2). More details of this geometry, along with kinematics and dynamics of the generated flow, have been described mathematically utilizing an orthogonal curvilinear coordinate system based on the shapes of the shearing surfaces defined by the parameter λ (Calvo and Campanella, 1990).

2.3.6 CAPILLARY VISCOMETRY

This technique is based on measurements of the pressures necessary to sustain the flow of a liquid through a capillary/tube. Although circular tubes are mostly utilized, other geometries, such as slits, have been used for on-line viscosity determination during extrusion operations due to practical considerations concerning pressure measurements (Li et al., 2004). Capillary viscometry can be used with two different approaches: (1) constant pressure and (2) constant rate. In the constant pressure approach the material is made to flow through a tube, or other geometry, by applying a constant pressure. The shear stress at the tube wall is determined from the applied pressure drop (ΔP) along the length of the capillary and the capillary dimensions, length L and radius R, whereas the shear rate is determined from the measured volumetric flow rate Q and the dimensions of the capillary L and R. For some liquids, notably power law liquids, the power law parameters are necessary to estimate the shear rate. A typical constant-pressure capillary viscometer consists of five components: (1) a test liquid reservoir, (2) a set of capillaries of known dimensions, (3) a means of controlling and measuring the applied pressure (generally done by pressurizing the liquid reservoir with an inert gas like air or nitrogen), (4) a means of measuring mass flow rate (often using a container and stopwatch method), and (5) a means of controlling temperature (e.g., by using a constant-temperature water bath). Mass flow rate is converted to volumetric flow rate by dividing the mass flow rate by the liquid density, ρ, that is, $Q = \dot{m}/\rho$. This type of viscometer can be easily assembled in a laboratory.

The constant-rate capillary viscometer is often used to measure the viscosity of liquids at extremely high shear rates. The reservoir that contains the testing liquid is a relatively large-bore cylinder with the capillary attached to one end. The other end is sealed by a close-fitting piston that can be moved at different speeds. The sample is forced through the capillary at a given rate, and the force required to move the piston is measured by a load cell attached to its end. Other designs have pressure transducers that measure the pressure at the capillary entrance. The working out of the experimental results to obtain the flow curve is the same for both the constant pressure and the constant rate approaches. Data analysis is described in specialized rheology books (Macosko, 1994; Steffe, 1996; Bourne, 2002) and briefly discussed below. The shear stress can be obtained from a balance between pressure forces and friction forces done between the extremes of the tube where the pressures are known. Calculation assumes that the pressure between the entrance and the exit of the capillary varies linearly. That is not the case due to entrance and exit effects, but they are corrected following the method described in several books (Macosko, 1994; Steffe, 1996) and publications (Wang et al., 2009). The shear stress at the capillary wall depends on the measured pressure drop and the dimensions of the capillary (Figure 2.14):

$$\sigma_w = \frac{\Delta p R}{2L}$$

(2.25)

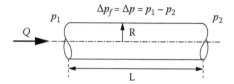

FIGURE 2.14 Schematic of the geometry used to describe capillary viscometers using cylindrical geometries.

The shear rate at the capillary wall, however, requires a more involved calculation because it needs to be calculated from the following equation:

$$\frac{Q}{\pi R^3} = \frac{1}{\sigma_w^3} \int_0^{\sigma_w} \sigma^2 f(\sigma) d\sigma \tag{2.26}$$

the function $f(\sigma)$ is the shear rate at the capillary wall, $\dot{\gamma}_w$.

There exist several approaches to estimate $\dot{\gamma}_w = f(\sigma)$ from Equation 2.26, which involves the inversion of an integral equation, often leading to multiple solutions due to the unavoidable noise that may be present in the experimental data (Berli and Deiber, 2001). The inversion of the integral to get the shear rate at the capillary wall $\dot{\gamma}_w$ is known as an ill-posed problem in rheology (Honerkamp, 1989). A simpler approache is to presume a rheological model for the liquid (i.e., $f(\sigma)$ is assumed) and solve the integral, which results in an explicit equation for the shear rate. This approach has been used successfully for Newtonian and non-Newtonian liquids and plastic foods, including power law, Bingham, and Herschel-Bulkley rheological models. The other approach has been to find an approximate solution of the shear rate from Equation 2.26 using regularization techniques like the Tikhonov regularization method (Yeow et al., 2003), which does not require the assumption of a rheological model. This approach allows one to include effects like wall slip or experimental data that exhibit the two Newtonian plateaus described before (Figure 2.5) at extremely low and high shear rates that are typical for macromolecule solutions (Berli and Deiber, 2001). Capillary viscometry has been largely used in food rheology as a tool to on-line characterize the steady shear viscosity of high-viscosity materials like extrudates (Campanella et al., 2002; Martin et al., 2003; Li et al., 2004) or high-viscosity materials like dough (Bagley et al., 1998). Although capillary viscometry data are often used to determine shear-based properties of liquid foods, pressure losses at the entrance and exit regions of the capillary, i.e., when the flow changes from pure shear flow to extensional flow, have been utilized for measuring the extensional properties of the liquid foods as well (Kwag and Vlachopoulos, 1991; Steffe, 1996).

2.4 OTHER TECHNIQUES

2.4.1 Squeezing Flow

In the methods described in the previous section food's properties are determined by using theoretical rheological models that combine the food rheological equation and the particulars of the sensor's geometry and test conditions. But most of the basic rheological methods that are used in food research were originally developed for synthetic polymeric melts, nonfood emulsions, and suspensions like paints. Their application to foods having a fragile gel structure like yogurt or tomato paste has two serious problems. First, almost without exception, the specimen has to be pressed into the narrow space of the sensor, e.g., the narrow gap of a coaxial viscometer, so mounting the specimen may disrupt or destroy its internal structure, and hence modify its rheological properties. The amount of damage is unknown, and there is always an element of uncertainty as to whether the measurements truly reflect the intact food's properties. The problem is further aggravated when the damage is irreversible, as in yogurt, for example, and hence cannot be reversed even by letting the specimen rest for a long time. Second, many foods exhibit what is known as wall slip due to the tendency of the particles in the shear field to migrate from the wall region to the tube center, leaving the suspension near the wall less concentrated. Hence, what is in contact with the measuring surface is not the original material but a layer of liquid with little or no suspended particles. This is equivalent to the presence of a thin film of a lubricant. The result is a plug flow instead of the expected fully developed shear flow on which the shear stress and shear rate calculations are based. A similar effect is produced when the food is self-lubricating through fat/oil exudation, e.g., peanut butter (Campanella and Peleg, 2002). A feasible solution to these two problems is squeezing flow viscometry, where the specimen, practically intact and with or without suspended particles, is squeezed between parallel plates. The outward flow pattern mainly depends on the friction between the fluid and plates or its absence (lubricated squeezing flow). Among the possible test geometries, the one of constant area and changing volume is the most practical for foods. The test can be performed at a constant displacement rate using common texture analyzers or under constant loads (using a creep array). Modern texture analyzers also allow a compression change with time in a manner that the extensional strain rate is constant (Pai et al., 2009). The test output is in the form of a force-height, force-time, or height-time relationship, from which several rheological parameters can be obtained (Campanella and Peleg, 2002). Despite the method's crudeness, its results are remarkably reproducible and sensitive to textural differences among soft viscoelastic food products. One of the main limitations of the squeezing flow method is that it is only restricted to small displacement rates; consequently, its data may not be useful to engineering design of operations that involve pumping, for example, where the strain rates are much higher. Even with these limitations, though, the method is a practical and inexpensive solution to the above-mentioned serious problems of slip and disturbance of the food fragile structure during handling. The method, as shown in a review by Campanella and Peleg (2002), can also help to solve a variety of problems, for which currently used shear-based viscometers are inadequate despite their high cost and sophisticated design.

The common feature of squeezing flow viscometry, in all its configurations, is the compression of the sample vertically between two plates to induce a horizontal flow. The flow pattern during the fluid squeezing can be classified as lubricated or frictional (Corvalan and Campanella, 2005). The first type of flow, lubricated squeezing flow, is produced when the plates are lubricated intentionally or by the sample itself due to slip. The flow generated is known as a biaxial elongational flow because the sample is stretched in two directions: the radial and azimuthal directions. The second type of flow is induced when there is friction, i.e., there is a good contact between the sample and the plates. This flow is a mix of shear and elongational flow produced by the stretching of the material when it is squeezed out of the plates. Lubricated and frictional squeezing flows have found large application to test a large number of foods that may exhibit self-lubrication. like peanut butter (Campanella and Peleg, 1987b) and melted cheese (Campanella et al., 1987), or that have fragile structure, such as yogurt, custard, mayonnaise (Suwonsichon and Peleg, 1999; Corradini et al., 2001; Raphaelides and Gioldasi, 2005; Janhoj et al., 2006; Terpstra et al., 2007), and tomato products (Lorenzo et al., 1997; Yan et al., 2010). Frictional squeezing flow on plastic materials was used to determine the yield stress of these materials with minimum disturbance of their structure (Campanella and Peleg, 1987c). Pai et al. (2009) used lubricated squeezing flow to study the influence of different modified fibers on the extensional properties of doughs prepared with cornmeal and up to 20% fibers. Extensional flow has a large influence in the expansion of extrudates, and these measurements were correlated to the expansion of extrudates along shear viscosities measured in a capillary viscometer at the extrusion shear and temperature conditions. Campanella and Peleg (2002) described different arrays used in squeezing flow viscometry along with the relevant working equations for each case.

2.4.2 Acoustic/Vibrational Methods

2.4.2.1 Introduction

Although, as discussed previously, many rheological methods can be used to characterize the rheological properties of foods, still there is a need to develop new rapid techniques that can perform precise evaluations of the properties and quality of foods without much disturbance of the samples. A good approach has been to use low-intensity acoustic/vibration waves within a wide range of frequencies that go from audible to ultrasound frequencies. As the energy in those waves is very low, they do not affect the structure of the food material. As they travel through the material, the waves are attenuated and the velocity of the waves changes in response to differences in the material physical properties.

2.4.2.2 Ultrasound Frequencies

Ultrasound is a novel and challenging method that is being developed for use with food products where nondestructive and noninvasive techniques are of great interest. Rheological properties of liquid and semiliquid foods have been determined by using ultrasound. Several approaches have been used. One of them, suitable for liquids, consists in determining, through ultrasound measurements, the velocity profile of the testing liquid while flowing in a tube or a part of an equipment (e.g., in an extruder die). By knowing the liquid velocity profile of the liquid and the geometric characteristics of the tube/equipment, the rheological properties of the material can be determined by methods such as those described in Section 2.3.6. Applications of this approach are described in a number of publications (Ouriev et al., 2003; Young et al., 2008; Wassell et al., 2010). One of the advantages of this approach is that it can be used with opaque liquids.

The other approach consists in measuring the ultrasound wave velocity, v, and the wave attenuation, α, which can be used to determine the rheological properties of the material (Figure 2.15). Two different techniques can be used. In one, the transmission technique, two transducers are used. One transducer is to emit the wave and the other to receive it. Wave velocity can be simply calculated from the sample thickness (path of the wave through the sample) and the time lag between the pulse emission and the pulse received by the receiver. Wave attenuation can be calculated from the decimal logarithm of the ratio between the amplitudes of the emitted and received waves (Figure 2.15). In the pulse echo technique the wave is reflected at the interface between the sample and the sample container and received by the emitter so the distance that the wave travels is twice the thickness of the sample. This technique was used by Kulmyrzaev and McClements (2000) to estimate the dynamic viscosity, $G''/\omega = \eta'$, of honey, with values that were comparable with those determined using classical concentric cylinder geometry. A similar approach was also used by Saggin and Coupland (2004) to measure the dynamic viscosity of xanthan gum containing different amounts of sucrose in a range of 5 to 20°C. Values obtained from the ultrasonic technique were compared with those measured by oscillatory viscometry. Good correlations were obtained among methods, but the numeric agreement was poor. The different frequencies used in both techniques could be one of the reasons for the divergences.

Both shear and longitudinal ultrasonic waves should be used to estimate the shear and bulk storage and loss moduli of the sample. The equations below illustrate how these viscoelastic properties can be calculated from velocity and attenuation measurements using both shear and longitudinal waves.

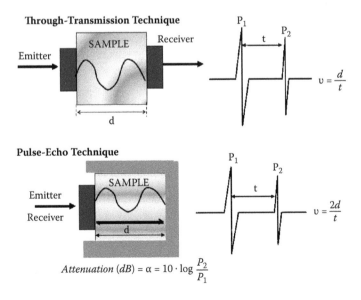

FIGURE 2.15 Different approaches to determine ultrasound velocity and ultrasound attenuation of the food materials with the aim of characterizing their rheological properties.

Elastic in-phase component:

$$M' = K' + \frac{4}{3}G' = \frac{\rho\upsilon^2\left(1 - \dfrac{\alpha^2\upsilon^2}{\omega^2}\right)}{\left(1 + \dfrac{\alpha^2\upsilon^2}{\omega^2}\right)^2}$$

(2.27)

Viscous out-of-phase component:

$$M'' = K'' + \frac{4}{3}G'' = \frac{2\rho\upsilon^2\dfrac{\alpha\upsilon}{\omega}}{\left(1 + \dfrac{\alpha^2\upsilon^2}{\omega^2}\right)^2}$$

(2.28)

where v and α are the velocity and attenuation of the ultrasound longitudinal wave measured during the test (Figure 2.15), ρ is the density of the material, and ω is the frequency of the ultrasound wave. K' and G' are the elastic in-phase bulk and shear moduli, whereas K'' and G'' are the viscous out-of-phase bulk and shear moduli, respectively. The application of a longitudinal wave gives the combined moduli M' and M'' calculated by Equations 2.27 and 2.28, respectively, whereas the application of a shear ultrasound wave gives the values of G' and G''. If both types of waves can be applied, the bulk moduli can be determined. However, shear waves often are highly attenuated, particularly for liquid-like foods, and the shear moduli cannot be determined. For these cases the combined moduli M' and M'' are reported (Lee et al., 2004).

Lee et al. (1992) evaluated the rheological properties of dough using ultrasonic waves and found a good agreement between the ultrasonic technique and conventional rheometry. Letang et al. (2001) studied the effects of the dough water content and mixing time and found that the ultrasonic properties of dough were dependent on frequency, which is the typical behavior of viscoelastic materials. Ultrasonic methods were used to monitor changes in the rheological properties of dough during mixing and fermentation (Lee et al., 2004; Ross et al., 2004). New studies on dough, other cereal foods, and batters using ultrasound dough and compared with classical viscoelastic measurements include the work of Gomez et al. (2008), Mehta et al. (2009), Bellido and Hatcher (2010), and Leroy et al. (2010).

Properties of gels and influence of process conditions on the gelling capacity of proteins and polysaccharide gums have been also studied using ultrasound-based rheological methods (Wang et al., 2005; Ross et al., 2006; Wang et al., 2007).

2.4.2.3 Vibrations at Audible Frequencies

Lower frequencies than those applied in ultrasound-based techniques have also been used to characterize the rheological properties of foods, notably fruits and vegetables. Butz et al. (2005) reviewed noninvasive methods, including rheological and other spectroscopic methods, to evaluate the quality of fresh fruits and vegetables. A recent review focuses mostly on vibration acoustic techniques to determine the quality of fruits and vegetables (Taniwaki and Sakurai, 2010). The technique is based on the generation of a vibration/acoustic impulse through a piezoelectric, a hammer (Wang et al., 2006), or a shaker moving with a controlled amplitude and frequency (Mert and Campanella, 2007). The response signal that moves through the sample can be captured with a microphone, a piezoelectric sensor, or accelerometers that are able to measure the characteristics of the received vibration wave in terms of force/pressure and velocity, whose ratio is defined as the mechanical impedance of the sample (Mert and Campanella, 2008).

Figure 2.16 illustrates a schematic of the principle utilized for the measurement. In this case the vibration, generated with a shaker, moves the can containing the liquid, and a wave is generated through the liquid and reflected back in the interface between the liquid and the headspace, forming standing waves. Properties of the standing waves can be measured in the frequency domain, in particular the resonance frequency and the amplitude of the wave at that resonant frequency, and related to the rheological properties of the liquid. These measurements do not provide the true viscosity of the testing liquid because the walls of the can are not rigid and deform significantly due to the vibration, and the development of theoretical equations to estimate the true viscosity becomes complicated. Despite that, properties of the standing waves were highly correlated with the rheology of the liquid tested off-line in a rheometer (Mert and Campanella, 2007). True measurements of the liquid rheological properties can be determined if the experiments are done in rigid tubes, as shown by Mert et al. (2004). Using a similar approach, the technique was extended to get more fundamental viscoelastic properties of food properties. The technique is known as oscillatory squeezing flow. One of the pioneer works on this technique was done by

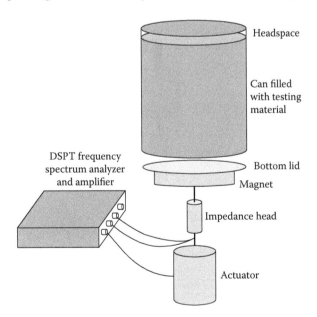

FIGURE 2.16 Schematic illustrating a system designed to measure the viscosity of liquids contained in a closed container using vibrations/acoustics at audible frequencies.

Field et al. (1996). It is based on the squeezing of the sample between a stationary lower plate and an oscillatory moving upper plate. The upper plate is moved by a linear motor that is electronically controlled to avoid overload and instrument inertia, which can be significant when the test is run at high frequencies. The approach used by Mert and Campanella (2008) is different; it consists of an impedance head, a step motor, a piezoelectric actuator, a load cell to monitor the pressure on the sample, and two parallel plates. The step motor provides a very sensitive gap thickness control with the help of a digital micrometer (Figure 2.17). The piezoelectric actuator and impedance head are used for the generation of a random frequency excitation and detection of the resulting force and velocity; the ratio of these Fourier-transformed variables is defined as the mechanical impedance of the sample. The main improvement of the measurement system is the low mass below the force gauge part of the impedance head. This enables the measurement of material properties at high frequencies, which are not possible with current commercial instruments. Physical dimensions and geometries of the top plate can be easily changed depending on the material; for instance, needles and spheres have been used to measure the viscoelastic properties of semisolid materials (Maher et al., 1999). The same technique and a round geometry were utilized by Mert et al. (2007) to determine the viscoelastic properties of corn grits during cooking and drying. Extension of the technique for large strains has been applied to soft materials such as dough and other biological materials (Phan-Thien et al., 2000a, 2000b). If the geometry of the sample is well

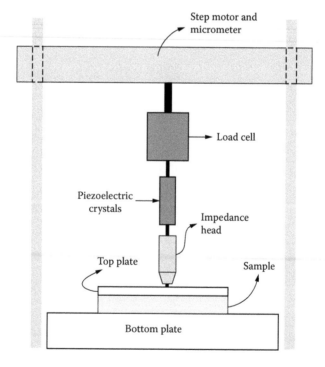

FIGURE 2.17 Schematic of the oscillatory squeezing flow test to measure the viscoelastic properties of soft viscoelastic foods.

defined, analytical equations can be obtained to estimate the storage and the loss moduli of the sample (defined in Section 2.2.3.1) when a shear strain is applied. If perfect lubrication between the plates and the samples can be achieved, the flow created would be extensional (Chatraei et al., 1981), and so the approach could be used to measure the extensional properties of soft viscoelastic materials as well. When the geometry of the sample is not well defined, the technique is still able to provide elastic and viscous components of viscoelastic samples such as corn grits during cooking and drying (Mert et al., 2007). As shown in Figure 2.18, the mechanical properties of the soft viscoelastic material can be described by an elastic component, the sample stiffness s, and a viscous component, the sample damping R. The equation of motion for the force-excited sample-probe system indicated in Figure 2.18 is

$$F(t) = m\dot{u}(t) + Ru(t) + s\int u(t)dt \tag{2.29}$$

where $\dot{u}(t)$, $u(t)$, and $\int u(t)dt$ are the acceleration, velocity, and displacement of the mass m. By Fourier-transforming the above equation to the frequency domain the following equation is obtained:

$$\hat{Z} = \frac{\hat{F}}{\hat{u}} = R + i\left(\omega m - \frac{s}{\omega}\right) \tag{2.30}$$

$\hat{Z}$ is the measured mechanical impedance, which is a complex number, where i is the standard imaginary unit with the property $i^2 = -1$. The inverse of the absolute value of the mechanical impedance, which is a real number defined as mobility, can be plotted as a function of the frequency, which for soft viscoelastic solids yields typical diagrams like the one depicted in Figure 2.19. The importance of these plots is that they contain relevant information to characterize the viscoelasticity of

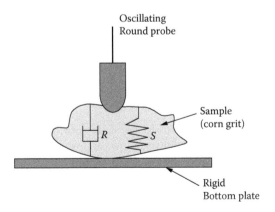

FIGURE 2.18 Schematic of a system used to estimate the viscoelastic properties of a material with an irregular shape (e.g., corn grits).

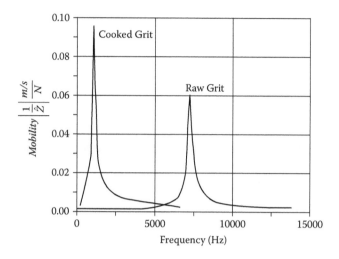

FIGURE 2.19 Typical data resulting from the acoustic technique oscillatory squeezing flow in terms of the function mobility as a function of the applied frequency.

semiliquid and semisolid foods. The frequency at which the resonance peak occurs can be used to calculate the elastic component of the sample s as

$$\omega_{peak} = \sqrt{\frac{s}{m}}$$

(2.31)

whereas the amplitude of the resonance peak can be used to estimate the viscous component R of the sample from the following equation:

$$Mobility = \left|\frac{1}{\hat{Z}}\right| = \frac{1}{R^2 + \left(\omega m - \dfrac{s}{\omega}\right)^2}$$

(2.32)

The technique has been successfully used to measure the glass transition of cereals aged at different conditions (Gonzalez et al., 2010).

2.5 FOOD TEXTURE EVALUATION

Bourne (2002) has described methods to assess the texture of foods and has classified them in two main categories: physical measurements and sensory assessment of food texture. While some of the rheological methods described in previous sections of this chapter are currently used to assess the texture of foods, a description of sensory assessments of foods is out of the scope of this chapter and will not be discussed in detail here. The reader can consult Chapter 7 of Bourne's book, which gives a detailed description of sensory methods of texture. Other useful references that cover this area closely related to methods used in food rheology are Guinard and

Mazzucchelli (1996) and Lawless and Heymann (1999). Although sensory evaluation of foods is an indication of quality, and it is associated to the information received by the consumer five senses—sight, smell, taste, touch, and hearing (Bourne, 2002)—food texture is mainly associated to the rheological and structural attributes of the food that are perceptible by only three senses: tactile, visual, and auditory receptors, with tactile being the more important.

Textural properties of food can be evaluated with the help of psychophysical tests, whereas other, different aspects of food texture can also be measured by physical and instrumental techniques. Thus, sensory assessment performed by psychological tests provides subjective information if these tests are not properly calibrated or performed with a trained panel, while physical measurements provide a more objective assessment of food texture. However, food is consumed by individuals, and ultimately their acceptability is determined by their sensory experience. Thus, there is an increasing interest in correlating physical measurements and sensory assessment of texture for reasons that include economical aspects (instrumental measurements are less expensive than a trained sensory panel); reproducibility of the information over time is better in instruments than among sensory panelists; data analysis of instrumental results is faster than sensory analysis; and ultimately, physical methods provide more quantitative information than sensory data, which tend to be more qualitative.

Visual, touch, and auditory signals are the three components of the sensory experience (Lawless and Heymann, 1999), and to be able to establish correlations between sensory data and physical measurements, rheology researchers should design tests to measure physical properties that are relevant to the sensory stimulus. Visual textures are associated with the condition of the food surface, which in many cases is related to the moistness of the sample. How pourable a liquid food is can be assessed visually by the movement of the liquid, which in turn is related to its viscosity.

The oral tactile texture or mouthfeel of liquids is closely related to its viscosity. There are notable examples that show good correlations between rheological properties, such as viscosity and sensory evaluation of milk, including the perception of fat (Mela, 1988). Correlations between sensory perception of soft viscoelastic semiliquid or semisolid foods and their viscoelastic properties are more difficult to establish given the rate and time dependence of these properties. To get better agreement between sensory and physical results, foods should be tested at strain rates and strains that are comparable with those that are applied when the food is consumed. Another aspect that must be taken into account is that while all rheological testing of viscoelastic materials is performed with geometries and equipment that are built with rigid materials, human tissues involved in the consumption and touching of foods are soft. It has been shown that forces and deformations vary significantly when the tests are performed using soft material fixtures instead of rigid ones (Campanella and Peleg, 1987a; Peleg and Campanella, 1988, 1989). Furthermore, consumption of a food involves several processes, which include deformation, flow, attrition, mixing, and hydration of the food by the saliva (Bourne, 2002), which often are not considered when the food is instrumentally tested. Despite the difficulties, there is a body of literature trying to establish correlations between sensory and physical assessment of texture of soft viscoelastic materials (Carson and Sun, 2001; Wright et al., 2001;

Goh et al., 2003; Barrangou et al., 2006; Tarrega and Costell, 2006; Bhattacharya and Jena, 2007; Sanz et al., 2008; Rogers et al., 2009).

Auditory texture is also an important property in food quality. It is associated with the crispness or crunchiness of the food and measured by the sound generated when the food is consumed. The sound generated could be due to the release of water from turgid cells in wet foods such as fruits and vegetables, whereas in dry foods it may be due to the release of air when the material is fractured. It has been reported that the structure of cellular foods (e.g., crackers, biscuits, and snacks) has a large influence on their textural perception, notably crispness and crunchiness (Duizer, 2001). Research has been conducted to determine the crunchiness and crispness of fruits and vegetables by either sensory evaluation (Fillion and Kilcast, 2002) or analysis of the sound generated by the chewing of these foods (De Belie et al., 2002). Concerning instrumental assessment of crispness and crunchiness, research has clearly shown that crispness and crunchiness of materials are associated with a combination of the force applied to break those materials and the sound generated by the fracture of the material itself (Duizer et al., 1998). Pioneer work on the sensorial assessment of crunchiness and crispness was started by Vickers and Bourne (1976a, 1976b). Dacremont et al. (1991) tried to imitate the attenuation of the sound generated during the consumption of crunchy foods, and Vickers (1991) related the perception of sound during food consumption with food quality. Two acoustic textural properties of foods, crunchiness and crispness, were able to be differentiated by defining some properties of the generated sound (Vickers, 1985). Although crunchiness and crispness of foods have been associated with the sound generated by their fracture during consumption, many studies have also focused on the applied force/stress to break the food (Suwonsichon and Peleg, 1998; Corradini and Peleg, 2006), a few studies on the sound generated during the fracture of the food in the oral cavity (Duizer et al., 1998), and a large body of work on the sound generated when the food is fractured in an instrument like a universal testing machine, a texture analyzer, or biting machines (Tesch et al., 1996a, 1996b; Duizer and Winger, 2006; Maruyama et al., 2008).

Many studies were aimed to correlate force measurement with the sound generated during the food fracture (Chen et al., 2005; Salvador et al., 2009), and others included the sensory evaluation of these foods (Bouvier et al., 1997; Chaunier et al., 2005; Varela et al., 2006). The analysis of the mechanical and sound signatures includes studies in the time and frequency domains, the latter obtained by the use of the Fourier transformation of the signatures in the time domain. The fractal dimension of the mechanical signature (Tesch et al., 1996b) was studied as a function of the water activity or correlated with the sensory assessment of extrudate samples prepared with different degrees of crispness/crunchiness (Duizer et al., 1998).

While most of the research has focused on the testing of the generated sound within an audible range of frequencies (Luyten et al., 2004; Chen et al., 2005) very little research has concentrated on other frequency ranges, specifically those that could include the ultrasound range. There is abundant literature related to the non-invasive characterization of nonfood materials using generated acoustics where the passive acoustic emission (AE) is analyzed to infer the structure of the material. Acoustic emission techniques are often used in the field of material sciences to monitor changes in the internal structure of materials caused by crack/fracture growth, crack opening, phase transformations, fiber breakage, and fiber-matrix debonding. An extensive literature review has shown only an application of AE to test micro-cracking of plant cellular materials (Zdunek and Konstankiewicz, 2004), and limited research has been applied to other foods (AlChakra et al., 1996).

2.6 CONCLUDING REMARKS

Rheological properties of foods cover a wide range of behaviors that require a large number of techniques for their characterization. Knowledge of these properties is very useful for design purposes, quality control and quality assessment, research development, and instrumental determination of food texture. However, defining an instrumental technique and the relevant food rheological properties to be deter-mined requires an analysis of their potential application. Concerning the rheologi-cal information necessary for the design of food processes, it is important to define the type of deformation to which the food material is subjected during the process, as well as the magnitude of the strain and strain rates applied. For amorphous soft viscoelastic materials, temperatures and moisture content are extremely important variables to consider during testing because both affect the glass transition of these foods. Glass transition is defined as either the temperature or the moisture at which a food material transitions from an extremely high-viscosity glassy material to a soft viscoelastic rubbery material (Abiad et al., 2009). Thus, at temperatures or mois-ture contents near the glass transition the rheological properties of the food may vary significantly and affect the consistency and reproducibility of the data if these variables are not well controlled. Rheological information of liquid foods has great value to the design piping and pump systems. Most liquid foods are non-Newtonian shear thinning, so their viscosities are affected by shear rate; thus, it is important to know the magnitude of the shear rate to be applied during the process. Decreases of viscosity with increases of temperature for liquid foods follow an Arrhenius type relationship. For Newtonian liquids the effect of temperature on viscosity can be studied by a straightforward analysis. For non-Newtonian liquids, it is required to define the shear rate at which the temperature changes are studied. Plastic materials exhibiting yield stress are more difficult to characterize, and use of data for process design, for example, piping and pumping systems, is more involved. A good descrip-tion of these applications is well covered by Steffe (1996). Viscoelastic properties of foods are mainly characterized by small strain or stress oscillation tests. They provide information useful to understand the effects of food composition, structure, and interactions between components of the food product quality and texture, but often it is difficult to associate these measurements with process parameters because during processing strain and deformations applied are very large. Small strain oscil-lation tests have also found great use in product development, in particular for foods that are considered soft viscoelastic matter, like gels, desserts, cheeses, and doughs. Ultimately, rheological properties of foods are associated with their texture, and research efforts are focusing on establishing relationships between these properties and texture parameters determined by sensory evaluation. Novel acoustic and ultra-sound techniques have potential to be adapted as on-line and at-line nondestructive and noninvasive techniques for controlling food processes.

REFERENCES

Abiad, M. G., Carvajal, M. T., and Campanella, O. H. 2009. A review on methods and theories to describe the glass transition phenomenon: Applications in food and pharmaceutical products. *Food Engineering Reviews* 1:105–32.

AlChakra, W., Allaf, K., and Jemai, A. B. 1996. Characterization of brittle food products: Application of the acoustical emission method. *Journal of Texture Studies* 27(3):327–48.

Andres, S. C., Zaritzky, N. E., and Califano, A. N. 2008. Stress relaxation characteristics of low-fat chicken sausages made in Argentina. *Meat Science* 79(3):589–94.

Arocas, A., Sanz, T., and Fiszman, S. M. 2009a. Influence of corn starch type in the rheological properties of a white sauce after heating and freezing. *Food Hydrocolloids* 23(3):901–7.

Arocas, A., Sanz, T., and Fiszman, S. M. 2009b. Clean label starches as thickeners in white sauces. Shearing, heating and freeze/thaw stability. *Food Hydrocolloids* 23(8):2031–37.

Bagley, E. B., Dintzis, F. R., and Chakrabarti, S. 1998. Experimental and conceptual problems in the rheological characterization of wheat flour doughs. *Rheologica Acta* 37(6):556–65.

Barnes, H. A. 1997. Thixotropy—A review. *Journal of Non-Newtonian Fluid Mechanics* 70(1):1–33.

Barrangou, L.A., Drake, M.A., Daubert, C.R., and Foegeding, E.A. 2006. Sensory texture related to large-strain rheological properties of agar/glycerol gels as a model food. *Journal of Texture Studies*. 37(3): 241–62

Bellido, G. G., and Hatcher, D. W. 2010. Ultrasonic characterization of fresh yellow alkaline noodles. *Food Research International* 43(3):701–8.

Berli, C. L. A., and Deiber, J. A. 2001. A procedure to determine the viscosity function from experimental data of capillary flow. *Rheologica Acta* 40(3):272–78.

Bhattacharya, S. and Jena, R. 2007. Gelling behavior of defatted soybean flour dispersions due to microwave treatment: Textural, oscillatory, microstructural and sensory properties. *Journal of Food Engineering* 78(4): 1305–14.

Biliaderis, C. G., and Tonogai, J. R. 1991. Influence of lipids on the thermal and mechanical properties of concentrated starch gels. *Journal of Agricultural and Food Chemistry* 39(5):833–40.

Bird, R. B., Armstrong, R. C., and Hassager, O. 1987. *Dynamics of polymeric liquids: Fluid dynamics*. Vol. 1. New York: John Wiley.

Bistany, K. L., and Kokini, J. L. 1983a. Dynamic viscoelastic properties of foods in texture control. *Journal of Rheology* 27(6):605–20.

Bistany, K. L., and Kokini, J. L. 1983b. Comparison of steady shear rheological properties and small amplitude dynamic viscoelastic properties of fluid food materials. *Journal of Texture Studies* 14(2):113–24.

Bourne, M. 2002. *Food texture and viscosity: Concept and measurement*. 2nd ed. New York: Academic Press.

Bouvier, J. M., Bonneville, R., and Goullieux, A. 1997. Instrumental methods for the measurement of extrudate crispness. *Agro Food Industry Hi-Tech* 8(1):16–19.

Butz, P., Hofmann, C., and Tauscher, B. 2005. Recent developments in noninvasive techniques for fresh fruit and vegetable internal quality analysis. *Journal of Food Science* 70(9):R131–41.

Byars, J. A., Carriere, C. J., and Inglett, G. E. 2003. Constitutive modeling of beta-glucan/amylodextrin blends. *Carbohydrate Polymers* 52(3):243–52.

Calvo, N., and Campanella, O. H. 1990. A novel geometry for rheological characterization of viscoelastic materials. *Rheologica Acta* 29(4):323–33.

Campanella, O. H., Li, P. X., Ross, K. A., and Okos, M. R. 2002. The role of rheology in extrusion. In *Engineering and food for the 21st century*, ed. J. Welti Chanes, G. V. Barbosa Canovas, and J. M. Aguilera. Boca Raton, FL: CRC Press, 393–413.

Campanella, O. H., and Peleg, M. 1987a. Analysis of the transient flow of mayonnaise in a coaxial viscometer. *Journal of Rheology* 31(6):439–52.

Campanella, O. H., and Peleg, M. 1987b. Squeezing flow viscosimetry of peanut butter. *Journal of Food Science* 52(1):180–84.

Campanella, O. H., and Peleg, M. 1987c. Determination of the yield stress of semiliquid foods from squeezing flow data. *Journal of Food Science* 52(1):214.

Campanella, O. H., and Peleg, M. 2002. Squeezing flow viscometry for nonelastic semiliquid foods—Theory and applications. *Critical Reviews in Food Science and Nutrition* 42(3):241–64.

Campanella, O. H., Popplewell, L. M., Rosenau, J. R., and Peleg, M. 1987. Elongational viscosity measurements of melting American process cheese. *Journal of Food Science* 52(5):1249–51.

Carson, L., and Sun, X. Z. S. 2001. Creep-recovery of bread and correlation to sensory measurements of textural attributes. *Cereal Chemistry* 78(1):101–4.

Chatraei, S., Macosko, C. W., and Winter, H. H. 1981. Lubricated squeezing flow—A new biaxial extensional rheometer. *Journal of Rheology* 25(4):433–43.

Chaunier, L., Courcoux, P., Della Valle, G., and Lourdin, D. 2005. Physical and sensory evaluation of cornflakes crispness. *Journal of Texture Studies* 36(1):93–118.

Chen, J. S., Karlsson, C., and Povey, M. 2005. Acoustic envelope detector for crispness assessment of biscuits. *Journal of Texture Studies* 36(2):139–56.

Cho, S. A., and Yoo, B. 2010. Comparison of the effect of sugars on the viscoelastic properties of sweet potato starch pastes. *International Journal of Food Science and Technology* 45(2):410–14.

Corradini, M. G., Engel, R., and Peleg, M. 2001. Sensory thresholds of consistency of semiliquid foods: Evaluation by squeezing flow viscometry. *Journal of Texture Studies* 32(2):143–54.

Corradini, M. G., and Peleg, M. 2006. Direction reversals in the mechanical signature of cellular snacks: A measure of brittleness? *Journal of Texture Studies* 37(5):538–52.

Corvalan, C. M., and Campanella, O. H. 2005. Squeezing and elongational flow. In *Food engineering, encyclopedia of life support systems*, 305–22. Norwich, UK: UNESCO Publishing.

Dacremont, C., Colas, B., and Sauvageot, F. 1991. Contribution of air- and bone-conduction to the creation of sounds perceived during sensory evaluation of foods. *Journal of Texture Studies* 22(4):443–56.

De Belie, N., Harker, F. R., and De Baerdemaeker, J. 2002. Crispness judgement of royal gala apples based on chewing sounds. *Biosystems Engineering* 81(3):297–303.

Del Nobile, M. A., Chillo, S., Mentana, A., and Baiano, A. 2007. Use of the generalized Maxwell model for describing the stress relaxation behavior of solid-like foods. *Journal of Food Engineering* 78(3):978–83.

Dickie, A. M., and Kokini, J. L. 1983. An improved model for food thickness from non-Newtonian fluid-mechanics in the mouth. *Journal of Food Science* 48(1):57.

Diez-Sales, O., Dolz, M., Hernandez, M. J., Casanovas, A., and Herraez, M. 2007. Rheological characterization of chitosan matrices: Influence of biopolymer concentration. *Journal of Applied Polymer Science* 105(4):2121–28.

Dolores Alvarez, M., Fernandez, C., and Canet, W. 2009. Enhancement of freezing stability in mashed potatoes by the incorporation of kappa-carrageenan and xanthan gum blends. *Journal of the Science of Food and Agriculture* 89(12):2115–27.

Dolz, M., Hernandez, M. J., and Delegido, J. 2006. Oscillatory measurements for salad dressings stabilized with modified starch, xanthan gum, and locust bean gum. *Journal of Applied Polymer Science* 102(1):897–903.

Duizer, L. 2001. A review of acoustic research for studying the sensory perception of crisp, crunchy and crackly textures. *Trends in Food Science and Technology* 12:17–24.

Duizer, L. M., Campanella, O. H., and Barnes, G. R. G. 1998. Sensory, instrumental and acoustic characteristics of extruded snack products. *Journal of Texture Studies* 29:397–411.

Duizer, L. M., and Winger, R. J. 2006. Instrumental measures of bite forces associated with crisp products. *Journal of Texture Studies* 37(1):1–15.

Eleya, M. M. O., and Turgeon, S. L. 2000. Rheology of kappa-carrageenan and beta-lactoglobulin mixed gels. *Food Hydrocolloids* 14(1):29–40.

Ferry, J. D. 1980. *Viscoelastic properties of polymers*. New York: John Wiley & Sons.

Field, J. S., Swain, M. V., and Phan-Thien, N. 1996. An experimental investigation of the use of random squeezing to determine the complex modulus of viscoelastic fluids. *Journal of Non-Newtonian Fluid Mechanics* 65:177–94.

Fillion, L., and Kilcast, D. 2002. Consumer perception of crispness and crunchiness in fruits and vegetables. *Food Quality and Preference* 13(1):23–29.

Goh, S. M., Charalambides, M. N., and Williams, J. G. 2003. Mechanical properties and sensory texture assessment of cheeses. *Journal of Texture Studies* 34(2):181–201.

Gomez, M., Oliete, B., Garcia-Alvarez, J., Ronda, F., and Salazar, J. 2008. Characterization of cake batters by ultrasound measurements. *Journal of Food Engineering* 89(4):408–13.

Gonzalez, D. C., Khalef, N., Wrigh, K., Okos, M. R., Hamaker, B. R., and Campanella, O. H. 2010. Physical aging of processed fragmented biopolymers. *Journal of Food Engineering* 100(2):187–93.

Guinard, J. X., and Mazzuchelli, R. 1996. The sensory perception of texture and mouthfeel. *Trends in Food Science and Technology* 7:213–19.

Gunasekaran, S., and Ak, M. M. 2000. Dynamic oscillatory shear testing of foods—Selected applications. *Food Science and Technology* 11(3):115–27.

Han, X. Z., Campanella, O. H., Guan, H. P., Keeling, P. L., and Hamaker, B. R. 2002. Influence of maize starch granule-associated protein on the rheological properties of starch pastes. Part I. Large deformation measurements of paste properties. *Carbohydrate Polymers* 49(3):315–21.

Holm, K., Wendin, K., and Hermansson, A. 2009. Sweetness and texture perception in mixed pectin gels with 30% sugar and a designed rheology. *LWT-Food Science and Technology* 42(3):788–95.

Honerkamp, J. 1989. Ill-posed problems in rheology. *Rheologica Acta* 28:263–371.

Janhoj, T., Petersen, C. B., Frost, M. B., and Ipsen, R. 2006. Sensory and rheological characterization of low-fat stirred yogurt. *Journal of Texture Studies* 37(3):276–99.

Joshi, N. S., Jhala, R. P., Muthukumarappan, K., Acharya, M. R., and Mistry, V. V. 2004. Textural and rheological properties of processed cheese. *International Journal of Food Properties* 7(3):519–30.

Kampf, N., and Nussinovitch, A. 1997. Rheological characterization of kappa-carrageenan soy milk gels. *Food Hydrocolloids* 11(3):261–69.

Kapri, A., and Bhattacharya, S. 2008. Gelling behavior of rice flour dispersions at different concentrations of solids and time of heating. *Journal of Texture Studies* 39:231–51.

Kaur, L., Singh, J., Singh, H., and McCarthy, O. J. 2008. Starch-cassia gum interactions: A microstructure—Rheology study. *Food Chemistry* 111:1–10.

Kokini, J. L., Bistany, K. L., and Mills, P. L. 1984. Guar and carrageenan using the Bird-Carreau constitutive model. *Journal of Food Science* 49(6):1569–71.

Kokini, J. L., and Cussler, E. L. 1983. Predicting the texture of liquids and melting semisolid food. *Journal of Food Science* 48:1221–25.

Kokini, J. L., and Dickie, A. 1981. An attempt to identify and model transient viscoelastic flow in foods. *Journal of Texture Studies* 12(4):539–57.

Kokini, J. L., and Dickie, A. 1982. A model of food spreadability from fluid-mechanics. *Journal of Texture Studies* 13(2):211–27.

Kulmyrzaev, A., and McClements, D. J. 2000. High frequency dynamic shear rheology of honey. *Journal of Food Engineering* 45(4):219–24.

Kwag, C., and Vlachopoulos, J. 1991. An assessment of Cogswell method for measurement of extensional viscosity. *Polymer Engineering and Science* 31(14):1015–21.

Lawless, H. T., and Heymann, H. 1999. *Sensory evaluation of food: Principles and practices.* Kluwer Academic Publishers: Dordrecht.

Lee, H. O., Luan, H. C., and Daut, D. G. 1992. Use of an ultrasonic technique to evaluate the rheological properties of cheese and dough. *Journal of Food Engineering* 16(1–2):127–50.

Lee, S. Y., Pyrak-Nolte, L. J., and Campanella, O. 2004. Determination of ultrasonic-based rheological properties of dough during fermentation. *Journal of Texture Studies* 35(1):33–51.

Leroy, V., Pitura, K. M., Scanlon, M. G., and Page, J. H. 2010. The complex shear modulus of dough over a wide frequency range. *Journal of Non-Newtonian Fluid Mechanics* 165(9–10):475–78.

Li, P. X., Campanella, O. H., and Hardacre, A. K. 2004. Using an in-line slit-die viscometer to study the effects of extrusion parameters on corn melt rheology. *Cereal Chemistry* 81(1):70–76.

Lorenzo, M. A., Gerhards, C., and Peleg, M. 1997. Imperfect squeezing flow viscosimetry of selected tomato products. *Journal of Texture Studies* 28(5):543–67.

Luyten, H., Lichtendonk, W., Castro, E.M., Visser, J. and van Vliet, T. 2004. Acoustic emission from crispy/crunchy foods to link mechanical properties and sensory perception. In *Food Colloids, Interactions, Microstructure and Processing* (E. Dickinson, ed.) Royal Soc. of Chemistry, Cambridge. 380–92.

Macosko, C. W. 1994. *Rheology, principles, measurements and applications.* New York: VCH Publishers.

Maher, A. M., Field, J., Pfister, B., Swain, M., and Phan-Thien, N. 1999. Measurement of the viscoelastic properties of bituminous materials using an oscillating needle technique. *Rheologica Acta* 38(5):443–50.

Mancini, F., Montanari, L., Peressini, D., and Fantozzi, P. 2002. Influence of alginate concentration and molecular weight on functional properties of mayonnaise. *Food Science and Technology* 35(6):517–25.

Mariotti, M., Sinelli, N., Catenacci, F., Pagani, M. A., and Lucisano, M. 2009. Retrogradation behaviour of milled and brown rice pastes during ageing. *Journal of Cereal Science* 49(2):171–77.

Markowski, M., Ratajski, A., Konopko, H., Zapotoczny, P., and Majewska, K. 2006. Rheological behavior of hot-air-puffed amaranth seeds. *International Journal of Food Properties* 9(2):195–203.

Martin, O., Averous, L., and Della Valle, G. 2003. In-line determination of plasticized wheat starch viscoelastic behavior: Impact of processing. *Carbohydrate Polymers* 53(2):169–82.

Maruyama, T. T., Arce, A. I. C., Ribeiro, L. P., and Costa, E. J. X. 2008. Time-frequency analysis of acoustic noise produced by breaking of crisp biscuits. *Journal of Food Engineering* 86(1):100–4.

Matalanis, A. M., Campanella, O. H., and Hamaker, B. R. 2009. Storage retrogradation behavior of sorghum, maize and rice starch pastes related to amylopectin fine structure. *Journal of Cereal Science* 50(1):74–81.

Mehta, K. L., Scanlon, M. G., Sapirstein, H. D., and Page, J. H. 2009. Ultrasonic investigation of the effect of vegetable shortening and mixing time on the mechanical properties of bread dough. *Journal of Food Science* 74(9):E455–61.

Mela, D. 1988. Sensory assessment of fat content in fluid dairy products. *Appetite* 10(1):37–44.

Mert, B., and Campanella, O. H. 2007. Monitoring the rheological properties and solid content of selected food materials contained in cylindrical cans using audio frequency sound waves. *Journal of Food Engineering* 79(2):546–52.

Mert, B., and Campanella, O. H. 2008. The study of the mechanical impedance of foods and biomaterials to characterize their linear viscoelastic behavior at high frequencies. *Rheologica Acta* 47(7):727–37.

Mert, B., Gonzalez, D., and Campanella, O. H. 2007. A new method to determine viscoelastic properties of corn grits during cooking and drying. *Journal of Cereal Science* 46(1):32–38.

Mert, B., Sumali, H., and Campanella, O. H. 2004. A new method to measure viscosity and intrinsic sound velocity of liquids using impedance tube principles at sonic frequencies. *Review of Scientific Instruments* 75(8):2613–19.

Morrison, F. A. 2001. *Understanding rheology*. New York: Oxford University Press.

Navarro, A. S., Martino, M. N., and Zaritzky, N. E. 1995. Effect of freezing rate on the rheological behavior of systems based on starch and lipid phase. *Journal of Food Engineering* 26(4):481–95.

Olivares, M. L., Zorrilla, S. E., and Rubiolo, A. C. 2009. Rheological properties of mozzarella cheese determined by creep/recovery tests: Effect of sampling direction, test temperature and ripening time. *Journal of Texture Studies* 40(3):300–18.

Ouriev, B., Windhab, E., Braun, P., Zeng, Y., and Birkhofer, B. 2003. Industrial application of ultrasound based in-line rheometry: Visualization of steady shear pipe flow of chocolate suspension in pre-crystallization process. *Review of Scientific Instruments* 74(12):5255–59.

Pai, D. A., Blake, O. A., Hamaker, B. R., and Campanella, O. H. 2009. Importance of extensional rheological properties on fiber-enriched corn extrudates. *Journal of Cereal Science* 50(2):227–34.

Peleg, M. 1980. Linearization of relaxation and creep curves of solid biological materials. *Journal of Rheology* 24(4):451–63.

Peleg, M., and Campanella, O. H. 1988. On the mathematical form of psychophysical relationships, with special focus on the perception of mechanical-properties of solid objects. *Perception & Psychophysics* 44(5):451–55.

Peleg, M., and Campanella, O. H. 1989. The mechanical sensitivity of soft compressible testing machines. *Journal of Rheology* 33(3):455–67.

Peleg, M., and Normand, M. D. 1983. Comparison of 2 methods for stress-relaxation data presentation of solid foods. *Rheologica Acta* 22(1):108–13.

Pereira, R. B., Bennett, R. J., Hemar, Y., and Campanella, O. H. 2001. Rheological and microstructural characteristics of model processed cheese analogues. *Journal of Texture Studies* 32(5–6):349–73.

Phan-Thien, N., Nasseri, S., and Bilston, L. E. 2000a. Oscillatory squeezing flow of a biological material. *Rheologica Acta* 39(4):409–17.

Phan-Thien, N., Newberry, M., and Tanner, R. I. 2000b. Non-linear oscillatory flow of a soft solid-like viscoelastic material. *Journal of Non-Newtonian Fluid Mechanics* 92(1):67–80.

Prokopowich, D. J., and Biliaderis, C. G. 1995. A comparative-study of the effect of sugars on the thermal and mechanical properties of concentrated waxy maize, wheat, potato and pea starch gels. *Food Chemistry* 52(3):255–62.

Purkayastha, S., Peleg, M., Johnson, E. A., and Normand, M. D. 1985. A computer-aided characterization of the compressive creep-behavior of potato and cheddar cheese. *Journal of Food Science* 50(1):45–47.

Purkayastha, S., Peleg, M., and Normand, M. D. 1984. Presentation of the creep curves of solid biological materials by a simplified mathematical version of the generalized Kelvin-Voigt model. *Rheologica Acta* 23(5):556–63.

Quemada, D. 1999. Rheological modelling of complex fluids. IV. Thixotropic and "thixoelastic" behaviour. Start-up and stress relaxation, creep tests and hysteresis cycles. *European Physical Journal—Applied Physics* 5(2):191–207.

Quintana, J. M., Califano, A. N., Zaritzky, N. E., Partal, P., and Franco, J. M. 2002. Linear and nonlinear viscoelastic behavior of oil-in-water emulsions stabilized with polysaccharides. *Journal of Texture Studies* 33(3):215–36.

Rao, M. A. 1999. *Rheology of fluid and semisolid foods.* Gaithersburg, MD: Chapman & Hall.

Raphaelides, S. N., and Gioldasi, A. 2005. Elongational flow studies of set yogurt. *Journal of Food Engineering* 70(4):538–45.

Rogers, N. R., Drake, M. A., Daubert, C., McMahon, D. J., Bletsch, T. K., and Foegeding, E. A. 2009. The effect of aging on low-fat, reduced-fat, and full-fat cheddar cheese texture. *Journal of Dairy Science* 92(10):4756–72.

Rolon-Garrido, V. H., and Wagner, M. H. 2009. The damping function in rheology. *Rheologica Acta* 48(3):245–84.

Ross, K. A., Pyrak-Nolte, L. J., and Campanella, O. H. 2004. The use of ultrasound and shear oscillatory tests to characterize the effect of mixing time on the rheological properties of dough. *Food Research International* 37(6):567–77.

Ross, K. A., Pyrak-Nolte, L. J., and Campanella, O. H. 2006. The effect of mixing conditions on the material properties of an agar gel—Microstructural and macrostructural considerations. *Food Hydrocolloids* 20(1):79–87.

Saggin, R., and Coupland, J. N. 2004. Rheology of xanthan/sucrose mixtures at ultrasonic frequencies. *Journal of Food Engineering* 65(1):49–53.

Salvador, A., Varela, P., Sanz, T., and Fiszman, S. M. 2009. Understanding potato chips' crispy texture by simultaneous fracture and acoustic measurements, and sensory analysis. *LWT-Food Science and Technology* 42(3):763–67.

Sanz, T., Salvador, A., Jimenez, A., and Fiszman, S. M. 2008. Yogurt enrichment with functional asparagus fibre. Effect of fibre extraction method on rheological properties, colour, and sensory acceptance. *European Food Research and Technology* 227(5):1515–21.

Song, K. W., Kuk, H. Y., and Chang, G. S. 2006. Rheology of concentrated xanthan gum solutions: Oscillatory shear flow behavior. *Korea-Australia Rheology Journal* 18(2):67–81.

Steffe, J. F. 1996. *Rheological methods in food process engineering.* 2nd ed. East Lansing, MI: Freeman Press.

Stern, P., Mikova, K., Pokorny, J., and Valentova, H. 2007. Effect of oil content on the rheological and textural properties of mayonnaise. *Journal of Food and Nutrition Research* 46(1):1–8.

Suwonsichon, T., and Peleg, M. 1998. Instrumental and sensory detection of simultaneous brittleness loss and moisture toughening in three puffed cereals. *Journal of Texture Studies* 29(3):255–74.

Suwonsichon, T., and Peleg, M. 1999. Rheological characterization of almost intact and stirred yogurt by imperfect squeezing flow viscometry. *Journal of the Science of Food and Agriculture* 79(6):911–21.

Tang, Q. N., Munro, P. A., and McCarthy, O. J. 1993. Rheology of whey-protein concentrate solutions as a function of concentration, temperature, pH and salt concentration. *Journal of Dairy Research* 60(3):349–61.

Taniwaki, M., and Sakurai, N. 2010. Evaluation of the internal quality of agricultural products using acoustic vibration techniques. *Journal of the Japanese Society for Horticultural Science* 79(2):113–28.

Tarrega, A., and Costell, E. 2006. Effect of composition on the rheological behaviour and sensory properties of semisolid dairy dessert. *Food Hydrocolloids* 20(6):914–22.

Terpstra, M. E. J., Janssen, A. M., and van der Linden, E. 2007. Exploring imperfect squeezing flow measurements in a Teflon geometry for semisolid foods. *Journal of Food Science* 72(9):E492–502.

Tesch, R., Normand, M. D., and Peleg, M. 1996a. Comparison of the acoustic and mechanical signatures of two cellular crunchy cereal foods at various water activity levels. *Journal of the Science of Food and Agriculture* 70(3):347–54.

Tesch, R., Normand, M. D., and Peleg, M. 1996b. On the apparent fractal dimension of sound bursts in acoustic signatures of two crunchy foods. *Journal of Texture Studies* 26(6):685–94.

van Soest, J. J. G., Benes, K., deWit, D., and Vliegenthart, J. F. G. 1996. The influence of starch molecular mass on the properties of extruded thermoplastic starch. *Polymer* 37(16):3543–52.

Varela, P., Chen, J., Fiszman, S., and Povey, M. J. W. 2006. Crispness assessment of roasted almonds by an integrated approach to texture description: Texture, acoustics, sensory and structure. *Journal of Chemometrics* 20(6–7):311–20.

Vickers, Z., and Bourne, M. C. 1976a. Crispness in foods—A review. *Journal of Food Science* 41(5):1153–57.

Vickers, Z. M., and Bourne, M. C. 1976b. A psychoacoustical theory of crispness. *Journal of Food Science* 41(5):1158–64.

Vickers, Z. M. 1985. Crispness and crunchiness—A difference in pitch? *Journal of Texture Studies* 15:157–63.

Vickers, Z. M. 1991. Sound perception and food quality. *Journal of Food Quality* 14(1):87–96.

Wang, B., Wang, L. J., Li, D., Bhandari, B., Wu, W. F., Shi, J., Chen, X. D., and Mao, Z. H. 2009. Effects of potato starch addition and cooling rate on rheological characteristics of flaxseed protein concentrate. *Journal of Food Engineering* 91(3):392–401.

Wang, J., Hernández Gómez, A., and Garcia Pereira, A. 2006. Acoustic impulse response for measuring the firmness of mandarin during storage. *Journal of Food Quality* 29:392–404.

Wang, L., and Zhang, L. M. 2009. Viscoelastic characterization of a new guar gum derivative containing anionic carboxymethyl and cationic 2-hydroxy-3-(trimethylammonio) propyl substituents. *Industrial Crops and Products* 29(2–3):524–29.

Wang, Q., Bulca, S., and Kulozik, U. 2007. A comparison of low-intensity ultrasound and oscillating rheology to assess the renneting properties of casein solutions after UHT heat pre-treatment. *International Dairy Journal* 17(1):50–58.

Wang, Q., Rademacher, B., Sedlmeyer, F., and Kulozik, U. 2005. Gelation behaviour of aqueous solutions of different types of carrageenan investigated by low-intensity-ultrasound measurements and comparison to rheological measurements. *Innovative Food Science & Emerging Technologies* 6(4):465–72.

Wassell, P., Wiklund, J., Stading, M., Bonwick, G., Smith, C., Almiron-Roig, E., and Young, N. W. G. 2010. Ultrasound Doppler based in-line viscosity and solid fat profile measurement of fat blends. *International Journal of Food Science and Technology* 45(5):877–83.

Wright, A. J., Scanlon, M. G., Hartel, R. W., and Marangoni, A. G. 2001. Rheological properties of milkfat and butter. *Journal of Food Science* 66(8):1056–71.

Xu, Y. L., Xiong, S. B., Li, Y. B., and Zhao, S. M. 2008. Study on creep properties of indica rice gel. *Journal of Food Engineering* 86(1):10–16.

Yan, Y., Zhang, Z., Cheneler, D., Stokes, J. R., and Adams, M. J. 2010. The influence of flow confinement on the rheological properties of complex fluids. *Rheologica Acta* 49(3):255–66.

Yang, X. H., and Zhu, W. L. 2006. Modeling the rheological properties of the solution of konjac glucomannan and milk powder. *Food Science and Technology International* 12(2):127–32.

Yeow, Y. L., Lee, H. L., Melvani, A. R., and Mifsud, G. C. 2003. New method of processing capillary viscometry data in the presence of wall slip. *Journal of Rheology* 47(2):337–48.

Young, N. W. G., Wassell, P., Wiklund, J., and Stading, M. 2008. Monitoring structurants of fat blends with ultrasound based in-line rheometry (ultrasonic velocity profiling with pressure difference). *International Journal of Food Science and Technology* 43(11):2083–89.

Yuan, R. C., and Thompson, D. B. 1998. Rheological and thermal properties of aged starch pastes from three waxy maize genotypes. *Cereal Chemistry* 75(1):117–23.

Zdunek, A., and Konstankiewicz, K. 2004. Acoustic emission in investigation of plant tissue micro-cracking. *Transactions of the ASAE* 47(4):1171–77.

3 Sensory Evaluation Using Artificial Intelligence

Seung Ju Lee

CONTENTS

3.1 INTRODUCTION

Implementing a food business consists of several factors, such as consumer surveys, product development, quality control, marketing, etc. The success of a food business is eventually determined by whether or not consumers buy the food being sold. So

no matter how high quality a food is in terms of nutritional and functional properties, storage stability, safety, and so forth, the key issue is how consumers respond
to the food product. A scientific approach has been utilized to measure consumers'
responses. It encompasses sensory evaluation in which food qualities are tested by
human subjects or consumers according to a well-established experimental design,
and the sensory data are statistically analyzed. Sensory evaluations, therefore, are
being applied to quality control (QC) and product development in the industrial
field.

Sensory tests require subjects to score the intensity or preference for sensory
attributes such as taste, odor, mastication feel, chewing sound, and appearance on
a questionnaire, or just verbally on site. Dating from the late nineteenth century,
sensory evaluation has evolved to be incorporated into such scientific fields as
psychology, physics, dentistry, and biology (Moskowitz et al., 2006). Even *in vitro*
experiments have applied sensory analysis by using animal tongues (Settles and
Mierson, 1993).

Major food attributes on sensory tests include texture, appearance, taste, and
flavor, and these sensorial data are treated with more logical methods due to their
uncertainty and imprecision. The responses of subjects can be rather subjective
and inconsistent, depending on their testing ability or prejudice, testing environment, and any other factors causing disturbance. Sensory attributes can be characterized in a more systematic manner by a group of professional panelists in
quantitative descriptive analysis (QDA) (Meilgaard et al., 1999). Once the target
attributes are established, their intensities are analyzed. The integration of each
panelist's data ends up being the mean (average) or the total number of votes of
yes or no, which are further treated with statistical methods such as the F-test,
χ^2-test, and multiple comparison, on the basis of analysis of variance (Lawless
and Heymann, 1999).

The attribute intensity data are generally described in scores or a mark on a line
by selecting a previously determined scale or stage. Recently, there have been several attempts to transform this quantification into fuzzy sets, which are a scale much
closer to human expression (Lee, 2000). The attribute intensities can be expressed
as a fuzzy set to reflect human, vague judgments rather than a deterministic value.
The integration of each panelist's fuzzy data is manipulated by fuzzy mathematics (Ruan and Zeng, 2004). The resultant fuzzy sets, in turn, can be converted to a
deterministic value.

Physical and chemical properties associated with sensory attributes are also
measured by such instruments and means as texture analyzers, gas chromatography
(GC), GC-olfactometry (GCO), high-performance liquid chromatography (HPLC),
electronic noses, electronic tongs, etc. (Meullenet and Sitakalin, 2001; Peris and
Escuder-Gilabert, 2009). Electrical signals or measures from the instruments are
directly read or statistically treated further. The instrumental measures represent
only the differences in sensory attribute intensities, and do not provide consumers' likes or dislikes or human expression. Although instrumental analyses don't
reflect responses directly from consumers, they are commonly used in the industrial
field because their results are more objective and consistent than those from sensory
evaluations. There have only been a few attempts to develop mathematical functions

to predict sensory data from instrumental data by correlating them with multivariate regressions and artificial neural networks (ANNs) (Bomio, 1998). ANN is a well-known tool to manage any kind of nonlinear problem, like sensory data.

In this chapter, the applications of artificial intelligence techniques such as fuzzy reasoning and ANN to food sensory evaluation are discussed. Fuzzy theory is initially dealt with as a prerequisite to help understand the sections that follow. The definitions of fuzzy sets, fuzzy calculus, and fuzzy reasoning algorithms are covered. Then, applications follow to facilitate practical adaptation. ANN is also included, and the basic principles of ANN and its applications to sensory data are discussed.

3.2 FUZZY THEORY

3.2.1 Introduction

Sciences associated with humans have been developed based on language and letters, and transmitted historically. Unless there is no numerical expression, especially in science, engineering, and economics, there cannot be a current monument of success due to the lack of such parametric media. Quantification of any data around us, therefore, has been attempted even in the food science field. Linguistic types of data are difficult to convert to numerical values. To overcome such barriers to the quantification of human expression, fuzzy sets and their mathematical manipulation algorithms have been introduced and are successfully being applied in relevant fields.

Fuzzy set theory was proposed in 1965 by Professor Zadeh of the University of California in Los Angeles, whose major area of study was automatic control (Stefanini et al., 2006). He realized that physically established control algorithms at that time could not be effectively adapted to reflect human expert control knowledge in the industrial field. He finally created a new theory that could cover fuzzy variables with uncertainty and imprecision, like linguistic expressions and their relation. Eventually, he made a more practical control system just by transplanting relevant expert knowledge to a control algorithm, using fuzzy theory.

Recently, many successful cases of fuzzy set applications can be found, such as pattern recognition, classification, sensory evaluation, fuzzy control for electric home appliances, fermentation, chemical plants, elevators, automobiles, even fuzzy expert systems for stocks, etc. In the field of food science, several research papers have been published on food process control and food sensory evaluation (Lee et al., 1994, 1995, 2002; Angerosa et al., 1996; Kupongsak and Tan, 2006; Lee and Kwon, 2007; Zeng et al., 2008). Still, however, the understanding and application of fuzzy theory remains insufficient in the area of food.

In most food problems, an experimental approach is more common than a theoretical approach. The experimental results usually come out to be a database with chemical, physical, and biological data, or expert knowledge with relevant principles. They are provided as either crisp data or approximate data, so fuzzy theory would be worth applying to them. In the next section, the basic concepts needed to understand fuzzy theory and further knowledge on how to apply it to practical sensory evaluation are discussed.

3.2.2   What Is a Fuzzy Set?

First, consider how to express variables. In the case that you are asked to describe the optimum temperature for a particular fermentation, you might be discomforted that the temperature condition is usually an approximate number rather than a precise number. If around 36°C is the known condition, it may be perplexing which temperature to choose, such as 36.2, 36.4, or 35.9°C. Around 36°C can be described in a crisp set such as {35, 36, 37} or {34, 35, 36, 37, 38}. But the crisp set cannot provide that 36 is the prime element in the set. To supplement this defect, the concept of a fuzzy set was devised.

The word *fuzzy* means vagueness or uncertainty, and is basically compared to the word *deterministic*. A fuzzy set is created by adding a membership degree to each element in a crisp set. Around 36°C can now be expressed well in a fuzzy set. That is, the crisp set {35, 36, 37} is converted to a fuzzy set {0.5/35, 1/36, 0.5/37}, where 0.5, 1, and 0.5 are membership degrees for the elements 35, 36, and 37, respectively. Membership degrees range from 1 to 0.

It is ensured that fuzzy sets facilitate converting linguistic expressions with vagueness or uncertainty to mathematical data. It is more reasonable to provide in fuzzy sets food qualities such as sensory, chemical, physical, and biological properties. This is because foods are natural materials with varying compositions even when from the same kind of source. And in sensory evaluations, human subjects' judgments are likely to be subjective, inconsistent, and prejudicial.

Fuzzy sets are denoted in several ways. The first is with discrete elements such as the following:

$$A = \{(\mu_{A,\,i},\, x_i)\},\ \forall x \in X\ (i = 1, 2, 3, \ldots, n) \tag{3.1}$$

$$A = \mu_{A,\,1}/x_1 + \mu_{A,\,2}/x_2 + \ldots + \mu_{A,n}/x_n\,(i = 1, 2, 3, \ldots, n) \tag{3.2}$$

where x_i and $\mu_{A,i}$ are the set elements and their membership degrees, respectively, and in the second equation, plus (+) does not mean arithmetic addition. Second, fuzzy sets are denoted with continuous elements as follows:

$$A = \{(\mu_A(x),\, x)\},\ \forall x \in X \tag{3.3}$$

where $\mu_A(x)$ is the membership function and x is the element in continuous variables. Membership functions are typically given in three types: triangle, trapezoid, and bell shapes (Figure 3.1).

When designating fuzzy set partitions, the shape and location of the partitions should be carefully determined. Fuzzy sets with wide partitions imply more fuzziness, or uncertainty, and those with narrow partitions have a lesser degree of fuzziness. A fuzzy set with a membership degree of 1 at one element only and 0 at the other elements is equivalent to a deterministic value, and that with a membership degree of 1 at all the elements becomes a crisp set.

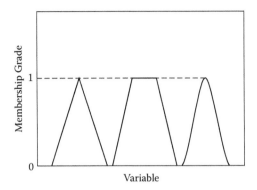

FIGURE 3.1   Patterns of membership functions for fuzzy set.

3.2.3 FUZZY RELATION

When there is a phenomenon that is described in a sentence with several words, its meaning is determined by the relationship of the words. The relationship can be described by a rule between the relevant fuzzy sets (words). For instance, in the case that a microorganism grows rapidly at about 36°C, a fuzzy rule can be built as follows: if the temperature is about 36°C, then the microorganism grows very fast, where the fuzzy sets are temperature and the microbial growth rate. Almost certainly most of the phenomena can be expressed by fuzzy rules or relations with fuzzy sets.

Fuzzy relations are denoted as if-then rules. In a sense, the fuzzy relation can be stated as an expert system with fuzzy set variables. Fuzzy relations can also be expressed like Table 3.1 rather than if-then rules. Table 3.1 represents three rules, for instance, if a patient has the symptom b_1 exceedingly, symptom b_2 moderately, and symptom b_3 slightly, then the outcome is disease a_1.

Compared to a function in mathematics, the fuzzy relation is a kind of function with fuzzy set variables. An independent variable of a fuzzy set is input into the relation, and then the dependent variable of a fuzzy set is estimated. In a regular function in mathematics, the above operation is easily executed, whereas the operation for the fuzzy relation may not be the same. Actually, the fuzzy relation is not like a mathematical function, but just expert knowledge in if-then rules. So substituting the input fuzzy set and estimating the output fuzzy set is called fuzzy reasoning, which is similar to terminology related to artificial intelligence. A series of operations, which are based on Boolean logic, have been successfully established for fuzzy reasoning. They include fuzzification, building fuzzy expert rules, the composition of expert rules and input fuzzy sets, and defuzzification. This will be covered in detail in the latter portion of the chapter.

3.2.4 MATHEMATICAL OPERATIONS OF FUZZY SETS

A large number of mathematical methods to apply to fuzzy sets have been established. Several methods that can be applied to sensory analysis are dealt with in the next section.

TABLE 3.1

Fuzzy Rules and Relation Matrix for Fuzzy Relations

	Symptom		
Disease	b1	b2	b3
a1	1	0.3	0.8
a2	0.2	0.1	1
a3	0.1	0.8	0.91

3.2.4.1 Basic Fuzzy Calculus

With sensory data, one usually needs to add and average them. In this part, several basic operations are discussed, such as addition, subtraction, multiplication, division, and averaging.

In order to facilitate arithmetic operations with fuzzy sets, the triangular and trapezoidal functions and α-cut method are applied. The first case is a fuzzy set X that has a triangular membership function and is defined by Equation 3.4:

$$X = <\text{minimum, apex, maximum}> \tag{3.4}$$

The membership grades of the minimum and maximum elements in the set are both 0, and that of the apex is 1. The triangular fuzzy set (TFS) is reexpressed as continuous intervals by α-cut (Figure 3.2). A and B, which are two TFSs by α-cut, are defined as follows:

$$A \rightarrow <a1, b1, c1>$$

$$\rightarrow \{A_{\alpha,\text{left}}, A_{\alpha,\text{right}}\}$$

$$\rightarrow \{(b1 - a1)\,\alpha + a1, (b1 - c1)\,\alpha + c1)\} \quad \forall \alpha \in [0,1]$$

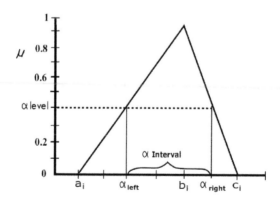

FIGURE 3.2   General form of a triangular fuzzy set and an alpha interval.

$B \rightarrow \langle a2, b2, c2 \rangle$

$\rightarrow \{B_{\alpha,\text{left}}, B_{\alpha,\text{right}}\}$

$\rightarrow \{(b2 - a2)\,\alpha + a2, (b2 - c2)\,\alpha + c2)\} \quad \forall \alpha \in [0,1]$ \hfill (3.5)

where $A_{\alpha,\text{left}}$ and $A_{\alpha,\text{right}}$ are the lower and upper branches on A, respectively.

The addition of two fuzzy sets is calculated by adding up all combinations of element values (Equation 3.6):

$C = A + B$ \hfill (3.6)

$\rightarrow \langle a1 + a2, b1 + b2, c1 + c2 \rangle$

$\rightarrow \{[(b1 + b2) - (a1 + a2)]\,\alpha + (a1 + a2), [(b1 + b2) - (c1 + c2)]\alpha + (c1 + c2)\}$

$\rightarrow \{A_{\alpha,\text{left}} + B_{\alpha,\text{left}}, A_{\alpha,\text{right}} + B_{\alpha,\text{right}}\}$

In the same manner, subtraction, multiplication, and division can be executed. The operations are as follows:

Subtraction: $C = A - B \rightarrow \{A_{\alpha,\text{left}} - B_{\alpha,\text{right}}, A_{\alpha,\text{right}} - B_{\alpha,\text{left}}\}$ \hfill (3.7)

Multiplication: $C = A \times B \rightarrow \{(AB)_{\alpha,\text{left}}, (AB)_{\alpha,\text{right}}\}$ \hfill (3.8)

where $(AB)_{\alpha,\text{left}} = \min\{A_{\alpha,\text{left}}\,B_{\alpha,\text{left}}, A_{\alpha,\text{left}}\,B_{\alpha,\text{right}}, A_{\alpha,\text{right}}\,B_{\alpha,\text{left}}, A_{\alpha,\text{right}}\,B_{\alpha,\text{right}}\}$, and $(AB)_{\alpha,\text{right}} = \min\{A_{\alpha,\text{left}}\,B_{\alpha,\text{left}}, A_{\alpha,\text{left}}\,B_{\alpha,\text{right}}, A_{\alpha,\text{right}}\,B_{\alpha,\text{left}}, A_{\alpha,\text{right}}\,B_{\alpha,\text{right}}\}$.

Division: $C = A/B \rightarrow \{(A/B)_{\alpha,\text{left}}, (A/B)_{\alpha,\text{right}}\}$ \hfill (3.9)

where $(AB)_{\alpha,\text{left}} = \min\{A_{\alpha,\text{left}}/B_{\alpha,\text{left}}, A_{\alpha,\text{left}}/B_{\alpha,\text{right}}, A_{\alpha,\text{right}}/B_{\alpha,\text{left}}, A_{\alpha,\text{right}}/B_{\alpha,\text{right}}\}$, and $(AB)_{\alpha,\text{right}} = \min\{A_{\alpha,\text{left}}/B_{\alpha,\text{left}}, A_{\alpha,\text{left}}/B_{\alpha,\text{right}}, A_{\alpha,\text{right}}/B_{\alpha,\text{left}}, A_{\alpha,\text{right}}/B_{\alpha,\text{right}}\}$.

The second case is where fuzzy set X has a trapezoidal membership function and is defined by Equation 3.10:

$A \rightarrow \langle a1, b1, c1, d1 \rangle$

$\rightarrow \{A_{\alpha,\text{left}}, A_{\alpha,\text{right}}\}$

$\rightarrow \{(b1 - a1)\,\alpha + a1, (c1 - d1)\,\alpha + d1)\} \quad \forall \alpha \in [0,1]$

$B \rightarrow \langle a2, b2, c2, d2 \rangle$

$\rightarrow \{B_{\alpha,\text{left}}, B_{\alpha,\text{right}}\}$

$\rightarrow \{(b2 - a2)\,\alpha + a2, (c2 - d2)\,\alpha + d2)\} \quad \forall \alpha \in [0,1]$ \hfill (3.10)

Addition, subtraction, multiplication, and division are executed the same as in Equations 3.6 to 3.9.

Averaging is slightly different than the other operations. Basically, averaging can be performed by the addition of all data and division of the summation by the total number of data. But if fuzzy sets for averaging have common discrete elements, another way is possible.

$$A = \left\{ \left(\sum_{k=1}^{m} \mu_{a,i,k} / n, x_i \right) \right\}, \ \forall x \in X \ (i = 1, 2, 3, \ldots, n) \qquad (3.11)$$

$$A = \sum_{k=1}^{m} \mu_{A,1,k} / x_1 + \sum_{k=1}^{m} \mu_{A,2,k} / x_2 + \ldots + \sum_{k=1}^{m} \mu_{A,n,k} / x_n \ (i = 1, 2, 3, \ldots, n) \qquad (3.12)$$

where m is the total number of data.

3.2.4.2 Fuzzy Reasoning

The procedure of fuzzy reasoning is established with the fuzzification of independent and dependent variables, building relationships between the fuzzy set variables as if-then rules, estimating an output fuzzy set for a given input fuzzy set by composition of the relation and input fuzzy sets, and defuzzification of the output fuzzy set if necessary. The individual steps are dealt with in more details, as follows:

1. *Fuzzification.* Before setting fuzzy relationships between variables, the variables should be converted to fuzzy sets. Select properties of interest should be targeted first, and then their fuzzy sets can be defined. A critical factor in the definition is the membership function. Partitions of fuzzy sets should be determined in terms of their number, shape, width, and location. The partition represents a graphical expression of a fuzzy set. The number of fuzzy sets determines how easy their relationships can be made in the next step. The shape, width, and location of a fuzzy set affect how accurately and reasonably the properties can be described in the fuzzy set.

2. *Fuzzy expert rules.* Setting fuzzy relationships as well as fuzzy sets should be a key part of or of some know-how in fuzzy reasoning, since the reasoning's accuracy and speed are determined by them. Constructing fuzzy relationships can start at expert knowledge or experiences. Expert knowledge is usually expressed in if-then rules:

If X1, then Y1 else

If X2, then Y2 else

If Xn, then Yn   (3.13)

where X and Y are fuzzy sets. The set rules are converted to a relationship matrix R, which is a Cartesian product. R is a binary fuzzy set with binary elements, x and y:

$$R = X \otimes Y \tag{3.14}$$

$$u_R(x, y) = \min [u_x(x), u_Y(y)] \tag{3.15}$$

where $u_R(x, y)$, $u_x(x)$, and $u_Y(y)$ are the membership degrees of fuzzy sets R, X, and Y, respectively.

3. *Composition.* The composition is just a mathematical operation to predict an output fuzzy set at an input fuzzy set from the fuzzy rules. Here, the composition by Mamdani's method is introduced:

$$Y_o = X_o \cdot R \tag{3.16}$$

$$u_{Y_o}(y) = \max [\min (u_{X_o}(x), u_R(x, y))] \tag{3.17}$$

where Y_o and X_o are output and input fuzzy sets, respectively, and $u_{Y_o}(y)$ and $u_{X_o}(x)$ are their membership degrees. If the input fuzzy set is just a deterministic value, which is common in fuzzy control applications, the membership degree should be one only at one element, but zero at all of the remaining elements.

4. *Defuzzification.* The output is obtained in the form of fuzzy sets from fuzzy reasoning. But if necessary, the fuzzy sets can be converted to a deterministic value again. Here, a defuzzification method for the center of gravity of a fuzzy set, which is most commonly used, is introduced:

$$C = \sum_{i=1}^{n} y_i \cdot u_{Y_o}(y_i) / \sum_{i=1}^{n} u_{Y_o}(y_i) \tag{3.18}$$

where C is the defuzzified deterministic value.

Consequently, it is realized that fuzzy reasoning may be more practical than regular mathematical functions of independent and dependent variables because it is based on practical expert knowledge or experience in the field.

3.3 APPLICATIONS OF FUZZY SETS

3.3.1 FUZZIFICATION OF SENSORY TEST DATA

The intensity or preference of attributes obtained from sensory tests can be expressed in fuzzy sets in two ways. One is to transform sensory data in numerical values to fuzzy values (fuzzification of rating), and the other is to transform sensory data in linguistic or hedonic expressions to fuzzy values (fuzzification of voting). In the next section, fuzzification of rating and fuzzification of voting are discussed.

3.3.1.1 Fuzzification of Rating

Fuzzy values can be defined and expressed as the following when sensory test data are obtained as attribute intensity scores (rating):

$$\text{Fuzzy value} = \{(\text{membership degree})_1/(\text{intensity level})_1, (\text{membership degree})_2/ \\ (\text{intensity level})_2, \ldots, (\text{membership degree})_n/(\text{intensity level})_n\} \quad (3.19)$$

where the intensity level is a sensory score, and its membership degree is the proportion of panelist votes to the total, who rated at the corresponding score.

There is another way to represent the results as fuzzy sets. A rated score reads the membership degrees from the membership functions of fuzzy values of false, borderline, and true, as shown in Figure 3.3. Compared to Equation 3.19, these three fuzzy values, so-called element fuzzy values, correspond to the intensity levels. When a rated score is substituted on the horizontal axis in Figure 3.3, the vertical axis values on the intersection points with each fuzzy value represent the membership degrees. For example, if the score is 50 out of 100, the fuzzy values will be {0.25/false, 0.5/ borderline, 0.25/true}. The element fuzzy values can vary in membership functions, as shown in Figures 3.4 and 3.5.

Figure 3.4 shows the case in which the membership functions of element fuzzy values are a triangle with five intensity levels. Also, Figure 3.5 shows the case of assuming only two intensity levels. This indicates that the shapes of the membership functions can be variously defined. The membership functions should vary with the specimen attributes, subjects' sensing abilities, the sensory test purpose, etc.

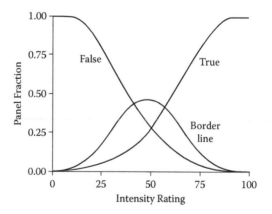

FIGURE 3.3 Membership functions as the panel fraction versus mean intensity rating.

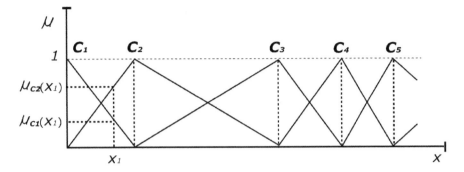

FIGURE 3.4   A battery of fuzzy sets, C_1 to C_n, with triangular membership functions.

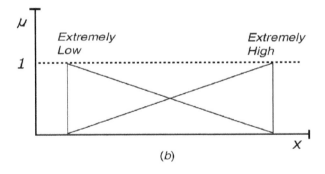

FIGURE 3.5   (a) A graphical line scale. (b) Two fuzzy sets representing the line scale.

3.3.1.2   Fuzzification of Voting

When attribute intensity is expressed at a linguistic level, panelists vote for the levels. The ratio of the votes for a linguistic level to the total is regarded as a membership degree of the linguistic level. This definition is similar to that for the fuzzification of rating. But the element here is just a linguistic level rather than the element fuzzy values. Table 3.2 shows an example; if the total vote for sample 1 is 100, and the votes for true (T), more or less true (MLT), neither true or false (NTF), more or less false (MLF), and false (F) are 40, 30, 15, 10, and 5, respectively, then the fuzzy set becomes {0.4/true, 0.3/more or less true, 0.15/neither true or false, 0.1/more or less false, 0.25/false}.

TABLE 3.2
Rice Stickiness: Panel Votes

	Subjective Factor				
Sample	T	MLT	NTF	MLF	F
Sample 1	40	30	15	10	5
Sample 2	30	40	20	8	2
Sample 3	10	20	55	10	5
Sample 4	3	7	15	30	45

The linguistic levels may not only be sensory attribute intensities, but also preferences on a hedonic scale, such as excellent, good, medium, fair, and not satisfactory.

3.3.2 Mathematics of Sensory Fuzzy Sets

There are several cases where the computation of fuzzy sets is needed. The mathematics of fuzzy sets associated with food sensory analysis is commonly limited to fuzzy set averaging, defuzzification of fuzzy sets, and fuzzy relationships. Although these principles were covered previously, further discussion with practical examples is given here.

3.3.2.1 Sensory Fuzzy Set Averaging

When fuzzy sets of individual panelists are combined as a representative fuzzy set of all panelists, the fuzzy sets are aggregated by averaging. The averaged fuzzy sets should have common elements to one another, and then the memberships can be averaged for each of the elements. Averaging is calculated by Equation 3.11, as mentioned previously.

3.3.2.2 Defuzzification of Sensory Fuzzy Sets

When comparing the magnitudes of sensory fuzzy sets, the fuzzy sets need to be defuzzified. The defuzzification of sensory fuzzy sets generally applies a specific formula or the center of gravity technique. First, consider the specific formula (Lincklaen Westenberg et al., 1989). This is the case of a fuzzy value with three element fuzzy values of true, borderline, and false (Figure 3.3).

$$q_L = \{1 + \mu_L(\text{true}) - \mu_L(\text{false})\}/2 \qquad (3.20)$$

where q_L is a normalized defuzzified value and μ_L, whose sum (μ_L (true), μ_L (borderline), μ_L (false)) is 1, are membership degrees. This equation implies that if the membership degrees of true, borderline, and false are 1, 0, and 0, respectively, then the value of q_L becomes 1, indicating the attribute intensity is the maximum.

Another defuzzification method commonly used is the center of gravity. For example, consider the fuzzy set $\{0.1/T_1, 0.6/T_2, 0.3/T_3, 0/T_4, 0/T_5\}$ in Figure 3.6. There are

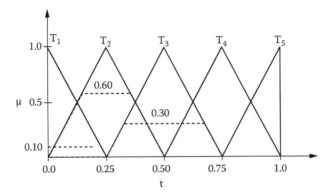

FIGURE 3.6 Triangular membership for linguistic truth values clipped at $\mu_{T1} = 0.1$, $\mu_{T2} = 0.6$, $\mu_{T3} = 0.3$, $\mu_{T4} = 0$, and $\mu_{T5} = 0$

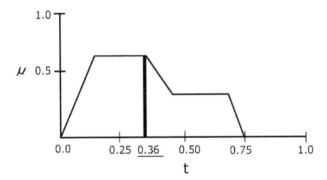

FIGURE 3.7 Center-of-gravity defuzzification for an aggregated truth value $t = 0.36$.

five element fuzzy values and their membership degrees are 0.1, 0.6, 0.3, 0, and 0, respectively. By applying the corresponding membership degrees to each fuzzy value, the aggregated fuzzy set can be obtained as shown in Figure 3.7. Eventually, the fuzzy sets can be transformed to a deterministic score of 0.36 through defuzzification by Equation 3.18.

For another simple example, the fuzzy elements in the previous example are simply replaced with a deterministic element. Figure 3.8 shows the fuzzy sets of preference for cooked rice. The fuzzy sets consist of hedonic scales and their membership degrees. But the hedonic scales are just substituted as EX = 5, GD = 4, FR = 3, PO =n2, and VP = 1. Then, the fuzzy set is defuzzified using Equation 3.18. The results are listed in Table 3.3.

3.3.2.3 Fuzzy Composition

Fuzzy composition is one phase in fuzzy reasoning, as mentioned previously. Consider some examples of applying fuzzy composition in a sensory test. Table 3.4

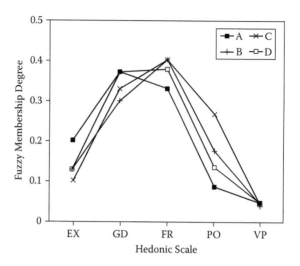

FIGURE 3.8 Fuzzy membership degrees for overall palatability of cooked rice samples A, B, C, and D. EX, excellent; GD, good; FR, fair; PO, poor; VP, very poor.

TABLE 3.3
Defuzzified Grades of Fuzzy Values of Overall Palatability of Cooked Rice

Variety	EX[a]	GD	FR	PO	VP	Grade
		Membership Degree				
A	0.20	0.37	0.33	0.09	0.05	3.6
D	0.13	0.37	0.38	0.14	0.04	3.4
B	0.13	0.30	0.40	0.18	0.04	3.3
C	0.10	0.33	0.40	0.27	0.05	3.2

[a] Magnitude of set elements: EX = 5, GD = 4, FR = 3, PO = 2, VP = 1.

shows a binary fuzzy relation obtained from a sensory test for cooked rice, which demonstrates the rules or relationship between fuzzy set A with the elements texture, taste, odor, and appearance, and fuzzy set B with the elements excellent, good, fair, poor, and very poor.

The membership degree for binary set elements means, for example, that with 0.18, 0.39, 0.30, 0.08, and 0.04 for texture, texture belongs to excellent, good, fair, poor, and very poor by weights of 0.18, 0.39, 0.30, 0.08, and 0.04, respectively. Therefore, texture is most likely to be good, but also fair, and subsequently excellent, poor, and very poor in decreasing order of preference. In the next step, if an input fuzzy set A_o {0.37/texture, 0.40/taste, 0.30/odor, 0.27/appearance} is composed with the relation fuzzy set, the output fuzzy set B_o is calculated by max-min operations

(Equation 3.17) as {0.20/excellent, 0.37/good, 0.33/fair, 0.09/poor, 0.05/very poor}. Here, A_0 is a fuzzy set representing the contributing weights of each attribute to the overall preference of cooked rice, and B_0 is the overall preference (Lee et al., 1994).

Fuzzy composition can also be applied to sample identification by the sensory test results. Table 3.5 shows an example of fuzzy rules demonstrating the relationships of sample preferences. In Table 3.5, the membership degree 1 in the fuzzy sets in the consequences of if-then rules means that the sample is identified as the one with a membership degree of 1. This rule or relationship can be transformed into a relation fuzzy set as shown in Table 3.6. Eventually, if an input fuzzy set for the preference of an unknown sample is composed with the relation fuzzy set, then the output fuzzy set of sample identification is calculated by max-min operations (Equation 3.17).

TABLE 3.4

Relation Fuzzy Set with a Matrix of Membership Degrees Corresponding to Binary Elements of Attributes and Preference of a Cooked Rice Sample

Attribute	Preference				
	Excellent	Good	Fair	Poor	Very poor
Texture	0.18	0.39	0.30	0.08	0.04
Taste	0.20	0.33	0.33	0.09	0.05
Odor	0.18	0.35	0.36	0.09	0.03
Appearance	0.17	0.36	0.33	0.09	0.05

TABLE 3.5

If-Then Rules for Sample Identification According to Sensory Preference

Rule #	If	Then
1	x is {0.2/good, 0.8/poor}	y is {1/sample1, 0/sample2, 0/sample3, 0/sample4}
2	x is {0.4/good, 0.6/poor}	y is {0/sample1, 1/sample2, 0/sample3, 0/sample4}
3	x is {0.7/good, 0.3/poor}	y is {0/sample1, 0/sample2, 1/sample3, 0/sample4}
4	x is {0.6/good, 0.4/poor}	y is {0/sample1, 0/sample2, 0/sample3, 1/sample4}

TABLE 3.6

Relation Fuzzy Set with a Matrix of Membership Degrees Corresponding to Binary Elements of Sensory Preference and Identification of Samples

Preference	Sample			
	Sample 1	Sample 2	Sample 3	Sample 4
Good	0.2	0.4	0.7	0.6
Poor	0.8	0.6	0.3	0.4

3.4   APPLICATIONS OF ANN WITH FUZZY SETS

3.4.1   INTRODUCTION

The basic concept of an artificial neural network (ANN) was proposed in 1943 by Warren McCulloch. Currently, ANN is known to be very effective for correlating variables with nonlinear relationships. Nonlinear relationships are often found in experimental data from food materials. If foods were a simple, pure material rather than a complex material, it would be possible to come up with theoretical mathematical functions for relevant relationships. Regression analyses, therefore, have been applied to food matters. Regression is a very useful and practical tool in the case where the mechanism for a relationship is not available. ANN is also included in the category of regression.

ANN has been utilized in food areas associated with sensory evaluation. ANN was applied to a sensory quality-based food process control, and control set points of processes were determined for sensory quality (Kupongsak and Tan, 2006). Generic authenticity can also be identified by ANN; Scotch whiskey or non-Scotch whiskey were correlated with thirteen sensory evaluation scores (Jack and Steele, 2002). Instrumental data can be correlated with sensory scores by ANN, especially when the data are complicated in number and nature like gas chromatographic data (Angerosa et al., 1996). ANN was applied for pattern recognition with an electronic nose and sensory properties (Han et al., 2001). And attempts have also been made to correlate whole texture profile measurements with sensory scores (Sitakalin and Meullenet, 2001). Overall, it is realized that most applications have not associated ANN with sensory evaluation methods methodologically, but correlated sensory scores with relevant data from other sources.

There is also a hybrid form of ANN and fuzzy sets, i.e., fuzzy neural networks (FNNs), which are known to be more powerful than ANN alone. FNN as well as response surface methodology (RSM) and multiple regression analysis (MRA) were applied to correlate food product formulations with sensory scores, where FNN proved the best in the predictions (Tominaga et al., 2001).

The procedures or mathematical algorithms to perform ANN or FNN have been well established. Only the basics of ANN theory are covered here since food scientists only need to know how to apply it to their particular cases. Some typical applications of ANN to sensory evaluation cases are discussed as well.

3.4.2   WHAT IS ANN?

The term *neural network* (NN) is used in two instances: biological NN (BNN) and artificial NN (ANN). BNN is a nerve system with a brain and biological neurons, whereas ANN is a calculation system with aggregated equations. The nerve system and biological neurons of BNN can be compared to the calculation system and aggregated equations of ANN. BNN and ANN work similarly. The neurons are educated from many and various experiences or occasions, and they work by transferring signals or information from one to another. In comparison, the equations are trained from many data, resulting in final set equations, and they work in a network manner for calculations. As a result, BNN can produce an output from some input information merely according to experiences, and ANN can also calculate output variables from input variables by the network with trained equations identical to the neurons.

ANN can predict relationships between independent (input) and dependent (output) variables with higher accuracy than other powerful methods, such as principal component regression (PCR) analysis and partial least squares regression (PLSR) analysis. Coverage can expand to even discontinuous functions with nonlinear relationships, which cannot be dealt with by PCR and PLSR. Nonlinear relationships are suggested to be far greater than linear ones.

3.4.3 How to Use ANN?

Before applying ANN, the relevant independent and dependent variables should be established. This aspect might be a key to enhance the prediction of the dependent variables at given independent variables. This is because ANN is only a versatile mathematical tool that is not associated with any environmental involvement in terms of physics, chemistry, biology, physiology, etc. Therefore, what variables may affect one another should be investigated very carefully.

Once the variable sets are established, the type of ANN should be selected. Many ANN types, such as radial basis function, Kohonen self-organizing, recurrent, stochastic, modular, and physical ANNs, are available, but a basic method will be covered here. It is back-propagation ANN also known as a feed-forward back-propagation ANN. ANN is composed of input, hidden, and output layers with neurons, as shown in Figure 3.9. Signals propagate from the input to output layers. In other words, input variables are given to the first layer, and then through the relevant calculation steps from one layer to another, output variables are finally obtained at the last layer. The number of hidden layers and neurons affect the accuracy and time of calculation,

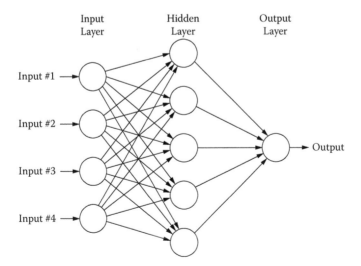

FIGURE 3.9 Structure of artificial neural network.

so they should reasonably be determined. More hidden layers and neurons may be necessary when there are more complicated correlations between the variables.

3.4.4 BACK-PROPAGATION ANN ALGORITHM

Those in food science may not be accustomed to calculating aggregated equations involved in ANN operations. Fortunately, there are many PC software tools available. Occasionally, these need to be modified or supplemented with relevant subroutines. Here, only the basics of an algorithm are covered to let the reader understand what occurs and to improve their application ability with ANN.

Figure 3.10 represents a sample mathematical process with input layers with several neurons and one hidden or output layer with only one neuron. This unit ANN extends to a complete ANN with more layers and neurons. The individual input values (x) are multiplied by their weights (w). The products are summed by Equation 3.21:

$$v_k = \sum_{j=1}^{P} w_{kj} x_j \tag{3.21}$$

The sum subtracted by a threshold (θ), which is Net in Figure 3.11, is input into several types of activation (transfer) functions, resulting in the output value from the neuron (Figure 3.11).

If there are more layers, this output value can consecutively be used as another input value to the next hidden layers, in turn producing final output values from the output layer.

The output value varies with the weights and thresholds of neurons for the linked neurons on the next layer, so there needs to be a process to optimize the weights to get the best predictions. It is the training or learning process that adjusts the weights by altering them until the errors between outputs given for the training and then calculated minimize, as shown in Figure 3.12. In the back-propagation ANN algorithm, the errors are propagated backwards, from the output layer toward the input layer. The time it takes to complete the corrections of the weights and thresholds is usually long.

3.4.5 APPLICATIONS OF ANN

Here, only two particular cases are dealt with. Other applications, if any, should be the same in principle, though. Relationships between sensory and instrumental data

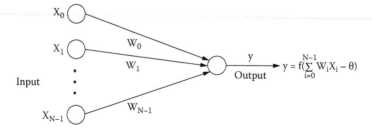

FIGURE 3.10   Mathematical process with transfer function.

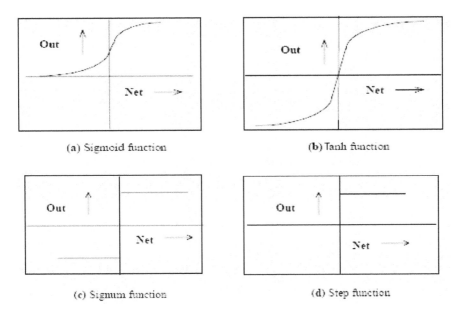

FIGURE 3.11 Types of transfer functions. (a, b) Less nonlinear. (c, d) More nonlinear. Net: v_k in Equation 3.21.

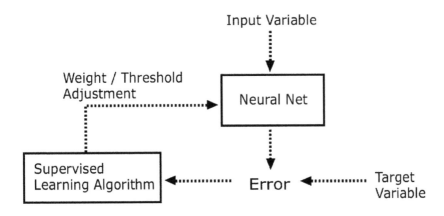

FIGURE 3.12 Back-propagation learning process to optimize the weights and thresholds.

can be established by ANN. Data analysis in sensory evaluation can be enhanced by ANN with fuzzy sets (FNNs).

3.4.5.1 Relating Instrumental Data to Sensory Fuzzy Sets by ANN

Sensory properties are the most important factors to determine product qualities, so it is very critical to measure them with objectivity, accuracy, and reproducibility. Sensory evaluation by human subjects can only be successful when there

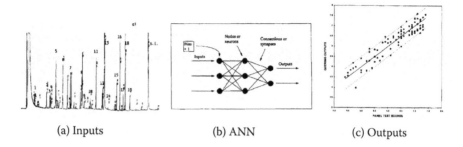

(a) Inputs   (b) ANN   (c) Outputs

FIGURE 3.13   Prediction of sensory properties from GC data. (a) GC data. (b) ANN learning and evaluation. (c) Sensory properties.

are well-trained panelists and a good testing environment available. If there are instrumental data matching sensory properties, they must be ideal, because instrumental data are more consistent and objective, and easily manageable. Therefore, to secure those instrumental data, the relationships between sensory and instrumental variables are examined by multivariate analyses. ANN is also employed as a good alternative to traditional statistical methods. Here two relevant examples will be discussed:

1. *Case for gas chromatographic data.* Gas chromatography (GC) is one of the most common methods to detect volatile compounds from food materials, so it can be a good alternative to sensory evaluations for flavor or taste. But GC data only contain the information of a large number of chemicals not providing any sensory attributes. To correlate GC data with sensory attributes, ANN has been utilized (Angerosa et al., 1996). GC and sensory data are regarded as input and output variables, respectively (Figure 3.13). After learning of the data, ANN can predict sensory properties from GC data, suggesting that sensory evaluation can be successfully replaced by GC analysis.

2. *Case for process control set points.* For process control, one needs to know the control set points. The processing conditions cause the qualities of products, so the former usually becomes the input and the latter the output. But to predict the set points from the sensory properties, their positions should be reversed. ANN has been applied to predict set points to obtain a particular sensory quality (Kupongsak and Tan, 2006). Sensory data were expressed as fuzzy sets and their membership degrees are assigned as input variables. A special approach was tried to obtain higher predictability. First, a test was performed to determine how the process variables influenced each of the sensory attributes, and vice versa. The former and latter steps were named the forward and backward mapping, respectively. After evaluation of their effects, a multiattribute model was developed by ANN. It was demonstrated that ANN with fuzzy sets is a promising means for sensory quality-based food process control.

3.4.5.2 Building Sensory Fuzzy Rules by ANN

From sensory evaluations, data for several attributes or preferences can be obtained to see the product qualities and preferences according to processing, storage, and any other treatment condition. But these data may not be efficient to show tendencies between conditions and product properties. It may be useful to use a graphical method with the data, so various regression methods have been utilized to draw graphs by their analytical equations. FNN has also been applied for the same purpose, but the way to show a tendency is not the same. Here, an example with fuzzy rules between sensory attributes and total evaluations instead of treatment conditions to characterize the essential attributes from sensory evaluations is introduced.

1. *Algorithm of FNN.* Figure 3.14 shows the structure of FNN. The difference from ANN is that is has specially designed paths between the neurons, and connection weights with particular meanings. Basically, a regular fuzzy reasoning procedure is used to calculate the output values from the input values. However, the membership functions of fuzzy sets are replaced with transfer functions used in ANN. To complete the adequate shapes of membership functions, one or two transfer functions are used, as shown in Equation 3.22 (Hanai et al., 1996):

$$v_k = \frac{1}{1+\exp(-w_g(x_1+w_c))} - \frac{1}{1+\exp(w_g(x_1,w_c))} \tag{3.22}$$

where the weights w_c and w_g are the parameters of transfer functions. The weight w_f, which is an interesting part, indicates the significance of the corresponding fuzzy rules. Thus, the optimization of w_c and w_g by the backpropagation algorithm of ANN is like the tuning of fuzzy sets, and that of w_f is like selecting the significant rules among all possible rules used in FNN. The positions and gradients of the input variables in the premise of if-then rules are determined by w_c and w_g. Consequently, the fuzzy rules with tuned fuzzy sets between sensory attributes and a total evaluation can be created from just regular sensory test data.

2. *Case for Ginjo sake.* Panelists tested the sensory properties of Ginjo sake according to evaluation criteria as shown in Table 3.7 (Hanai et al., 1996). Their averages were used for either the learning or evaluation of the FNN in Figure 3.14.

As a result, Table 3.8 can be summarized as three fuzzy rules as follows:

---

a. If color is small (clear) and aging is small (young), total evaluation is small (good).
b. If flavor base is small (high), total evaluation is small (good).
c. If aging is small (young), total evaluation is small (good).

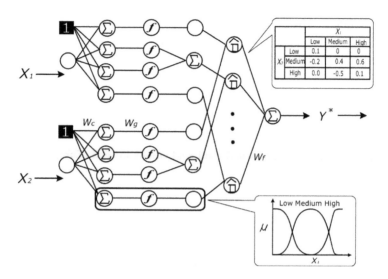

FIGURE 3.14 Structure of fuzzy neural network. x_1, x_2, inputs; y^*, output.

TABLE 3.7

Grades of Sensory Attributes of Ginjo Sake: Color, Flavor Base, Aging, and Total Evaluation

Attributes	Grade		
Color	Clear (S)	Ordinary (M)	Cloudy (B)
Flavor base	High (S)		Low (B)
Aging	Young (S)		Aged (B)
Total evaluation	Good (–1)	Ordinary (0)	Bad (1)

TABLE 3.8

Fuzzy Rules between Sensory Attributes and Total Evaluation Created by the Trained FNN

	Aging					
	S			B		
	Color			Color		
Flavor base	S	M	B	S	M	B
S	–0.40	–0.46	–0.09	0.30	–0.14	0.47
B	0.13	0.45	0.34	0.83	0.54	0.61

3.5 CONCLUSIONS

Currently, the quantification of food qualities is being addressed in food research and business practices. Food data are necessary to describe food qualities for product development, quality control, and process control. However, there is much difficulty in quantifying these data, particularly due to the intrinsic characteristics of food accompanied with uncertainty and imprecision. Thus, it is effective to apply fuzzy sets to such food matters. Particularly in relation to sensory tests, objective quantification is more difficult, as the evaluation subject is human rather than instrumental, making quantification more difficult combined with food data.

Statistical techniques for sensory tests have already been developed and continuously advanced. Statistical quantification is usually based on the assumption of error probability for the quantification. Linguistic expressions are also used, but they should be converted to numerical values for statistical treatment, where there is vagueness as an obstacle to assign a linguistic expression a numerical value. By contrast, fuzzy theory is a mathematical technique that can directly quantify human linguistic expression, and therefore is very practical to quantify even primary data obtained from sensory tests. In addition, it is possible to perform mathematical operations on fuzzy sets for a particular purpose by applying the fuzzy operation techniques that are currently available. Therefore, if fuzzy set theory is applied along with any other available techniques, it ensures that more effective performance and analysis of sensory tests is possible. ANN results in a good match. ANN is a versatile method for the prediction of nonlinear relationships, and can readily be adapted by learning from experiences. Most of the variables from sensory evaluation are in nonlinear relationships of one another. FNN is developed by combining the advantages of fuzzy sets with ANN. FNN is an effective means to handle sensory data with fuzziness and their nonlinear relationships from sensory evaluations. Consequently, it is thought that the integration of fuzzy set theory and ANN is continually required for more successful food sensory analysis.

REFERENCES

Angerosa, F., L. D. Giacinto, R. Vito, et al. 1996. Sensory evaluation of virgin olive oils by artificial neural network processing of dynamic head-space gas chromatographic data. *Journal of the Science of Food and Agriculture* 72:323–28.

Bomio, M. 1998. Neural networks and the future of sensory evaluation. *Food Technology* 52:62–63.

Han, K. Y., J. H. Kim, and B. S. Noh. 2001. Identification of the volatile compounds of irradiated meat by using electronic nose. *Food Science and Biotechnology* 10:668–72.

Hanai, T., A. Kakamu, H. Honda, et al. 1996. Modeling of total evaluation process of Ginjo sake using a fuzzy neural network. *Transaction of the Society of Instrument and Control Engineers* 32:1113–20.

Jack, F. R., and G. M. Steele. 2002. Modelling the sensory characteristics of Scotch whisky using neural networks—A novel tool for generic protection. *Food Quality and Preference* 13:163–72.

Kupongsak, S., and J. Tan. 2006. Application of fuzzy set and neural network techniques in determining food process control set points. *Fuzzy Sets and Systems* 157:1169–78.

Lawless, H. T., and H. Heymann. 1999. *Sensory evaluation of food: Principles and practices.* New York: Aspen Publishers.

Lee, S. J. 2000. Introduction of fuzzy theory. *Food Science and Industry* 33:20–26.

Lee, S. J., C. G. Hong, T. S. Han, et al. 2002. Application of fuzzy control to start-up of twin screw extruder. *Food Control* 13:301–6.

Lee, S. J., and Y. A. Kwon. 2007. Study on fuzzy reasoning application for sensory evaluation of sausages. *Food Control* 18:811–16.

Lee, S. J., W. S. Noh, and Y. C. Choi. 1994. Sensory evaluation of cooked rice with fuzzy reasoning. *Korea Journal of Food Science and Technology* 26:776–80.

Lee, S. J., C. S. Won, O. Han, et al. 1995. Simulation of fuzzy logic controller for food extrusion process. *Korea Journal of Food Science and Technology* 27:164–69.

Lincklaen Westenberg, H. W., S. D. Jong, D. A. Meel, et al. 1989. Fuzzy set theory applied to product classification by a sensory panel. *Journal of Sensory Studies* 4:55–72.

Meilgaard, M. C., B. T. Carr, and G. V. Civille. 1999. *Sensory evaluation techniques.* Boca Raton, FL: CRC Press.

Moskowitz, H. R., J. H. Beckley, and A. V. A. Resurreccion. 2006. *Sensory and consumer research in food product design and development.* IFT Press Series. Hoboken, NJ: Wiley-Blackwell.

Peris, M., and L. Escuder-Gilabert. 2009. A 21st century technique for food control: Electronic noses. *Analytica Chimica Acta* 638:1–15.

Ruan, D., and X. Zeng. 2004. *Intelligent sensory evaluation: Methodologies and applications.* New York: Springer.

Settles, A. M., and S. Mierson. 1993. Ion transport in rat tongue epithelium in vitro: A developmental study. *Pharmacology, Biochemistry and Behavior* 46:83–88.

Sitakalin, C., and J. C. Meullenet. 2001. Prediction of cooked rice texture using an extrusion test in combination with partial least squares regression and artificial neural networks. *Cereal Chemistry* 78:391–94.

Stefanini, L., L. Sorini, and M. L. Guerra. 2006. Parametric representation of fuzzy numbers and application to fuzzy calculus. *Fuzzy Sets and Systems* 157:2423–55.

Tominaga, O., F. Ito, T. Hanai, et al. 2001. Sensory modeling of coffee with a fuzzy neural network. *Journal of Food Science* 67:363–68.

Zeng, X., D. Ruan, and L. Koehl. 2008. Intelligent sensory evaluation: Concepts, implementations, and applications. *Mathematics and Computers in Simulation* 77:443–52.

4 Application of Computer Vision Systems for Objective Assessment of Food Qualities

Patrick Jackman and Da-Wen Sun

CONTENTS

4.1   INTRODUCTION

Food products have varied commercial value due to attributes expected to be present by consumers. Food products rich in the expected attributes will be of high commercial value. Beef, for example, is judged by consumers on its tenderness, juiciness, and flavor (Warriss, 2000). A reliable means of estimating consumer response before distribution to wholesalers is essential, as presenting high-quality food products in the low-end market is wasteful, and presenting low-quality food products in the high-end market is extremely damaging to a retailer's reputation. As food products have some observable quality indicators that can be measured before distribution, a good estimate of quality is thus possible. Among these quality indicators, color, texture, shape, and size are the most common and important.

Color immediately stands out as quality indicator for foods, as color provides much information on the condition of both the external surface and the whole food product. For example, beef muscle color is an indicator of final palatability (USDA, 1997a). Color is also a very convenient quality indicator that is very easy to observe with a computer vision system, and it can also be reliably independently verified with a colorimeter.

In the meantime, surface texture is also recognized as being necessary for many food items, as it can reveal information on the food structure. For example, beef muscle texture is an indicator of final palatability (USDA, 1997a). It is not as easy to observe as color, as there can be very fine texture details that might only be visible under high magnification or even nonvisible wavelengths. In addition, texture can be difficult to describe succinctly as the patterns can be highly irregular.

In addition, shape can carry useful quality information because if the product deviates strongly from the normal shape, it might indicate a product malaise. Abnormal growths, both internal and external, can alter the product shape. Shape can also indicate the maturity and breed. For example, the shape of pickling cucumbers forms part of the USDA grading system, where substantially odd shapes require downgrading (USDA, 1997b). Exact acceptable limits for food shape can, however, be subjective.

Finally, size can reveal information on vital food qualities, as substantial deviations from the normal size might also be an indicator of malaise. Abnormal growth, both internal and external, can cause deviations in the product size. Similar to shape, size can also be an indicator for the maturity and breed of the food. For example, the USDA (2006) sets size limits for grading strawberries and rules on how many strawberries in a batch can be off-size.

It is possible for highly trained and experienced personnel in the food industry to recognize such quality indicators in a food product before it is distributed to wholesalers and make an "expert" judgment as to the likely consumer response. In the light of this, the USDA has published a vast array of expert grading standards (USDA Agricultural Marketing Service, 2009) to allow food quality to be assessed on a range of criteria, such as color, shape, size, and texture. A prime example of these standards is the beef carcass grading standard (USDA, 1997a).

Such expert grading has a number of strong advantages. First, the intuition and experience of a sufficiently well-trained grader should prevent gross misclassifications.

Second, the manual grading process is nondestructive (Jackman et al., 2008). However, there remain some fundamental problems with manual grading. First, human judgments are subjective; that is, any two manual graders may reach different opinions. Second, human judgments are inconsistent; that is, the same grader may reach a different opinion on any different occasion. Third, human grading can be expensive and slow. This is succinctly described by Du and Sun (2004).

One of the solutions to the above problems is a computer vision system (Sun, 2008), which can deliver the required objective, consistent, speedy, and affordable quality judgments, while still being nondestructive and avoiding gross misclassifications. A computer vision system is a simulation of the human assessment process (Du and Sun, 2006a), with the camera replacing the eye and a learning algorithm replacing the brain. The camera records objective and consistent quality indication data free of confounding noise. Then a trained learning algorithm matches the quality indication data to the correct quality class or level.

Using computer vision, quality assessment steps can be fully automated, with the resulting quality classifications being fully objective and consistent. Other benefits of using computer vision include improved speed and efficiency, increased accuracy, as well as reduced costs (Brosnan and Sun, 2004). Especially with the rapid development in computer vision hardware and software, up-to-date computer vision equipment is normally affordable. For example, a three-CCD (charged-coupled device) digital camera will cost approximately \$1,000 to \$5,000, and a frame grabber approximately \$500 to \$2,000 (Edmund Optics, 2009). However, some highly advanced three-dimensional imaging systems, such as magnetic resonance imaging, are prohibitively expensive.

There have been very many successful transpositions of expert grading methods to computer vision systems, thus providing rapid, consistent, and objective classification without any product damage. Many of these are detailed by Aguilera and Briones (2005), Brosnan and Sun (2004), Du and Sun (2006a), Sun (2008), and Tan (2004). Thus, computer vision has demonstrated a considerable pedigree in automatic food quality evaluation and grading. With increasing computational power and hardware advancements in recent years, there have been new opportunities to both widen and deepen the application of computer vision systems.

4.2 PRINCIPLE OF COMPUTER VISION

A computer vision system acquires digital image data for providing useful information on food quality. As foods are widely diverse, their features that reveal quality information can be very different. Similarly, the point on the production line where the food needs to be observed can also be changeable, as can the design of the production line itself. Thus, computer vision systems should be tailored to the task at hand rather than prefabricated. Fortunately for the vast majority of foods, the tailoring is not overly elaborate or difficult.

Ideally the basic CCD camera acquiring visible wavelength data will suffice. This is by far the least expensive option and precise color data can be acquired (Du and Sun, 2004). Only where this fails should the more expensive options be considered. Nonvisible wavelengths can illuminate surface regions of interest, making them easier to identify, or allow the internal regions of the food to be seen.

There are instances where the object surface does not contain enough quality information. In these cases attempts should be made to find more quality information. In many foods the quality can be determined by what is on the inside as well as on the outside. Thus, penetration imaging can be used to observe the internal regions of the food. The use of penetration imaging is not useful for finding any color information but is usually very useful for finding internal object outlines for shape and size evaluation. All penetration technologies do suffer from the same flaw to one degree or another, which is that deeper penetration comes at the cost of lower resolution. As might be expected, the equipment that best retains resolution will also be the most expensive.

In order to obtain consistent image data, consistent illumination conditions are required, especially if color is a key quality indicator. Thus, data acquisition should take place under a fixed and appropriate illumination regime (Brosnan and Sun, 2004). Color perception is dependent on lighting; therefore, if absolute color measurement is required, the illumination regime should be tailored so that the camera will record red, green, and blue (RGB) data that can be immediately converted to device-independent color data without the introduction of substantial error (Valous et al., 2009a). As imaging is most typically recorded in amounts of red, green, and blue, assuming a three-channel digital camera is used, the image acquisition environment is designed to minimize noise to a level where it has a confounding or corrupting effect on the image data.

The recorded data are first preprocessed to eliminate any noise and distortions (Brosnan and Sun, 2004); then a segmentation algorithm tries to isolate the region of interest, as there is most likely to be a background or subregions containing no useful data (Brosnan and Sun, 2004). The purpose of segmentation is to identify the useful parts of the image, i.e., region of interest (ROI). The first step is to remove the redundant background. With the background removed, the next task is to identify the useful parts of the remaining object. The whole object may be useful or only certain parts may be useful. Segmenting the object into useful and nonuseful regions can be far more difficult. A fully reproducible and robust automatic segmentation process is desirable to ensure objective and consistent data. There are three major approaches to segmentation: thresholding, gradient-based segmentation, and cluster-based segmentation (Jackman et al., 2009b). Thresholding seeks a universal limit that all pixels of interest either lie above or below. Gradient-based segmentation assumes that abrupt changes in pixel values constitute an object edge. Cluster-based segmentation presumes that a pixel belongs to the most similar cluster. As autonomous segmentation is one of the greatest and most difficult challenges in computer vision (Gunasekaran, 1996), segmentation usually involves fine-tuning an established algorithm or writing a customized algorithm. Occasionally the segmentation process will be so difficult that manual segmentation is required to complete segmentation (Jackman et al., 2008, 2009a). Hence, aspirations toward a universal segmentation algorithm with general application are lofty, to say the least.

With the useful regions successfully isolated, the region should then be described in a meaningful and succinct way. A region of interest might have 100,000 pixels, so describing individual pixels is unrealistic. Thus, effective data compression is required. Condensing the data into a reasonable number of representative features is essential to allow a model linking the features to independent quality data to be built with training and testing data. A model with too many variables carries many difficulties with training and testing, and thus a simple model is fundamentally more desirable (Jackman et al., 2008). With a well-trained and tested model the quality of future samples can be estimated reliably.

Finally, the image data should be used to correlate with independent quality data. These independent quality data can be true public opinion (consumer panel), a subjective approximation (trained sensory panel), or an objective approximation (e.g., Instron) in the order of usefulness (Jackman et al., 2008). A trained expert opinion can also be used as the independent quality data, which would be akin to a subjective approximation. The well-established correlations can then be used for food quality prediction and evaluation.

4.3   COMPUTER VISION HARDWARE AND SOFTWARE

All computer vision devices involve a food image being recorded by an array of light-sensitive devices, with a CCD camera as the most common one (a CCD scanner operates on the same principles). A CCD camera receives light reflected off the object surface or light from behind the object. The lighting arrangement can even be specially structured to help reveal surface features (Du and Sun, 2004). This illustrates the need for sufficient and consistent lighting as well as consistent camera placement and settings. The received light is passed through a special prism to separate out the red, green, and blue elements (Canon, 2009). These are then directed onto photosensitive capacitors. The charge can then be converted into a voltage, and hence an electrical signal (Kodak, 2009). The strength of the signal is proportionate to the strength of the light in a highly linear way. For most problems this will allow clear object outlines to be calculated, leading to reliable size and shape assessment. It will also allow reliable color and surface texture measurement.

The most basic nonvisible wavelength approach is ultraviolet illumination. This is useful for contrast enhancement, as it can cause some substances, such as fat, to glow or fluoresce. This can be a valuable tool in assisting image segmentation and object identification (Basset et al., 2000). Ultraviolet light has the weakness of visible light being unable to penetrate beneath the food surface. Furthermore, ultraviolet light does not yield any color information.

Hyperspectral imaging and multispectral imaging are other popular nonvisible imaging technologies that record images under a range of light frequencies one by one. They follow the principle of seeing how a pixel changes under light of different frequencies (Chen, 1999). The difference is that hyperspectral involves much higher band density. A food product can thus be scanned at various wavelengths, and a three-dimensional hyperspectral "image cube" is formed, which is denser than the multispectral one. The image cube can then be analyzed for useful information. Many substances have a "spectral signature" in which they are illuminated strongly at particular wavelengths.

For penetrative imaging the low-cost option of ultrasound is fine for measuring parameters like fat thickness (Du and Sun, 2004); however, for a highly detailed internal examination very powerful and very expensive equipment such as magnetic resonance imaging (MRI), computed tomography (CT), or ET (electrical tomography) is required, the details of which are discussed by Du and Sun (2004). Of these, MRI is the most powerful, as a dense MRI scan can reveal fine details in three dimensions. Unfortunately, MRI is also the most expensive and is very slow; thus, it remains restricted mainly to medical applications where cost and throughput are less of a concern. This makes a three-dimensional measurement of texture or a two-dimensional measurement far below the food surface very difficult at any kind of reasonable price. Typical costs for nonvisible wavelength imaging devices are approximately $2,000 for a near-infrared (NIR) camera adapter (Edmund Optics, 2009). For penetrative imaging with x-rays the equipment cost increases dramatically.

Regardless of what imaging equipment is used, there is an important question of image precision. This is the number of rows and columns of pixels. The finer the precision, the more details that can be observed. However, the finer the image precision the greater the equipment cost and image processing computational load. Hence, a compromise is required. Image precisions of approximately 500 to 1,000 pixels in each row and column would be a typical compromise.

Also, regardless of the imaging equipment chosen, the field of vision may well extend beyond the food boundaries. In this case, segmentation will be required to exclude pixels that are not of interest. Once the region of interest has been isolated, its color, texture, shape, or size can be quantified. The simplest and most desirable means of doing so is with a contrasting background, allowing robust thresholding to separate out the background. This was successfully used by Barni et al. (1997) and Borggaard et al. (1996). In addition, segmentation can be assisted with secondary equipment, such as an ultraviolet flasher, to illuminate the region of interest (Basset et al., 2000).

Image processing can be very elaborate, requiring highly complex and detailed data processing. Thus, there is a need for specialized software and persons capable of using it correctly and efficiently. Matlab version 7 (The Mathworks, Natick, Massachusetts) is highly suited to such image processing, as it has many custom-built programs and functions included. With efficient data processing, classification and predictive models can be developed. Both explicit (regression, discriminant, etc.) and implicit (neural network, fuzzy inference, etc.) models can be used.

A predictive model's reliability and utility go beyond simply good validation statistics. There are main model features that can be used to indicate the reliability of a model. For example, for regression models there are a number of very simple tests to determine a model's utility (Jackman et al., 2009c). First, a regression matrix will have a condition number that is easily computed. This number is a measurement of the fragility or vulnerability to small perturbations of the regression equation. If this number is high, then a small amount of error could substantially change the regression prediction. Second, a partial least squares regression (PLSR) model will have a root mean square error of prediction (RMSEP) curve. This should smoothly decline to a minimum point without obvious local maxima. Local maxima along the way to the minimum point raise doubts about the number of genuinely useful model factors. Third, if the quality measurement is replaced by a random variable, the model should collapse; on the other hand, if it does not, the model is essentially capable of producing any result.

4.4 METHODS FOR FEATURE EXTRACTION AND ANALYSIS

4.4.1 EXTRACTING COLOR FEATURES

Color is the simplest property to measure but is also the property that requires the most care if absolute rather than relative color measurement is required. Measuring color can be done with a three-CCD digital camera, which yields a red, green, and blue value for each pixel. After segmentation the food color can easily be expressed with histograms of red, green, and blue and compared. The color can be used as a feature for a predictive model (Li et al., 1999) or for comparison against a standard (Quevedo et al., 2008a). Simple summary properties such as mean, standard deviation, skewness, kurtosis, and interquartile range are easily calculated (Jackman et al., 2008, 2009a). Local variations in the color of the food can be critical, as they may indicate stages of ripening or a malaise. For example, most of the product flaws described by the USDA standards for peaches (USDA, 2004) could be observed as local variations in color. These local variations can be found by noting peaks and troughs in the color channels across the image.

In the food industry color is normally expressed in the device-independent L*a*b* colorspace developed by the Commission Internationale d'Eclairage (CIE) in 1976 (HunterLab, 2008). This avoids the peculiarities and quirks of various brands of digital camera and variations in lighting. Thus, for absolute color measurement an image acquisition environment (including the camera) must be specially calibrated to a standard red, green, and blue colorspace (sRGB). This allows direct comparison of image data with colorimeter readings. A format for sRGB was developed by Hewlett-Packard© and Microsoft© (Stokes et al., 1996).

While RGB is the source data format, it is not always the most convenient for data analysis. A variety of colorspaces exist that might express the color information more efficiently, such as hue, saturation, and intensity (HSI). HSI has the advantage of expressing the color type, dilution, and brightness in separate variables. It was found by Li et al. (2001) that the saturation color channel expresses beef surface texture better than the RGB color channels. In addition, a number of colorspaces allow data compression without substantial loss of information, such as the National Television Services Committee YIQ colorspace or the digital video YCbCr colorspace. As data compression is not ordinarily required, the colorspaces of RGB, sRGB, and HSI are the most popular. These alternative colorspaces can be used to find features with greater discriminant capacity (Du and Sun, 2005; Jackman et al., 2009e).

Pixel data precision like image precision is a compromise between maximizing image information with cost and computational load. A good compromise is 8-bit data. For three-channel color data this leads to 24-bit color data. Each of three channels has 256 increments, from no exposure (0) to full exposure (255). This leads to the need for correct camera exposure, as overexposure will lead to clipped data and underexposure will lead to poor precision. Where color features other than the mean are sought, the need for correct camera focus becomes clear. Any blurring affects the other color histogram features and makes local variations harder to identify. Blurring makes segmentation more difficult and may lead to corruption of the region of interest, and thus skewed data.

Color data accuracy can be enhanced with the most modern cameras, such as PIXIS (Princeton Instruments, Trenton, New Jersey), with "dark current" control; i.e., they read consistently when faced with a light trap. However, these cameras are very expensive. Older and less expensive cameras will have some dark-current fluctuation and will have slight inconsistencies in readings when faced with a light trap.

4.4.2 Extracting Texture Features

Texture is the localized spatial variation in the region of interest. Computer vision data can be single channel, but usually RGB data are used. When color data are used, a greyscale should be chosen to best express the texture information. With a wide variety of colorspace transformations available, there are many choices of greyscales. The most typical choices would be saturation and intensity. If RGB data are not used (nonvisible wavelengths), the greyscale should have already been fixed as noncolor data and only have one channel. Unlike with color, there is no possibility of condensing texture information into a few simple summary properties, as the patterns of these spatial variations can be highly elaborate and convoluted. Even looking for local extrema in the channel histogram is of little or no benefit, as a large number of variables are required to describe the subtle patterns that exist between neighboring pixels.

The classic texture algorithm of Haralick et al. (1973) is the original and still one of the best algorithms for expressing texture. This is a pixel pair co-occurrence algorithm. For a given pair of pixels a certain distance apart, a comparison is made, and the two pixel values are interpreted as coordinates in a matrix. A set of matrix features can then be calculated. The co-occurrence matrix algorithm is discussed in more detail by Zheng et al. (2006a). Another classic texture algorithm is pixel run lengths. This is a pixel chain algorithm. For a given pixel a journey is made across neighboring pixels until an abrupt change is encountered, which is the run length. The distribution of run lengths gives the texture features. This is also discussed in more detail by Zheng et al. (2006a). On the other hand, a simpler and less computationally demanding algorithm is pixel difference histograms (PDHs), which record the difference between neighboring pixels and build up a histogram. This is discussed in more detail by Chandraratne et al. (2006a).

The classical approaches have some key strength in that the derived texture features immediately make sense and are easy to program into a computer file without specialist software. Conversely, they have some key weakness. Mainly, the pixel co-occurrence algorithm in particular is very slow, and can create many redundant features. The run length algorithm can also be very slow. In addition, the PDH does not record the actual pixel values, and thus does miss some information.

The above weakness with the classical algorithms has led to the development of alternative texture characterizations such as the Fourier transform, the wavelet transform, and fractal dimensions. Each of these seeks to condense texture information into fewer variables, thus considerably speeding up the computation process without overlooking any significant texture information in the image.

A means of characterizing highly complex patterns is the Fourier transform. This treats the pattern as a convolution of sinusoidal waves. The higher frequencies are most likely noise and can be discarded without loss of information, leading to substantial data compression. This is explained in more detail by Stein and Shakarchi (2003). In the case of image texture the Fourier transform imagines the greyscale image as a two-dimensional wave with the dark pixels as the troughs and the bright pixels as the crests. The Fourier transform can therefore break down the two-dimensional wave into a number of important sinusoidal components. Those component sinusoids are characteristic of the texture.

The wavelet transform follows similar ideas of decomposing the two-dimensional wave into component elements, but in this case the component elements are localized in both frequency and time. Applying the wavelet transform at dyadic scales allows for effective data compression (Jackman et al., 2009d). The wavelet transform has a key edge over the Fourier transform in that it is better placed to account for local fluctuations in texture. A comprehensive discussion of the utility of the wavelet transform is given by Kaiser (1994).

Fractal dimensions seek to compress texture data by identifying recurring self-similar patterns at various modalities in the image. By observing such repeating patterns, we can compute characteristic dimensions that closely describe the texture patterns. This is quite an ambitious form of texture analysis but is highly effective if such repeating patterns are at least closely approximated. As can be expected, such a simple approach often fails, and a more realistic approach of multifractal dimensions is required. In this case a single exponent is not used; instead, a spectrum of exponents is used. The utility of fractal dimensions in image analysis is explained by Soille and Rivest (1996).

As with color, pixel data precision is vital. Too much data precision will slow down the computational process, and too little will cause a loss of information. While data precision of 8 bits is most common, greyslicing is often required for pixel co-occurrence. Correct camera exposure and focus are essential. In particular, blurring must be avoided, as it will make correct texture measurement impossible.

4.4.3 Extracting Shape and Size Features

While visible wavelength imaging is often sufficient, nonvisible wavelength equipment is more frequently used if an object shape or size is desired. This is mainly due to the need to highlight a surface object that is obscured or to find an object beneath the food surface. Therefore, costly penetrative imaging may be the only option. To measure shape and size, first an object outline must be found with image segmentation. With a clear object outline, various features can be calculated, depending on the food object. Shape and size features chosen must be indicative of quality. This can be a feature that indicates positive qualities or a feature that indicates the presence of some kind of defect or malaise. An array of important shape and size features are discussed by Du and Sun (2004). Exact size and shape may not always be critical, but approximate size and shape should be. For example, classes of fruit are expected to lie within particular size ranges (Moreda et al., 2009). Further discussion of main size parameters is detailed by Moreda et al. (2009). The USDA (2009) lists a selection of key size and shape features for foods.

The critical importance of image segmentation comes to the fore when shape and size features are required. Unlike developing color and texture features where slight underestimation of the region of interest might not substantially skew the resulting features, the object outline must be reliably developed. Without a reliable outline, the extracted features could be heavily overestimated or underestimated. This can be seen when considering the shape and size features listed by Du and Sun (2004).

Pixel data precision is imperative only for the segmentation step. Once the outline has been found, the image can be binarized without loss of information. Correct camera exposure is important, like with color and texture, to ensure no data clipping. Camera focus is of utmost importance, as any blurring will make accurate identification of the object boundary more difficult, or even impossible.

4.4.4 Modeling and Dimensionality Reduction

Once a series of image features have been extracted from the image, an attempt should be made to link the image features with independent food quality data. A robust classification or prediction model that can estimate quality accurately the vast majority of the time is required. Ideally, a predictive model could be created to predict a particular food quality value; however, this will be more difficult than just a classification model that assigns a sample to one of a small number of classes. Any model should be calibrated and validated with a representative number of samples before it can be used on new samples with confidence.

Some form of dimensionality reduction or feature selection is required in order to make the model tractable. This is because images can generate huge amounts of data (Du and Sun, 2006a). There are two ways of doing this: one way is to transform the feature data into a more efficient set of vectors, and the other way is to attempt to exclude weak or unimportant features.

The first way can involve methods that only analyze the feature data, such as principle component analysis (PCA), which transforms the features into a new space where most of the variability lies in a few vectors and allows the later vectors (which are essentially noise) to be excluded (Du and Sun, 2006a). This approach has the advantage of creating new variables that are orthogonal; however, it makes an assumption that what explains most of the variability in features should also explain most of the variability in the food quality. This assumption can be flawed. An alternative to PCA is partial least squares regression (PLSR), which searches for vectors in the feature space that explain both feature variability and food quality variability. This approach has the advantage of finding vectors suited to explaining the response variability. The PLSR method is explained in depth by Wold et al. (2001).

The second way uses what is known as a best subset method. Each feature is assessed to see if it is of value, and then included or excluded. Like with data transformation, the food quality data may or may not be part of the inclusion/exclusion decision-making process. Approaches such as stepwise selection or genetic selection can assess image features for their utility in explaining the response variability.

Stepwise selection is a very popular approach (Du and Sun, 2006a), which cycles through the image features to see if adding or subtracting a feature offers an improvement with a certain degree of certainty. This is explained in more detail by Han and Kamber (2001). Genetic selection follows the same idea of searching for the ideal subset but has an additional advantage of making no assumptions about the data. Thus, with appropriate fitness function definition and algorithm parameters, a close convergence on the ideal subset is ensured. This was successfully applied by Jackman et al. (2009a).

4.4.5 CLASSIFICATION AND PREDICTION

With the issue of dimensionality reduction settled, a classification or prediction model can be built. Building a classification model will require us to use the selected image features to decide which class the food should belong to. This can be done with both explicit modeling and implicit modeling. Explicit modeling involves methods such as linear discriminant analysis, k-nearest neighbors, and principle component analysis. Implicit modeling involves artificial intelligence methods such as neural networks, fuzzy logic, or support vector machines.

Explicit or statistical modeling is very much preferred, as there is a clearly understood model present. There are three types of statistical models: Bayesian, discriminant, and nearest neighbor (Du and Sun, 2006a). Implicit models such as artificial neural networks (ANNs) are essentially a black box into which feature data enter and a predicted class emerges (Du and Sun, 2006a). Only if explicit models fail to perform adequately should an implicit model be considered.

Predictive modeling is far more difficult, as an exact value is to be predicted, not a discrete class (within a range of values). Predictive modeling follows a similar line to classification modeling in that explicit models are very much preferred. The main types of explicit models are multilinear regression (MLR), partial least squares regression (PLSR), and principle component regression (PCR).

Any model built should be tested for robustness. Thus, comprehensive calibration and validation is required. This entails a substantial data set spanning a large range. Calibration is the easy part, where image and food quality data are used to compute the model parameters. This carries a danger of the noise in the data being integrated into the model. To protect against this, validation must take place after calibration. There are three major forms of validation: train and test, cross-validation, and segmented cross-validation (Harrell, 2002).

Train and test excludes some data, and the model is calibrated only with the training data and then validated with the test data. This has the advantage of being a truly independent test and is usually preferred when there is a large amount of data. However, it has the weakness that the test data are wasted, as the model does not get to learn anything from the test data set. Cross-validation cycles through the data set, excluding one sample at a time. The excluded sample is predicted from the rest. The predictions are then collected at the end and compared to the original measured response data. This has the advantage of making full use of the data and is a fully reproducible answer; however, it has the drawback of lacking a fully independent test data set. A compromise between the above is segmented cross-validation, which divides the data into, say, ten segments and cross-validates between the ten segments. This keeps the advantages of cross-validation by using all of the data while effectively amounting to ten training-testing splits. After validation has been completed, the final model can be applied to new data with confidence that the true relationship between the image features and the food quality data has been absorbed into the model parameters, while simultaneously being confident that the effects of the noises of the training and testing data have been excluded from the model parameters.

4.5   APPLICATIONS

Computer vision has followed trends in the last few years quite similar to previous research, as there have been few major technological advances in either hardware or software to facilitate radical change. The cost of penetrative imaging devices remains prohibitive; thus, every effort has been made to find solutions with visible and nonvisible surface examinations, with penetrative imaging used as a last resort.

The most notable change in recent years in terms of hardware is the considerable increase in the use of image scanners, rather than a camera with independent lighting. This offers a considerable advantage of eliminating worries about the consistency of lighting, as external lighting is completely excluded. Consistent lighting makes developing sRGB data easier, allowing for immediate comparison with L*a*b* data. The images of many food products acquired by scanners include potato chips (Romani et al., 2009), cakes (Wilderjans et al., 2008), bread (Shittu et al., 2008), fish (Kause et al., 2008), almonds (Varela et al., 2008a), cheese (Jelinski et al., 2007), seeds (Venora et al., 2007), hazelnuts (Pallotino et al., 2009), and apples (Menesatti et al., 2009).

While new and more efficient image hardware and software packages continue to be developed, their fundamental principles and applications have not substantially changed; thus, in the last few years there has been little noticeable change. Image data are still recorded and saved in the same manner.

Image segmentation has drawn on previous successful techniques with few segmentation algorithms standing out as being particularly different. The most obvious exceptions to this trend would be the region growing algorithm of Blasco et al. (2007a) and the active contour method of Antequera et al. (2007).

An image feature description of color, shape, and size has already been almost fully explored. The only area where there have been some substantially new approaches is for the expression of image texture features, which will be discussed in greater detail in later sections. An exception to this trend was the customized colorspace developed by Lee et al. (2008a) for examining dates. This availed of the fact that dates vary from yellow to dark red; thus, blues and greens are not required for consideration. Hence, a customized color channel, as shown in Figure 4.1, was developed that greatly simplified the image processing.

Similarly, image data processing has not changed notably in the last few years, with the problems still needing to be solved being palatability prediction, quality classification, or defect detection. Thus, the tried and trusted approaches of multivariate statistics and artificial intelligence continue to be applied, and simple and easily tractable predictive models are still desired.

(a)

(b)

FIGURE 4.1   The customized color channel of Lee et al. (2008a): (a) in color and (b) in grey-scale. (From D. J. Lee et al., *Journal of Food Engineering*, 88(3):364–372, 2008. Elsevier. With permission.)

4.5.1 EVALUATION OF COLOR

Color continues to be recorded in the usual way of a pixel's intensity in red, green, and blue, whether with a digital camera or a scanner. Conversion to L*a*b* is sometimes necessary for comparison with colorimeter data or for convenience, as the food science community is long accustomed to expressing color information as L*a*b*.

4.5.1.1 Meats

Meat can be split into two *ad hoc* categories of color: "red" meat, which is beef, pork, lamb, and other such large quadrupeds, and "white" meat, which is chicken, turkey, fish, and other birds. Color is a more ergonomic indicator for red meats, as larger animal carcasses need to be cut into pieces for further processing. The white meat animals are usually processed whole, and thus the muscle color cannot be observed. Some fish such as salmon have colored flesh, and the color is a reflection of diet, which is expected to affect quality. Also, some fish have distinctively colored scales and fins allowing them to be identified. An example of meat color is shown in Figure 4.2, indicating the color of beef muscle after carcass cutting.

Considerable efforts were made by Jackman et al. (2008, 2009a, 2009e) to characterize the color of beef muscle. Muscle color is important, as it reflects the muscle pH, which in turn reflects the muscle fiber spacing. In these studies (Jackman et al., 2008, 2009a, 2009e), simple histograms of red, green, and blue were efficiently used to condense the color information into fifteen features. Color on its own was ineffective, but it made a valuable contribution to the final models. Similarly, Folkestad et al. (2008) used RGB values to characterize salmon muscle color, but only the mean of each color channel was used, in which the predictions of fat and pigment content from color correlated very strongly (r^2 up to 0.84) with the independent chemical measurements. The muscle of salmon fillets was also very effectively characterized with RGB values by Stien et al. (2007), which allowed strips of fat in between the muscles to be identified, and a strong correlation ($r = 0.84$) was developed between fat area by image analysis and fat content by chemical analysis.

More convenient means of expressing color were used by Zion et al. (2008), as the color of fish skin differs greatly in the hue channel of the hue, saturation, and intensity (HSI) colorspace. Hue angle profiles proved to be a powerful indicator of gender, allowing for 96% correct classification. The RGB colorspace was not found

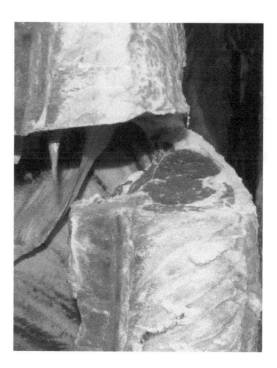

FIGURE 4.2   The exposed muscle after beef carcass cutting.

to be suitable by Kause et al. (2008) for expressing the color of trout flesh; instead, the L*C*h* (luminance, chromacity, and hue) colorspace was used. This keeps lightness from the L*a*b* colorspace but expresses the color information in a manner conceptually similar to that of the HSI colorspace. The mean of the hue channel discriminated well between high-quality red flesh and low-quality yellow flesh, and the mean of the lightness channel discriminated well between high-quality dark flesh and low-quality bright flesh. Similarly, Stien et al. (2006a) used both L*a*b* and L*C*h* to characterize fish flesh color with mean values. This allowed fat to be identified and correlated with chemically analyzed fat (r = 0.78).

The importance of being able to accurately convert raw color data to L*a*b* was illustrated by Quevedo et al. (2008a) as expert grader's color cards for assessing salmon flesh are manufactured according to set L*a*b* values; thus, the RGB values derived from image analysis must be convertible to L*a*b* values without substantial error. A selection of algorithms for reliably converting RGB data into L*a*b* data is discussed by Leon et al. (2006). It was shown to be possible to minimize data conversion error to less than 1% with the neural network model. Any system that converts RGB data to L*a*b* data must not introduce errors of greater than 2.2 color units; a robust system was thus developed by Valous et al. (2009a) to perform this.

At very high magnifications shadows can be a substantial hazard in attempting to develop meaningful images of meat. Thus, a means of retrospectively eliminating the shadows without loss of image information is required. Such a method was

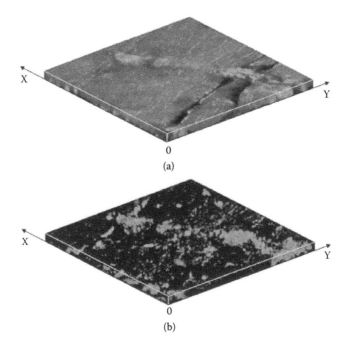

(a)

(b)

FIGURE 4.3 Shading corrected images of beef muscle: (a) unsegmented and (b) segmented (From Du and Sun, *Food and Bioprocess Technology*, 2(2):167–76, 2009. With permission.)

developed by Du and Sun (2009). This allowed shadow removal from beef confocal microscope images, allowing for high-quality imaging of beef muscle with a modest three-dimensional component. An example is shown in Figure 4.3.

4.5.1.2 Fruit and Vegetables

Fruit color reflects quality and maturity. Deviation from normal color values and distributions shows the presence of a malaise or advancement of ripening beyond an acceptable or expected level. Expression of color data in L*a*b* form is still preferred by the food science community. For fruits, color and color change are something that can be predicted. For example, the mean color features of banana cuts were expressed in L*a*b* form by Quevedo et al. (2009a), as were pear cuts by Quevedo et al. (2009b), pumpkin cuts by Zenoozian and Devahastin (2009), and kiwifruit cuts by Fathi et al. (2009).

Citrus fruit color was characterized with a wide variety of colorspaces by Blasco et al. (2007b) to find the color channels with the strongest discriminating capacity for separating good and defect pixels. The HSI colorspace was used by Abdullah et al. (2006) to assist in the segregation of starfruits by maturity based on the hue histograms, which change strongly with maturity. Very accurate classification (up to 100% correct classifications) was possible. Fruit juice color is also important, as it can be related to rheological properties. The rheology of grape juice was related to juice brightness by Kaya et al. (2008), as the addition of whitener powder altered both the color and viscosity.

Much like fruit, vegetable color similarly indicates quality and maturity. Deviation from normal values will show a malaise or advancement of maturity beyond an acceptable or expected level. The color of mushrooms, for example, degrades and dark brown spots appear with advancing maturity. This alters the mean and standard deviation of the mushrooms' brightness in a logistic manner (Aguirre et al., 2009). As process conditions such as temperature and relative humidity heavily influence the rate of maturation, vegetable malnutrition can be observed by color changes, as with legumes (Wiwart et al., 2009), in which the mean HSI and L*a*b* values of the leaves changed noticeably as the supply of essential nutrients of nitrogen, phosphorous, magnesium, and potassium was varied.

Color is an essential quality factor of pomegranate arils, as color varies from white to dark red/brown, depending on the degree of maturity. Using color features, accurate sorting of the arils (up to 90% correct classification) was developed by Blasco et al. (2009a) using RGB thresholds. A variety of arils that were classified are shown in Figure 4.4.

Charles et al. (2008) observed how the leaf greening and base browning on endives is delayed by various active packaging arrangements. The mean L*a*b* color change for the endive surface reflects the progression of maturity, and hence the effect of the active packaging.

Like other fruit and vegetables, tomatoes degrade upon exposure to oxygen, which triggers a simultaneous color change. The change in the mean RGB values can thus be a proxy for degradation, as studied by Lana et al. (2006a). Simultaneously, tomato

FIGURE 4.4 A selection of pomegranate arils and loose material. (From Blasco et al., *Journal of Food Engineering*, 90(1):27–34, 2009a. With permission.)

firmness decreases exponentially over time; thus, a strong exponential model was found for color changes over time with an r^2 of over 0.95. In addition, the fading of tomato color as well as the changing of color can be a significant indicator of maturity. Once cut, fresh tomatoes begin to degrade, and the translucency of tomato flesh increases. Thus, by filming tomato slices under both a black and white background, the translucent effect can be observed, which was investigated by Lana et al. (2006b).

Color changes do not just happen naturally. Food processing such as frying also makes large changes to color properties of vegetable-like potatoes. Development of brown areas during frying indicates acrylamide, and development of clear areas indicates excessive fat absorption. Potato color is normally expressed with L*a*b* data, and the change in the mean of the three color channels can be observed as the frying advances, allowing characterization of these phenomena (Romani et al., 2009). Investigation of the presence of acrylamide in fried potato chips was also conducted by Viklund et al. (2007). The browning of the chips due to acrylamide was reflected in a decline in lightness (L*), while green to red (a*) and blue to yellow (b*) increased. Strong correlations (from 0.9 to 0.92) between mean color values and chemically evaluated acrylamide content were found. A similar study was done by Gokmen et al. (2007) that assessed homemade french fries as well as potato chips, and found strong correlations between acrylamide concentration and the brown area extracted from the potato image based on RGB thresholds. Frying temperature is a fundamental process parameter in the development of browning. As might be expected, higher temperature will increase browning. This was investigated by Pedreschi et al. (2006) where strongly linear relationships were found between total mean L*a*b* color change and frying time for a selection of frying temperatures. Figure 4.5 shows the browning of a potato chip.

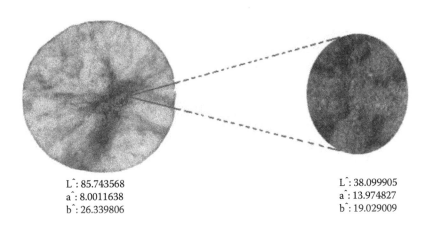

L^: 85.743568
a^: 8.0011638
b^: 26.339806

L^: 38.099905
a^: 13.974827
b^: 19.029009

FIGURE 4.5 The browning of a potato chip. (From Pedreschi et al., *Food Research International*, 39(10):1092–98, 2006. With permission.)

4.5.1.3   Other Foods

Color changes from white to yellow are regular occurrences in food processing. In this case advancement of the yellow color can be characterized by the decline in blue, and therefore the RGB data do not need to be converted to L*a*b*. Such changes are seen in yeast fermentation (Acevedo et al., 2009).

During baking, the color of bread sees a change from a very pale brown to a much darker brown as the crust develops. Purlis and Salvadori (2009) were able to monitor the browning of bread during crust development so the crust color could be related to the progression of baking. The change of the mean L*a*b* values was fed into an exponential model as crust formation follows first-order kinetics. A small relative error (<4%) was observed. The water activity of the bread and the baking temperature were important factors in the browning rate.

The roasting of coffee beans must be stopped before excessive burning occurs. Thus, Hernandez et al. (2008) monitored the color change of green raw coffee beans as they turned dark brown during roasting and related it to process parameters very accurately ($r^2 > 0.98$). It was then possible to determine the correct time to stop roasting using the predicted bean color. Like coffee beans, most beans have essential color features. A selection of beans was conducted by Kilic et al. (2007) based on color features. With the mean, variance, skewness, and kurtosis of the RGB values, an overall bean classification rate of greater than 90% was possible.

Moisture content is a critical process variable in food production. It is possible for some foods like cereal grains to predict the moisture content from the color properties. This was investigated by Tahir et al. (2007), who indicated that color features contributed to classification models capable of up to 98% correct classifications.

Some foods have a sheddable skin, such as hazelnuts. This skin can be a different color than the pulp or flesh. Thus, peeling of the skin will show up as a color change. For example, after roasting, hazelnuts shed some or all of their skin. As the skin is a dark brown while the pulp is almost white, RGB values were used by Pallotino et al. (2009) to allow exposed and unexposed pulp surface to be identified.

4.5.2   Evaluation of Texture

4.5.2.1   Meats

For red meats texture has proven to be a very powerful quality indicator, as carcass cutting exposes a fairly flat muscle surface, allowing for the analysis of high-quality two-dimensional surface texture. The surface texture reflects the connective tissue distribution and the geometry of muscle fibers that feed closely into important quality features when cooked, such as acceptability, tenderness, juiciness, and flavor. A comprehensive investigation into the predictive power of beef surface texture features was undertaken by Jackman et al. (2008, 2009a, 2009c, 2009d, 2009e), who found that it was possible to develop some accurate models ($r^2 > 0.8$, standard error < 0.333) using texture features alone or in concert with color and marbling features. Texture was normally expressed with wavelet transforms rather than classic algorithms. An image of beef texture is shown in Figure 4.6.

FIGURE 4.6 Saturation channel image of beef surface texture. (From Jackman et al., *Meat Science*, 80(4):1273–81, 2008. With permission.)

Very similar work on lamb surface texture was performed by Chandraratne et al. (2006a, 2006b, 2007), who relied on the classic texture algorithms of grey level co-occurrence, grey level run lengths, and grey level difference histograms. Multiclassification of carcasses into six grades was possible with up to 97% accuracy in concert with some geometrical features (Chandraratne et al., 2007). The tenderness of lamb could be predicted with a reasonably high level of accuracy ($r^2 = 0.75$) from a very similar approach with texture and geometric features (Chandraratne et al., 2006a).

Processed meat such as ham retains useful texture features, as the ham is composed of at least one muscle that is cut in a manner similar to a butcher cut of a carcass. Work by Valous et al. (2009b) on pork ham found that texture features expressed with fractal dimensions and lacunarity functions yielded substantial information on overall ham roughness, porosity, and heterogeneity. These are required, as high-quality ham has a smoother surface and low-quality ham has a rougher surface.

Further use of fractal dimensions was found by Niamnuy et al. (2008) to observe how shrimp microtexture changes in concert with protein content due to various boiling conditions. Strong correlations were found between the normalized fractal dimension and chemically measured protein content.

4.5.2.2 Fruit and Vegetables

Fractal dimensions were used by Quevedo et al. (2009a) to describe banana browning. By noting that browning advances with the creation and growth of brown spots on the flesh, it was possible to imagine the pattern of yellow and brown regions as a form of roughness. As such roughness can be suited to expression with a fractal dimension, the fractal dimension therefore mirrors the advancement of browning.

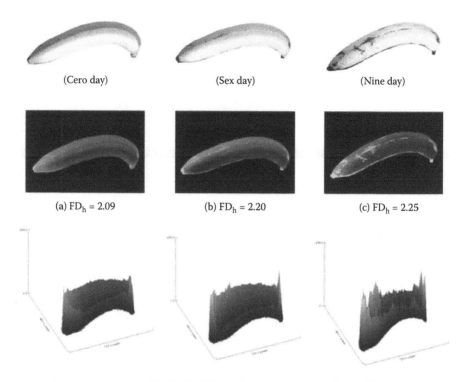

(Cero day)   (Sex day)   (Nine day)

(a) FD$_h$ = 2.09   (b) FD$_h$ = 2.20   (c) FD$_h$ = 2.25

FIGURE 4.7 Images of ripening banana and the changes of its fractal dimension. (From Quevedo et al., *Journal of Food Engineering*, 84(4):509–15, 2008b. With permission.)

A very similar approach was applied by Quevedo et al. (2009b) on pears and on the browning of banana skin as overripening sets in (Quevedo et al., 2008b). The fractal dimension changes as the browning spots are created and grow. Images of ripening bananas are shown in Figure 4.7.

The wavelet transform was used by Zenoozian and Devahastin (2009) to extract textural features from dried pumpkin flesh. The resulting wavelet coefficients contributed to very accurate models (r^2 up to 1) of important quality parameters such as moisture ratio, Heywood shape factor, and overall color change.

A selection of fruit diseases and malaises can be seen in their accompanying leaves. For example, citrus fruits can suffer from greasy spot, melanose, and scab (Pydipati et al., 2006). It was possible to use color co-occurrence matrices from the HSI colorspace to extract useful texture features, allowing for up to 95 to 100% correct classification of normal, greasy spot, melanose, and scab.

Automatically separating walnut shell fragments from walnut pulp is important to the development of the walnut industry due to the high costs of manual removal of residual shell fragments (Jin et al., 2008). As pulp and shell display different texture properties when subjected to strong backlighting, accurate classification (up to 98%) of objects as shell or pulp is possible (Jin et al., 2008). Simple textural features of local binary pattern and local variance measurement proved adequate. Figure 4.8 shows some examples of walnut images.

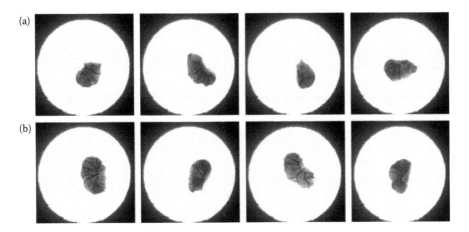

FIGURE 4.8 Backlight images of walnuts. (From Jin et al., *Journal of Food Engineering*, 88(1):75–85, 2008. With permission.)

4.5.2.3 Other Foods

Aside from the above, image texture properties proved useful for a wide range of other food types. Baked products develop a matrix of holes due to gas evolution during cooking. The density and distributions of the holes are a simple means of expressing texture. The texture of various baked products is caused by their ingredients and baking process parameters. Thus, texture data can feed back into the ingredient and baking parameter selection process. Bread (Shittu et al., 2008) is the most typical baked product as well as cake (Wilderjans et al., 2008). The use of different ingredients was shown to be reflected in the surface texture feature by Gonzales-Barron and Butler (2008), who used three principal components derived from classic co-occurrence features to account for 93% of bread crumb variance.

The features of the bread bubble matrix were also used by Lassoued et al. (2007) in both two-dimensional CCD scanners and three-dimensional x-ray images. Crumb fineness and heterogeneity could be extracted from the two-dimensional images, and cell size from the three-dimensional images. This allowed the recipe and cooking method to be identified. Crumb texture can also be expressed with simple features such as cell area, density, and shape distribution (Brescia et al., 2007). These features, however, proved weak indicators of the type of bread prepared. Similarly, Rosales-Juarez et al. (2008) used cell density, mean cell area, and shape factors to characterize the crumb texture in an attempt to distinguish between bread without soybean and bread with either germinated or nongerminated soybean. Germinated soybean did cause a coarser crumb texture.

If the bubble matrix properties could be approximated by a faster technique such as ultrasound, then crumb texture analysis would become a more viable option for quality control. This was investigated by Lagrain et al. (2006). The ultrasound allowed estimation (via the Biot-Aillard model) of flow resistivity, open porosity, size of the intersections in the crumb cell walls, and tortuosity, which are related to crumb texture. Similarly for cake, the bubble size and size distribution are heavily dependent on the gluten content of the cake mix (Wilderjans et al., 2008). A typical crumb texture is shown in Figure 4.9.

Surface texture can be intimately related to color and gloss. This was investigated for chocolate by Briones et al. (2006). Surface roughness was simulated by applying sandpaper, and the resulting change in brightness and gloss was noted. There

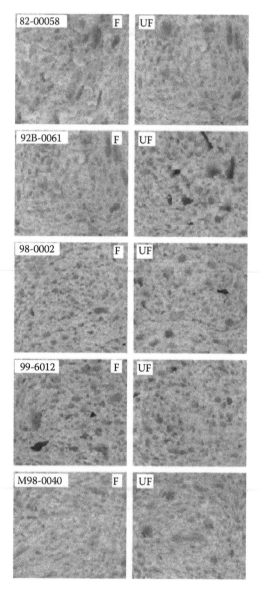

FIGURE 4.9 Typical textures produced during bread baking. (From Shittu et al., *Food Research International*, 41(6):569–78, 2008. With permission.)

was an exponential decline in gloss ($r^2 = 0.96$) with increasing roughness. Lightness decreased linearly with increasing roughness ($r^2 = 0.95$).

Occasionally an internal inspection is required. This can be the case for examining wheat kernels (Neethirajan et al., 2006). Both transmitted light and soft x-rays were used to develop texture features. Vitreous and nonvitreous kernels could be correctly classified using the transmitted light features to approximately 90% using texture as well as other features.

Not only solids have important textural features, but liquids such as salad dressing can also have important textural features (Christiansen et al., 2006). The important property of viscosity could be related to image texture properties derived from scanning electron microscope (SEM) images. Features of average absolute difference in contrast vectors as well as box counting vectors were used.

Texture features can also have practical engineering application, as in the case of milk fouling in a Couette. Fouling is particularly challenging, as it is temperature dependent; at low temperatures weaker van der Waals bonding is dominant, while at higher temperatures stronger covalent bonding is dominant. Fouling is also dependent on shear rate, as a greater shear rate will accelerate the formation of aggregates (Simmons et al., 2007). The difference in fouling can be observed with image texture, as at higher temperatures larger aggregate particles form, leading to a coarser texture. Addition of minerals to the whey protein concentrate successfully reduced particle size, but at the price of increased deposition onto the Couette.

4.5.3   EVALUATION OF SHAPE AND SIZE

4.5.3.1   Meats

For ham, porosity is an essential quality indicator as the development of pores indicates a high yield or high level of brine injection. Pore density and area are thus parameters that need to be evaluated. A means of evaluating pores was developed by Valous et al. (2009b). Figure 4.10 shows some porosity in ham. Further work on characterizing ham pores was carried out by Du and Sun (2006b) to develop correlations with processing time, moisture content, and texture profile analysis. A series of significant correlations were found. Estimation of weight loss in beef joints was studied by Zheng et al. (2006b). Ellipsoid curves were fitted to images of the joints and very strong predictions ($r^2 > 0.99$) were developed for the shrinkage of volume, surface area, and joint axes.

For uncut meat muscle, the shape and size of the muscle can be characterized with B-spline curves (Goni et al., 2008) with magnetic resonance imaging (MRI), thus allowing muscle assessment in a truly nondestructive way. For carcass pork ham it is possible to hang the ham over a period of time to develop palatability attributes. With MRI the ripening process can be followed by using active contour algorithms (Antequera et al., 2007) to identify the ham muscle. The muscle shrinking identified with MRI correlates extremely well with weight loss ($r^2 = 0.99$).

For fish, shape and size are a much more immediately obvious significant parameter, as different types of fish have different shapes and sizes. For example, the

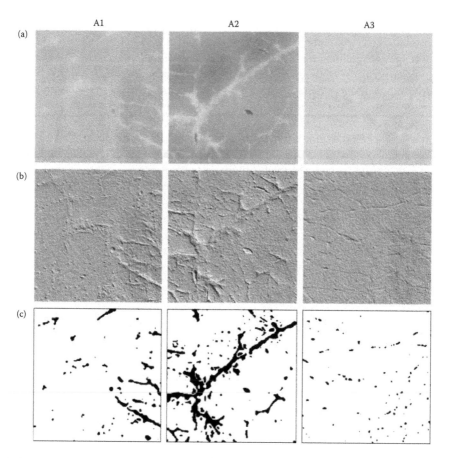

FIGURE 4.10   Ham images showing some porosity. (From Valous et al., *Food Research International*, 42(3):353–62, 2009b. With permission.)

gender of guppies could be classified with up to 90% accuracy using shape features alone by Zion et al. (2008), such as tail area-to-body area or tail length-to-body length ratio. Three species of fish were almost perfectly correctly classified (up to 99% correct classifications) by Zion et al. (2007) using shape features of the back-light fish, such as tail area, fish perimeter, and body length. Figure 4.11 shows the shape outlines.

Rigor contraction is a common phenomenon in fish processing. Attempts to relate overall length shrinkage to contraction and isometric tension throughout the fillet were successfully performed by Stien et al. (2006b). The ability of fish to recover their shape following a compression is a strong indicator of firmness. Thus, Quevedo and Aguilera (2008) characterized the curvature of salmon fillets when compressed with a weight. This was compared with direct firmness measurements of the fillet muscle.

The morphology of clams is crucial to identify their different breeds. The shape of clams can be awkward; therefore, Costa et al. (2008) used Fourier harmonics to

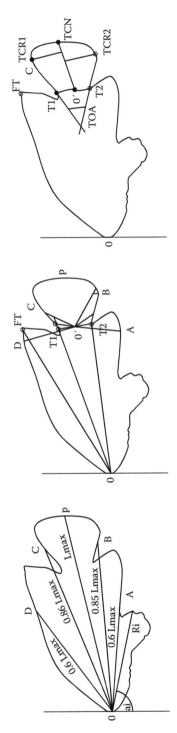

FIGURE 4.11 Shape analysis of fish breed. (From Zion et al., *Computer and Electronics in Agriculture*, 56(1):34–45, 2007. With permission.)

characterize the clam morphology. This allowed a partial least squares discriminant analysis model to correctly classify over 96% of clams.

4.5.3.2 Fruit and Vegetables

An odd or eccentric shape in fruits points to poor quality or some kind of malaise. Satsuma segments were analyzed by Blasco et al. (2009b) for deviation from expected shape. Using backlighting silhouettes from the object, shapes were analyzed, and good segments, broken segments, loose skin, and other raw material were identified with basic morphological features. The results showed that identification of good segments at high accuracy (93% correct classifications) was possible.

Olive defects can be easily characterized with morphological features as applied by Riquelme et al. (2008). Defects of mussel scale or serpeta, hail damaged or granizo, mill or rehús, wrinkled olive or agostado, purple olive and undefined damage or molestado could be classified up to 100% correctly, although there was large variability in the success rate.

The size of dates and the fraction of delaminated skin are essential to their quality. High-quality dates are larger, with a low fraction of skin delaminated. Size and delaminated area can be measured with near-infrared (NIR) spectroscopy as described by Lee et al. (2008b) as natural variation in date color, making color a poor choice as a discriminator. Some good classification rates were possible offline (up to 87%), but this declined when the system went online (up to 79%). Example images are given in Figure 4.12.

Apple stem and calyx need to be removed before packing. Their removal procedures can be imperfect, and thus the presence of stems and calyxes must be monitored. Simple shape factors can identify a stem or calyx, including area perimeter and circularity. A support vector machine was capable of perfect classification of calyx (100% correct classification) and almost perfect (99% correct classification) of stems (Unay and Gosselin, 2007). Furthermore, deformation of starfruits can be characterized with the Fourier transform (Abdullah et al., 2006). This proved to be

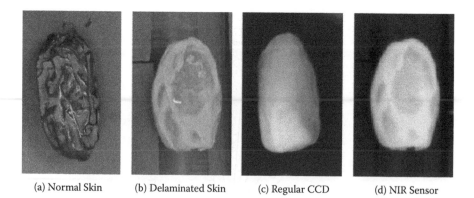

(a) Normal Skin   (b) Delaminated Skin   (c) Regular CCD   (d) NIR Sensor

FIGURE 4.12 Normal and delaminated dates under visible and near-infrared lighting. (From Lee et al., *Journal of Food Engineering*, 86(3):388–98, 2008b. With permission.)

highly successful, leading to perfect classification of good, slightly deformed, and badly deformed starfruits.

Food shape and surface area are of great interest to food microbiologists, as greater area implies greater bacterial contamination. Thus, Eifert et al. (2006) used surface fitting and three-dimensional wire frame models to estimate fruit surface area from weight for a selection of fruits that have reasonably spherical shapes. Results were mixed with some excellent predictions (r^2 up to 0.96 for strawberries) and some poorer ones.

Shape and size features are of similar importance to vegetables as fruit. Tamarind pods, for example, have expected shapes and sizes; the size of the surface defect also affects expected quality. Thus, Jarimopas and Jaisin (2008) used pod curvature and defect size to classify pods as being of acceptable quality or not with high rates of correct classification (94% for Srichompoo pods). Likewise, for chicory the importance of shape and size is well known. A comparison of classic and Fourier-based descriptors was made by Lootens et al. (2007). Both types had advantages and disadvantages, leading to the conclusion that a combination of both would be ideal. The utility of the size features of beans is described by Kilic et al. (2007). The size measured by image analysis methods was able to correlate very well with independent size measurement with callipers (r up to 0.98).

4.5.3.3 Other Foods

Many other food products also have important shape and size properties. For example, the surface area of coffee beans must be accurately estimated, as heat transfer is dependent on heat transfer area. A model for estimating bean surface area was therefore developed by Hernandez et al. (2008), with which bean surface area could be predicted with very high accuracy ($r^2 = 0.99$).

Rice kernels will expand upon water absorption in cooking. The extent of the expansion was studied by Yadav and Jindal (2007). With basic dimensions like length, width, perimeter, and projected area, a satisfactory exponential model and a satisfactory Peleg model relating water absorption and morphological changes were developed. Meanwhile, excessive moisture absorption into rice kernels can cause the kernels to crack and irreversibly damage the product. It proved possible to correlate the elongation of rice kernels with the degree of fissuring (Shimizu et al., 2008). In addition, wheat kernels' vitreousness is related to the kernel morphology (Neethirajan et al., 2006).

The cracking of almonds under compression has a direct effect on human taste sensation. Thus, nut compression fragment size distribution can be compared to sensory panel results (Varela et al., 2008a). This allows image analysis of the size distributions of the fragments in machine testing to be used for classification. Further work by Varela et al. (2008b) used similar fracture pattern analysis to predict the crunchiness and crispiness of almonds.

Cereals, like many foods, dilate upon absorption of water. The extent of that dilation is reflected in changes of shape and size. This was investigated by Tahir et al. (2007) using wheat and barley grain kernels. Shape and size were characterized with simple dimensions, invariant shape moments, and Fourier transform properties.

These morphological features proved less powerful than color and texture features for predicting the moisture content.

The expansion and subsequent contraction of hot extruded bread flour is related to the viscosity of the dough, and therefore is a fundamental process parameter. Image analysis was used by Arhaliass et al. (2009) to develop a model of expansion indices and velocities based on the bubble growth model.

4.6   CONCLUSIONS

Based on the above discussion, it can be seen that computer vision technology continues to be a very useful tool in tackling a wide variety of food classification and quality prediction problems. Good predictions and accurate classifications have proven possible on a large number of occasions. Color, texture, and morphological features continue to be the bedrock of computer vision models, as these are the discriminating features that are usually found in the expert grading manuals issued by competent authorities on food quality assessment. These features remain the most convenient to record, characterize, and process.

In future perspectives, computer vision will continue to be successful, and that success will follow similar patterns. This is due to the fact that aside from faster computers with greater processing memory, the fundamentals of image acquisition, image recording, image data handling, and processing have changed little and are unlikely to change substantially in the near future. The only area remaining for real innovation is data generation, in particular very high-resolution two-dimensional data or affordable three-dimensional data that can retain resolution with deep penetration. However, both of these will continue to be mercurial, as the equipment cost cannot be expected to fall substantially in a short time. In addition, models developed should maintain the simplicity and ease of traction, with good accuracy.

REFERENCES

Abdullah, M. Z., Mohamad-Saleh, J., Fathinul-Syahir, A. S., and Mohd-Azemi, B. M. N. 2006. Discrimination and classification of fresh-cut starfruits (*Averrhoa carambola* L.) using automated machine vision system. *Journal of Food Engineering* 76(4):506–23.

Acevedo, C. A., Skurtys, O., Young, M. E., Enrione, J., Pedreschi, F., and Osorio, F. 2009. A non-destructive digital imaging method to predict immobilized yeast-biomass. *LWT-Food Science and Technology* 42(8):1444–49.

Aguilera, J. M., and Briones, V. 2005. Computer vision and food quality. *Food Australia* 57(3):79–87.

Aguirre, L., Frias, J. M., Barry-Ryan, C., and Grogan, H. 2009. Modelling browning and brown spotting of mushrooms (*Agaricus bisporus*) stored in controlled environmental conditions using image analysis. *Journal of Food Engineering* 91(2):280–86.

Antequera, T., Caro, A., Rodriguez, P. G., and Perez, T. 2007. Monitoring the ripening process of Iberian ham by computer vision on magnetic resonance imaging. *Meat Science* 76(3):561–67.

Arhaliass, A., Legrand, J., Vauchel, P., Fodil-Pacha, F., Lamer, T., and Bouvier, J. M. 2009. The effect of wheat and maize flours properties on the expansion mechanism during extrusion cooking. *Food and Bioprocess Technology* 2(2):186–93.

Barni, M., Cappellini, V., and Mecocci, A. 1997. Colour-based detection of defects on chicken meat. *Image and Vision Computing* 15(7):549–56.

Basset, O., Buquet, B., Aboulekaram, S., Delachartre, P., and Culioli, J. 2000. Application of texture image analysis for the classification of bovine meat. *Food Chemistry* 69(4):437–45.

Blasco, J., Aleixos, N., Gomez, J., and Molto, E. 2007b. Citrus sorting by identification of the most common defects using multispectral computer vision. *Journal of Food Engineering* 83(3):384–93.

Blasco, J., Aleixos, N., and Molto, E. 2007a. Computer vision detection of peel defects in citrus by means of a region oriented segmentation algorithm. *Journal of Food Engineering* 81(3):535–43.

Blasco, J., Aleixos, N., Cubero, S., Gomez-Sanchis, J., and Molto, E. 2009b. Automatic sorting of satsuma (*Citrus unshiu*) segments using computer vision and morphological features. *Computers and Electronics in Agriculture* 66(1):1–8.

Blasco, J., Cubero, S., Gomez-Sanchis, J., Mira, P., and Molto, E. 2009a. Development of a machine for the automatic sorting of pomegranate (*Punica granatum*) arils based on computer vision. *Journal of Food Engineering* 90(1):27–34.

Borggaard, C., Madsen, N. T., and Thodberg, H. H. 1996. In-line image analysis in the slaughter industry, illustrated by beef carcass classification. *Meat Science* 43(S1):151–63.

Brescia, M. A., Sacco, D., Sgaramella, A., Pasqualone, A., Simeone, R., Peri, G., and Sacco, A. 2007. Characterisation of different typical Italian breads by means of traditional, spectroscopic and image analyses. *Food Chemistry* 104(1):429–38.

Briones, V., Aguilera, J. M., and Brown, C. 2006. Effect of surface topography on color and gloss of chocolate samples. *Journal of Food Engineering* 77(4):776–83.

Brosnan, T., and Sun, D.-W. 2004. Improving quality inspection of food products by computer vision—A review. *Journal of Food Engineering* 61(1):3–16.

Canon USA. 2009. *GL2 user guide.* Lake Success, NY.

Chandraratne, M. R., Kulasiri, D., Frampton, C., Samarasinghe, S., and Bickerstaffe, R. 2006b. Prediction of lamb carcass grades using features extracted from lamb chop images. *Journal of Food Engineering* 74(1):116–24.

Chandraratne, M. R., Kulasiri, D., and Samarasinghe, S. 2007. Classification of lamb carcass using machine vision: Comparison of statistical and neural network analyses. *Journal of Food Engineering* 82(1):26–34.

Chandraratne, M. R., Samarasinghe, S., Kulasiri, D., and Bickerstaffe, R. 2006a. Prediction of lamb tenderness using image surface texture features. *Journal of Food Engineering* 77(3):492–99.

Charles, F., Guillaume, C., and Gontard, N. 2008. Effect of passive and active modified atmosphere packaging on quality changes of fresh endives. *Postharvest Biology and Technology* 48(1):22–29.

Chen, C.-H. 1999. *Information processing for remote sensing.* Hackensack, NJ: World Scientific.

Christiansen, K. F., Krekling, T., Kohler, A., Vegarud, G., Langsrud, T., and Egelandsdal, B. 2006. Microstructure and sensory properties of high pressure processed dressings stabilized by different whey proteins. *Food Hydrocolloids* 20(5):650–62.

Costa, C., Menesatti, P., Aguzzi, J., D'Andrea, S., Antonucci, F., Rimatori, V., Pallotino, F., and Mattoccia, M. 2008. External shape differences between sympatric populations of commercial clams *Tapes decussatus* and *T. philippinarum. Food and Bioprocess Technology,* in press. doi: 10.1007/s11947-008-0068-8.

Du, C.-J., and Sun, D.-W. 2004. Recent developments in the applications of image processing techniques for food quality evaluation. *Trends in Food Science and Technology* 15(5):230–49.

Du, C.-J., and Sun, D.-W. 2005. Comparison of three methods for classification of pizza topping using different colour space transformations. *Journal of Food Engineering* 68(3):277–87.

Du, C.-J., and Sun, D.-W. 2006a. Learning techniques used in computer vision for food quality evaluation: A review. *Journal of Food Engineering* 72(1):39–55.

Du, C.-J., and Sun, D.-W. 2006b. Automatic measurement of pores and porosity in pork ham and their correlations with processing time, water content and texture. *Meat Science* 72(2):294–302.

Du, C.-J., and Sun, D.-W. 2009. Retrospective shading correction of confocal laser scanning microscopy beef images for three-dimensional visualization. *Food and Bioprocess Technology* 2(2):167–76.

Edmund Optics. 2009. *Online catalogue.* Barrington, NJ.

Eifert, J. D., Sanglay, G. C., Lee, D. J., Sumner, S. S., and Pierson, M. D. 2006. Prediction of raw produce surface area from weight measurement. *Journal of Food Engineering* 74(4):552–56.

Fathi, M., Mohebbi, M., and Razavi, S. M. A. 2009. Application of image analysis and artificial neural network to predict mass transfer kinetics and color changes of osmotically dehydrated kiwifruit. *Food and Bioprocess Technology,* in press. doi: 10.1007/s11947-009-0222-y.

Folkestad, A., Wold, J. P., Rorvik, K. A., Tschudi, J., Haugholt, K. H., Kolstad, K., and Morkore, T. 2008. Rapid and non-invasive measurements of fat and pigment concentrations in live and slaughtered Atlantic salmon (*Salmo salar* L.). *Aquaculture* 280(1–4):129–35.

Gokmen, V., Senyuva, H. Z., Dulek, B., and Cetin, A. E. 2007. Computer vision-based image analysis for the estimation of acrylamide concentrations of potato chips and french fries. *Food Chemistry* 101(2):791–98.

Goni, S. M., Purlis, E., and Salvadori, V. O. 2008. Geometry modelling of food materials from magnetic resonance imaging. *Journal of Food Engineering* 88(4):561–67.

Gonzales-Barron, U., and Butler, F. 2008. Discrimination of crumb grain visual appearance of organic and non-organic bread loaves by image texture analysis. *Journal of Food Engineering* 84(3):480–88.

Gunasekaran, S. 1996. Computer vision technology for food quality assurance. *Trends in Food Science and Technology* 7(8):245–56.

Han, J., and Kamber, M. 2001. *Data mining: Concepts and techniques.* San Francisco: Morgan Kaufmann Publishers.

Haralick, R. M., Shanmugam, K., and Dinstein, I. 1973. Textural features for image classification. *IEEE Transactions on Systems, Man & Cybernetics* 6(3):610–21.

Harrell, F. E. 2002. *Regression modelling strategies: With applications to linear models, logistic regression, and survival analysis.* New York: Springer.

Hernandez, J. A., Heyd, B., and Trystram, G. 2008. Prediction of brightness and surface area kinetics during coffee roasting. *Journal of Food Engineering* 89(2):156–63.

HunterLab. 2008. *Insight on color: CIE L*a*b* color scale.* Reston, VA.

Jackman, P., Sun, D.-W., and Allen, P. 2009b. Automatic segmentation of beef *longissimus dorsi* muscle and marbling by an adaptable algorithm. *Meat Science* 83(2):187–94.

Jackman, P., Sun, D.-W., and Allen, P. 2009c. Comparison of various wavelet texture features to predict beef palatability. *Meat Science* 83(1):82–87.

Jackman, P., Sun, D.-W., and Allen, P. 2009d. Comparison of the predictive power of beef surface wavelet texture features at high and low magnification. *Meat Science* 82(3):353–56.

Jackman, P., Sun, D.-W., and Allen, P. 2009e. Prediction of beef palatability from colour, marbling and surface texture features of *longissimus dorsi. Journal of Food Engineering* 96(1):151–65.

Jackman, P., Sun, D.-W., Du, C.-J., and Allen, P. 2009a. Prediction of beef eating qualities from colour, marbling and wavelet surface texture features using homogenous carcass treatment. *Pattern Recognition* 42(5):751–63.

Jackman, P., Sun, D.-W., Du, C.-J., Allen, P., and Downey, G. 2008. Prediction of beef eating quality from colour, marbling and wavelet texture features. *Meat Science* 80(4):1273–81.

Jarimopas, B., and Jaisin, N. 2008. An experimental machine vision system for sorting sweet tamarind. *Journal of Food Engineering* 89(3):291–97.

Jelinski, T., Du, C.-J., Sun, D.-W., and Fornal, J. 2007. Inspection of the distribution and amount of ingredients in pasteurized cheese by computer vision. *Journal of Food Engineering* 83(1):3–9.

Jin, F., Qin, L., Jiang, L., Zhu, B., and Tao, Y. 2008. Novel separation method of black walnut meat from shell using invariant features and a supervised self-organizing map. *Journal of Food Engineering* 88(1):75–85.

Kaiser, G. 1994. *A friendly guide to wavelets*. Boston: Birkhauser.

Kause, A., Stien, L. H., Rungruangsak-Torrissen, K., Ritola, O., Ruohonen, K., and Kiessling, A. 2008. Image analysis as a tool to facilitate selective breeding of quality traits in rainbow trout. *Livestock Science* 114(2–3):315–24.

Kaya, A., Ko, S., and Gunaskaran, S. 2008. Viscosity and color change during *in situ* solidification of grape pekmez. *Food and Bioprocess Technology*, in press. doi: 10.1007/s11947-008-0169-4.

Kilic, K., Boyaci, I. H., Koksel, H., and Kusmenoglu, I. 2007. A classification system for beans using computer vision system and artificial neural networks. *Journal of Food Engineering* 78(3):897–904.

Kodak. 2009. *Technical overview: CCD technology*. Rochester, NY.

Lagrain, B., Boeckx, L., Wilderjans, E., Delcour, J. A., and Lauriks, W. 2006. Non-contact ultrasound characterization of bread crumb: Application of the Biot–Allard model. *Food Research International* 39(10):1067–75.

Lana, M. M., Tijskens, L. M. M., and van Kooten, O. 2006a. Effects of storage temperature and stage of ripening on RGB colour aspects of fresh-cut tomato pericarp using video image analysis. *Journal of Food Engineering* 77(4):871–79.

Lana, M. M., Tijskens, L. M. M., and van Kooten, O. 2006b. Modelling RGB colour aspects and translucency of fresh-cut tomatoes. *Postharvest Biology and Technology* 40(1):15–25.

Lassoued, N., Babin, P., Della-Valle, G., Devaux, M. F., and Regurre, A. L. 2007. Granulometry of bread crumb grain: Contributions of 2D and 3D image analysis at different scale. *Food Research International* 40(8):1087–97.

Lee, D.-J., Archibald, J. K., Chang, Y.-C., and Greco, C. R. 2008a. Robust color space conversion and color distribution analysis techniques for date maturity evaluation. *Journal of Food Engineering* 88(3):364–72.

Lee, D.-J., Schoenberger, R., Archibald, J. K., and McCollum, S. 2008b. Development of a machine vision system for automatic date grading using digital reflective near-infrared imaging. *Journal of Food Engineering* 86(3):388–98.

Leon, K., Mery, D., Pedreschi, F., and Leon, J. 2006. Color measurement in $L^*a^*b^*$ units from RGB digital images. *Food Research International* 39(10):1084–91.

Li, J., Tan, J., Martz, F. A., and Heymann, H. 1999. Image texture features as indicators of beef tenderness. *Meat Science* 53(1):17–22.

Li, J., Tan, J., and Shatdal, P. 2001. Classification of tough and tender beef by image texture analysis. *Meat Science* 57(4):341–46.

Lootens, P., van Waes, J., and Carlier, L. 2007. Description of the morphology of roots of *Chicorium intybus* L. partim by means of image analysis: Comparison of elliptic Fourier descriptors and classical parameters. *Computers and Electronics in Agriculture* 58(2):164–73.

Menesatti, P., Zanella, A., D'Andrea, S., Costa, C., Paglia, G., and Pallotino, F. 2009. Supervised multivariate analysis of hyper-spectral NIR images to evaluate the starch index of apples. *Food and Bioprocess Technology* 2(3):308–14.

Moreda, G. P., Ortiz-Canavate, J., Garcia-Ramos, F. J., and Ruiz-Altisent, M. 2009. Non-destructive technologies for fruit and vegetable size determination—A review. *Journal of Food Engineering* 92(1):119–36.

Neethirajan, S., Karunakaran, C., Symons, S., and Jayas, D. S. 2006. Classification of vitre-ousness in durum wheat using soft x-rays and transmitted light images. *Computers and Electronics in Agriculture* 53(1):71–78.

Niamnuy, C., Devahastin, S., and Soponronnarit, S. 2008. Changes in protein compositions and their effects on physical changes of shrimp during boiling in salt solution. *Food Chemistry* 108(1):165–75.

Pallotino, F., Menesatti, P., Costa, C., Paglia, G., De Salvador, F. R., and Lolletti, D. 2010. Image analysis techniques for automated hazelnut peeling determination. *Food and Bioprocess Technology*, 3(1):155–159.

Pedreschi, F., Leon, J., Mery, D., and Moyano, P. 2006. Development of a computer vision system to measure the color of potato chips. *Food Research International* 39(10):1092–98.

Purlis, E., and Salvadori, V. O. 2009. Modelling the browning of bread during baking. *Food Research International* 42(7):865–70.

Pydipati, R., Burks, T. F., and Lee, W. S. 2006. Identification of citrus disease using color texture features and discriminant analysis. *Computers and Electronics in Agriculture* 52(1–2):49–59.

Quevedo, R. A., and Aguilera, J. M. 2010. Computer vision and stereoscopy for estimating firmness in the salmon (*Salmon salar*) fillets. *Food and Bioprocess Technology*, 3(4):561–567.

Quevedo, R. A., Aguilera, J. M., and Pedreschi, F. 2010. Color of salmon fillets by computer vision and sensory panel. *Food and Bioprocess Technology,* in press. doi: 10.1007/s11947-008-0106-6.

Quevedo, R. A., Diaz, O., Caqueo, A., Ronceros, B., and Aguilera, J. M. 2009b. Quantification of enzymatic browning kinetics in pear slices using non-homogenous L^* color information from digital images. *LWT-Food Science International* 42(8):1367–73.

Quevedo, R. A., Diaz, O., Ronceros, B., Pedreschi, F., and Aguilera, J. M. 2009a. Description of the kinetic enzymatic browning in banana (*Musa cavendish*) slices using non-uniform color information from digital images. *Food Research International* 42(9):1309–14.

Quevedo, R. A., Mendoza, F., Aguilera, J. M., Chanona, J., and Gutierrez-Lopez, G. 2008b. Determination of senescent spotting in banana (*Musa cavendish*) using fractal texture Fourier image. *Journal of Food Engineering* 84(4):509–15.

Riquelme, M. T., Barreiro, P., Altisent, M. R., and Valero, C. 2008. Olive classification according to external damage using image analysis. *Journal of Food Engineering* 87(3):371–79.

Romani, S., Rocculi, P., Mendoza, F., and Rosa, M. D. 2009. Image characterization of potato chip appearance during frying. *Journal of Food Engineering* 93(4):487–94.

Rosales-Juarez, M., Gonzalez-Mendoza, B., Lopez-Guel, E. C., Lauzano-Bautista, F., Chanona-Perez, J., Gutierrez-Lopez, G., Farrera-Rebollo, R., and Calderon-Dominguez, G. 2008. Changes on dough rheological characteristics and bread quality as a result of the addition of germinated and non-germinated soybean flour. *Food and Bioprocess Technology* 1(2):152–60.

Shimizu, N., Haque, M. A., Andersson, M., and Kimura, T. 2008. Measurement and fissuring of rice kernels during quasi-moisture sorption by image analysis. *Journal of Cereal Science* 48(1):98–103.

Shittu, T. A., Dixon, A., Awonorin, S. O., Sanni, L. O., and Mazia-Dixon, B. 2008. Bread from composite cassava–wheat flour. II. Effect of cassava genotype and nitrogen fertilizer on bread quality. *Food Research International* 41(6):569–78.

Simmons, M. J. H., Jayaraman, P., and Fryer, P. J. 2007. The effect of temperature and shear rate upon the aggregation of whey protein and its implications for milk fouling. *Journal of Food Engineering* 79(2):517–28.

Soille, P., and Rivest, J. F. 1996. On the validity of fractal dimension measurements in image analysis. *Journal of Visual Communication and Image Representation* 7(3):217–29.

Stein, E. M., and Shakarchi, R. 2003. *Fourier Analysis: An introduction.* Princeton, NJ: Princeton University Press.

Stien, L. H., Kiessling, A., and Manne, F. 2007. Rapid estimation of fat content in salmon fillets by colour image analysis. *Journal of Food Composition and Analysis* 20(2):73–79.

Stien, L. H., Manne, F., Ruohonene, K., Kause, A., Rungruangsak-Torrinsen, K., and Kiessling, A. 2006a. Automated image analysis as a tool to quantify the colour and composition of rainbow trout (*Oncorhynchus mykiss* W.) cutlets. *Aquaculture* 261(2):695–705.

Stien, L. H., Suontama, J., and Kiessling, A. 2006b. Image analysis as a tool to quantify rigor contraction in pre-rigor-filleted fillets. *Computers and Electronics in Agriculture* 50(2):109–20.

Stokes, M., Anderson, M., Chandrasekar, S., and Motta, R. 1996. *A standard default color space for the Internet: sRGB.* Reston, VA: International Color Consortium.

Sun, D.-W. (ed.). 2008. *Computer vision technology for food quality evaluation.* San Diego: Academic Press/Elsevier.

Tahir, A. R., Neethirajan, S., Jayas, D. S., Shahin, M. A., Symons, S. J., and White, N. D. G. 2007. Evaluation of the effect of moisture content on cereal grains by digital image analysis. *Food Research International* 40(9):1140–45.

Tan, J. 2004. Meat quality evaluation by computer vision. *Journal of Food Engineering* 61(1):27–35.

Unay, D., and Gosselin, B. 2007. Stem and calyx recognition on 'Jonagold' apples by pattern recognition. *Journal of Food Engineering* 78(2):597–605.

USDA. 1997a. *United States standards for grades of carcass beef.* Washington, DC: USDA.

USDA. 1997b. *United States standards for grades of pickling cucumbers.* Washington, DC: USDA.

USDA. 2004. *United States standards for grades of peaches.* Washington, DC: USDA.

USDA. 2006. *United States standards for grades of strawberries.* Washington, DC: USDA.

USDA Agricultural Marketing Service. 2009. *Grading, certification and verification: USDA quality standards.* Washington, DC: USDA.

Valous, N. A., Mendoza, F., Sun, D.-W., and Allen, P. 2009a. Colour calibration of a laboratory computer vision system for quality evaluation of pre-sliced hams. *Meat Science* 81(1):132–41.

Valous, N. A., Mendoza, F., Sun, D.-W., and Allen, P. 2009b. Texture appearance characterization of pre-sliced pork ham images using fractal metrics: Fourier analysis dimension and lacunarity. *Food Research International* 42(3):353–62.

Varela, P., Aguilera, J. M., and Fiszman, S. 2008b. Quantification of fracture properties and microstructural features of roasted *Marcona* almonds by image analysis. *LWT-Food Science and Technology* 41(1):10–17.

Varela, P., Salvador, A., and Fiszman, S. 2008a. On the assessment of fracture in brittle foods: The case of roasted almonds. *Food Research International* 41(5):544–51.

Venora, G., Grillo, O., Shahin, M. A., and Symons, S. J. 2007. Identification of Sicilian landraces and Canadian cultivars of lentil using an image analysis system. *Food Research International* 40(1):161–66.

Viklund, G., Mendoza, F., Sjoholm, I., and Skog, K. 2007. An experimental set-up for studying acrylamide formation in potato crisps. *LWT-Food Science and Technology* 40(6):1066–71.

Warriss, P. D. 2000. *Meat science: An introductory text.* Wallingford, UK: CABI Publishing.

Wilderjans, E., Pareyt, B., Goesaert, H., Brijs, K., and Delcour, J. A. 2008. The role of gluten in a pound cake system: A model approach based on gluten–starch blends. *Food Chemistry* 110(4):909–15.

Wiwart, M., Fordonski, G., Golaszewska, K. Z., and Suchowilska, E. 2009. Early diagnostics of macronutrient deficiencies in three legume species by color image analysis. *Computers and Electronics in Agriculture* 65(1):125–32.

Wold, S., Sjostrom, M., and Eriksson, L. 2001. PLS–regression: A basic tool of chemometrics. *Chemometrics and Intelligent Laboratory Systems* 58(2):109–30.

Yadav, B. K., and Jindal, V. K. 2007. Dimensional changes in milled rice (*Oryza sativa* L.) kernel during cooking in relation to its physicochemical properties by image analysis. *Journal of Food Engineering* 81(4):710–20.

Zenoozian, M. S., and Devahastin, S. 2009. Application of wavelet transform coupled with artificial neural network for predicting physicochemical properties of osmotically dehydrated pumpkin. *Journal of Food Engineering* 90(2):219–27.

Zheng, C., Sun, D.-W., and Du, C.-J. 2006b. Estimating shrinkage of large cooked beef joints during air-blast cooling by computer vision. *Journal of Food Engineering* 72(1):56–62.

Zheng, C., Sun, D.-W., and Zheng, L. 2006a. Recent applications of image texture for evaluation of food qualities—A review. *Trends in Food Science and Technology* 17(3):113–28.

Zion, B., Alchanatis, V., Ostrovsky, V., Barki, A., and Karplus, I. 2007. Real-time underwater sorting of edible fish species. *Computers and Electronics in Agriculture* 56(1):34–45.

Zion, B., Alchanatis, V., Ostrovsky, V., Barki, A., and Karplus, I. 2008. Classification of guppies' (*Poecilia reticulata*) gender by computer vision. *Aquacultural Engineering* 38(2):97–104.

5 NIR Spectroscopy for Chemical Composition and Internal Quality in Foods

Sukwon Kang

CONTENTS

5.1 INTRODUCTION

Absorption of near-infrared (NIR) radiation, particularly by CH, NH, and OH bonds commonly present in components of food materials, has been used for determining chemical compositions and internal quality in foods and food products as a nondestructive analytical technique. Its response time is fast, its sample preparation is easy, multiple components can be analyzed, anyone can operate the instrumentation, and the instrumentation cost is lower than that for ultraviolet, visible, mid-infrared, Raman, and other spectral techniques.

The spectral region of the NIR radiation spans the wavelength range from 750 to 2,500 nm (14,300 ~ 4,000 cm^{-1}). NIR measurements can reflect significant complexity as a result of weak or highly overlapped absorptions. The practical application of the study of NIR radiation and its interaction with matter, i.e., near-infrared spectroscopy (NIRS), for agricultural and food purposes started in the 1960s with Karl Norris of the U.S. Department of Agriculture (USDA). Since then, NIRS has been a major analytical tool for determining the physicochemical properties of agricultural and food products.

5.2 PRINCIPLE OF THE TECHNIQUE

5.2.1 Physical and Spectroscopic Principles

Light (electromagnetic radiation) from the sun appears white, but if the light passes a prism (matter), it shows colors from violet to red. The human eye can recognize this visible light, but other types of light also dispersed by the prism are not visible to the human eye. Spectroscopy is a useful tool by which to measure the interaction between the electromagnetic radiation and matter that is composed of molecules, atoms, or ions, and may exist in gaseous, liquid, or solid form.

Electromagnetic radiation behaves like both a wave and a composition of particles. The properties of electromagnetic waves can be represented as oscillating perpendicular electric and magnetic fields (Figure 5.1). These fields are at right angles to each other and to the direction of propagation of the light. The oscillation shape appears sinusoidal. The crest-to-crest distance between two successive maxima is defined as the wavelength, λ, and the maximum of the vector from the origin to a point displacement of the oscillation is defined as the amplitude of the wave. The

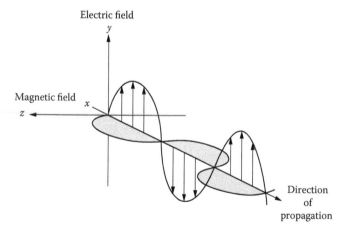

FIGURE 5.1   Single-frequency electromagnetic radiation.

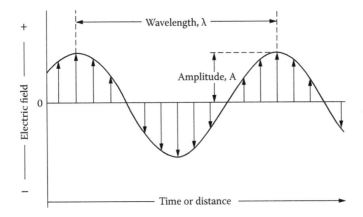

FIGURE 5.2   The electric field oscillations.

number of crests passing a fixed point per second is the frequency, ν, of the wave (Figure 5.2).

The speed of light can be presented by the wavelength of light, λ, and frequency, ν:

$$c = \lambda \nu \qquad (5.1)$$

where c is the speed of light in a vacuum, 2.997×10^8 m/s, ν is the frequency of the light in inverse seconds (Hz), and λ is the wavelength in meters. In a vacuum, the speed of light is at a maximum and does not depend on the wavelength. The frequency of light is determined by the source and does not vary. When light passes through matter, its speed is decreased. Because the frequency remains invariant, the wavelength must decrease. The speed, frequency, and wavelength of light are always positively signed.

The properties of a stream of photons, i.e., discrete quantum particles of light, can be used to describe phenomena in electromagnetic radiation, such as a permanent energy transfer from an emitting object or to an absorbing medium. These emission and absorption processes have been explained by quantum theory based on the following two important postulates. First, atoms, ions, and molecules can exist only in certain specific discrete states, characterized by defined discrete amounts of energy. If the state is changed from one specific state to another, the amount of energy involved in the emission and absorption processes is equal to the energy difference between the two states. Second, the frequency and wavelength of the radiation absorbed or emitted when a particle makes the transition from one energy state to another is related to the energy difference between the states by the equation

$$\Delta E = h \nu \qquad (5.2)$$

where E is the energy in joules (J), h is Planck's constant, 6.626×10^{-34} J s, and ν is the frequency in inverse seconds (Hz).

From Equations 5.1 and 5.2, we can deduce that

$$\Delta E = \frac{hc}{\lambda} \qquad (5.3)$$

Interaction between matter and radiation provides information about the matter. When radiation strikes a sample of matter, the radiation energy can be absorbed by the sample, transmitted through the sample, reflected off the sample surface, or scattered by the sample. If absorption occurs, it may be followed by an emission of radiation by the sample, such as luminescence.

Heat, electrical energy, chemical reactions, light, and particles can be used to stimulate a sample. Before the stimulus, a sample is in its lowest possible energy state, also called the ground state. After the stimulus, if the sample shows a transition to a higher energy state, then it is now in an excited state. Information about the sample matter is acquired by measuring the amount of electromagnetic radiation emitted by the sample as it returns from the excited state to the ground state, or by measuring the amount of electromagnetic radiation that was absorbed or scattered when the sample was excited from the ground state. If absorption happens, then the energy of sample increases and ΔE is positive. However, if emission happens, then the energy of sample decreases and ΔE is negative.

Radiant power is produced by the emission of excess energy in the form of photons while the excited particles (atoms, ions, or molecules) return to the ground state. This can provide identification and concentration information about the sample matter. An emission spectrum is a plot form of the relative power of the emitted radiation as a function of wavelength or frequency.

When the electromagnetic radiation passes through a layer of matter, energy at selected frequencies may be removed via absorption—that is, the energy is transferred to the atoms, ions, or molecules composing the matter. The amount of light absorption that occurs can be described by a function of wavelength and provides both qualitative and quantitative information about the matter. The absorption spectra for monatomic particles can be plotted for a few well-defined frequencies. Absorption spectra for polyatomic molecules can be described by the electronic energy of the molecule that arises from the energy states of its several bonding electrons, $\Delta E_{electronic}$; vibrational energy from the molecule's various atomic vibrations, $\Delta E_{vibrational}$; and rotational energy by the rotational motions within a molecule, $\Delta E_{rotational}$:

$$\Delta E = \Delta E_{electronic} + \Delta E_{vibrational} + \Delta E_{rotational} \tag{5.4}$$

The NIR radiation does not provide enough energy for the electronic transitions of polyatomic molecules, but can explain small energy differences between various vibrational and rotational states. If a molecule has a net change in dipole moment as it vibrates or rotates, it can absorb the NIR radiation. Homonuclear species such as O_2, N_2, or Cl_2 do not exhibit any net change in dipole moment during vibration or rotation, and thus cannot absorb NIR radiation.

Due to many different types of vibration, the relative positions of atoms in a molecule fluctuate continuously (Szymanski, 1964; Colthup et al., 1990). A simple diatomic or triatomic molecule can be easily described by the number and nature of such vibrations and its absorption energy. For polyatomic molecules of more than three atoms, it is more difficult due to the large number of vibrating centers and interactions among several centers. Stretching and bending are the basic categories of vibrations. A stretching vibration is a continuous variation in the interatomic distance along the axis of the bond between two atoms. A bending vibration is a change in the angle between two bonds and includes rocking, scissoring, wagging, and twisting (Figure 5.3).

Radiant radiation observed after an excited species returns to the ground state may be fluorescence or phosphorescence relaxation. When the radiation is scattered, if the wavelength of the scattered radiation is the same as that of the source radiation, it is called elastic scattering and can be used for measurements in particle sizing and concentration, such as for *nephelometry* and *turbidimetry*. Inelastic scattering produces a vibrational spectrum of sample molecules and is called Raman scattering.

The wave properties of electromagnetic radiation are diffraction, coherent, transmission, scattering, refraction, reflection, and polarization.

Diffraction happens when a parallel beam of radiation is bent as it passes by a sharp opening or through a narrow opening such as a slit. If the opening area of the slit is the same order of magnitude of wavelength, diffraction becomes pronounced. On the other hand, if the slit is wide to the wavelength, diffraction is slight and difficult to detect.

The coherent radiation pattern is the definite phase relationships with time between different points in a cross-section of the beam. It happens when the two

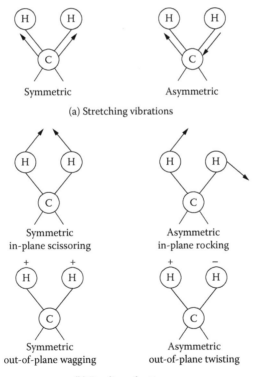

(a) Stretching vibrations

(b) Bending vibrations

FIGURE 5.3   Types of molecular vibrations.

sources of radiation have identical frequencies and the phase relationship between the two beams remains constant with time.

After the radiation passes a transparent substance, the velocity in the matter is slower than that in a vacuum, and rate radiation depends on the kinds and concentrations of atoms, ions, or molecules in the matter. If the particles in a medium are small, a significant amount of radiation will be in the original light path. On the other hand, if the particles are large, the light beam will be scattered in all directions. Among the scattering, *Rayleigh scattering* is by molecules or aggregates of molecules with dimensions significantly smaller than the wavelength of the radiation. Its intensity is proportional to the inverse fourth power of wavelength, the square of the polarizability of the particles, and the dimensions of the scattering particles. If the particle is too big, there are differences in scattering intensity for the different directions, and this characteristic can be used to determine the size and shape of large molecules and colloid particles. If there is a vibrational energy level transition, it makes the quantized frequency change and there is a scatted radiation called Raman scattering.

Refraction happens when a light beam passes at an angle through the interface between the two transparent media that have different densities. The difference in density changes the velocity of the light as it travels through the two media, and the direction of the beam is bent. Bending toward the normal to the interface occurs when the beam passes from a less dense to a more dense medium. If the beam passes from a more dense to a less dense medium, the bending is away from the normal.

The extent of refraction is given by Snell's law:

$$\frac{Sin\theta_1}{Sin\theta_2} = \frac{n_1}{n_2} = \frac{v_1}{v_2} \tag{5.5}$$

where θ_1 and θ_2 are the angle from the normal, n_1 and n_2 are the refractive indexes, and v_1 and v_2 are the velocities of the two media (Figure 5.4).

Reflection of radiation always occurs when radiation crosses an interface between media that have different refraction indexes. As the difference between the refraction indexes increases, the fraction of reflected radiation becomes greater. For the

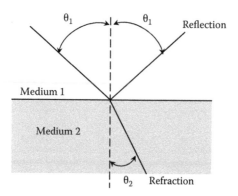

FIGURE 5.4   Refraction of light passing from medium 1 into medium 2.

solid samples, two types of reflectance may occur. One is diffuse or body reflectance and the other is specular reflectance. If radiation impacts on a solid sample and the reflected radiation travels equally in all directions, this is called diffuse reflectance. Specular reflectance happens when the angle of incidence is the same as the angle of reflection; this type of reflectance imparts little or no information about properties of the sample material.

Polarization looks like a bundle of electromagnetic waves that are equally distributed among a hung number of planes centered along the path of the beam. The passage of radiation through media that selectively absorb, reflect, or refract radiation that vibrates in only one plane produces polarized ultraviolet and visible radiation.

5.2.2 Measurements of Spectrum

Radiant energy converted by a radiation detector into an electrical signal or intensity, I, has been used to determine the radiant power. In emission, fluorescence, and scattering, the power of the radiation emitted by an analyte after excitation is proportional to the analyte concentration, c:

$$I = kc \tag{5.6}$$

where k is a constant determined by measuring I for the excitation of the analyte material in one or more reference standards of known concentration.

Absorption methods require two power measurements: one measurement of the light source energy before it falls on the surface of the medium containing the analyte (I_0), and the other measurement after the energy has passed through the analyte (I).

From the Beer-Lambert law, the absorption is proportional to the concentration, c, of the absorbing species (Figure 5.5).

$$I = I_0 \, 10^{-\varepsilon bc} \tag{5.7}$$

where ε is an absorption coefficient, and b is the thickness of the sample.

The transmittance T of the medium is the fraction after the radiation passes through a medium:

$$T = \frac{I}{I_0} = 10^{-\varepsilon bc} \tag{5.8}$$

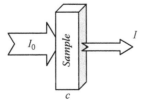

FIGURE 5.5   The Beer-Lambert law for absorption.

The absorption A of a medium is defined by the equation

$$A = -\log_{10} T = \log\frac{I_0}{I} = \varepsilon bc \tag{5.9}$$

If the sample absorbs all of the initial energy, the I becomes zero and absorption increases to infinity. For the absorption measurement, the homogeneous and non-scattering sample is required, but it is impossible to make, even for very pure liquids. Because of the index of refraction at the interface between the cuvette walls and the liquid, energy losses happen. To minimize the losses, the double-beam instrument was developed.

For diffuse reflectance in NIRS, some reflected energy passed through the sample and returned to the detector is a function of the change in path length b, and the reflectance R and absorption A may be written as

$$R = \frac{I}{I_0} = 10^{-\varepsilon bc} \tag{5.10}$$

$$A = -\log_{10} R = \log\frac{I_0}{I} = \varepsilon bc \tag{5.11}$$

There are five configurations to obtain NIR spectra: transmittance, specular reflectance, transflectance, diffuse reflectance, and interaction (Figure 5.6). If the sample medium is clear and exhibits no scattering, some electromagnetic energy is absorbed by the atoms, ions, or molecules that compose the matter. After absorption, transmitted energy can be used to obtain spectra of a medium, such as liquids or clear solids. Specular reflectance can be used to evaluate the roughness of a surface. If the angle of incidence is the same as the angle of repose, and the surface is smooth, specular reflectance happens and there is little or no information from the sample. The transflectance can be used to make measurements of reflectance spectra. The energy crosses the sample twice, and it is not good for a high optical density sample. The diffuse (or body) reflectance arrangement was designed for obtaining NIR spectra from small particles such as dry and ground samples. The interaction configuration is to obtain diffusely reflected energy from deep within a sample without specular or reflected energy from the surface. This method can be used to obtain the information or spectra from deep within the flesh of fruits and vegetables.

5.2.3 MATHEMATICAL AND CHEMOMETRIC PRINCIPLES

Unlike raw spectra from the mid-infrared spectral regions, the raw spectra measured in the near-infrared spectral region straightforwardly cannot provide physical and chemical information about molecular structures without any pretreatment or data transformation (Martens and Næs, 1991; Mark, 1991; Kramer, 1998; Vandenginster et al., 1998; Næs et al., 2002; Ozaki et al., 2001, 2003). NIR spectral bands are broad, ill defined, and weak, arising from overtones, and can exhibit heavy overlap with each other, with band shifting due to hydrogen bonding. This makes difficult

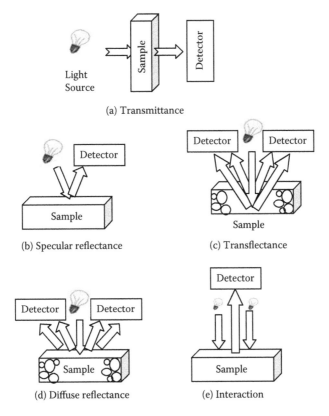

(a) Transmittance

(b) Specular reflectance

(c) Transflectance

(d) Diffuse reflectance

(e) Interaction

FIGURE 5.6   Transmittance, specular reflectance, transflectance, diffuse reflectance, and interaction.

the extraction of useful information from simple visual examination of the spectra. Generally, NIRS deals with samples that have a variety of components, such as carbon-carbon, carbon-oxygen, carbon-nitrogen, peroxide, and other bonds, consequently resulting in a poor signal-to-noise ratio, baseline fluctuations, and severe overlapping of bands. Light from solid samples or cloudy liquids can be scattered. Variations in NIR spectra can be affected by the path length, temperature, density, and particle size of samples. Spectral distortions can be caused by spectrometer hardware, such as baseline drift, wavelength shifts, nonlinearity of the detector, and noise from the hardware. These interferences affect the raw spectral responses such that predictions cannot be guaranteed from model equations with spectral responses. Proper pretreatment methods must be used to improve the signal-to-noise ratio and the linear relationship between analyte concentration and spectral signals, or to correct baseline fluctuations (Ozaki et al., 2007). There are two major groups of pretreatment methods. One group consists of reference-independent pretreatments that do not use available reference values, and another one is for a reference-dependent pretreatment. The latter group uses the reference to orthogonalize the data (Karstang and Manne, 1992; Goicoechea and Olivieri, 2001; Westerhuis et al., 2001). Because the response variables are used actively in the modeling, this method cannot be applicable generally. Methods in the former group are good for exploratory studies where reference values are unavailable. These methods can be divided into two subgroups: derivation methods and scatter correction methods. Also, pretreatment methods can be categorized by the purpose of pretreatment, such as noise reduction, baseline correction, resolution enhancement, and centering and normalization. These methods are introduced in the following sections.

5.2.3.1 Noise Reduction Methods

A variety of noise can be encountered while collecting NIR spectra data from instruments. High-frequency noise is associated with instrument hardware and electronic circuits. Low-frequency noise happens due to instrument drift during scanning measurements. Because low-frequency noise patterns can be similar to useful spectra, it is not as easy to reduce low-frequency noise as it is high-frequency noise. The general noise reduction method for high-frequency noise is accumulating the spectra data and calculating the average of accumulated data, but this method requires longer measurement times, greater data storage capacity, and computational speed. If the noise is not reduced sufficiently, smoothing methods, wavelets, eigenvector reconstruction, and artificial neural networks can also be used to address high-frequency noise (Chau et al., 1996; Stork et al., 1998). The most commonly used smoothing methods are the moving-average method and the Savitzky-Golay method (Savitzky and Golay, 1964; Hruschka, 2001; Steiner et al., 1972). The results of the moving-average method, which calculates the arithmetic mean of all original spectra signal values within a corresponding spectral window, can be affected by the height and area of spectral peaks. The Savitzky-Golay method fits the spectrum with low-degree polynomials through data points within the local spectral window by using least squares regression. Wavelet transform, which can be used to reduce both high- and low-frequency noise, changes a spectrum into the wavelet domain and then returns it to the spectral domain (Chau et al., 1996; Stork et al., 1998). Wavelets can denoise the data by using a user-selected threshold such that the noise of reconstructed data is under the threshold level. Wavelet transform produces much better results for signals exhibiting narrow lines and small or irregular features, compared to traditional Fourier transformation used to reduce noise (McClure and Davies, 1998; Schulze et al., 2005; Davis et al., 2007; Misiti et al., 2008).

5.2.3.1.1 Baseline Correction Methods

There are multiplicative and additive scatter factors in observed NIR spectra that affect baseline variations of the spectra. One of the oldest methods is the derivative method that can be used for baseline correction and resolution enhancement (Griffiths, 1987; Griffiths et al., 1987). A derivative spectrum is expressed as derivative values as a function of wavelength. Use of the second derivative is popular and changes the maxima peaks in an original spectrum to minima peaks in the second derivative spectrum and vice versa. As a resolution enhancement method, it can identify weak peaks that were not clear in the original spectrum. However, the more the spectrum is differentiated, the worse the signal-to-noise ratio becomes. Generally, spectral data are composed of discrete values and derivatives are calculated by algebraic differences between data taken at closely spaced wavelengths. If the local filters for smoothing are used with a modified coefficient, this can be used as a derivative-like pretreatment. The technique of using local filters for the derivatives was first introduced by Karl Norris. The Fourier domain filtering technique and the Savitzky-Golay method can also be used for derivative-like calculations. The derivatives from the Savitzky-Golay method are much closer to the true derivative than those from the Norris method, but there is no significant impact on the quality of subsequent multivariate calibrations.

Multiplicative deviations are produced from light scattering from solid samples, emulsions, and dispersions. The vertical variations and inclination of the baseline can be corrected by the multiplicative scatter correction (MSC) method (Martens and Jensen, 1983; Geladi et al., 1985). This method corrects each spectrum by shifting and scaling to fit a given standard spectrum, which may be a spectrum of a specific sample or a mean spectrum of the calibration set. Martens and Stark (1991) include wavelength dependency in their MSC method, called expanded MSC (EMSC). Unlike the MSC method, the standard normal variate (SNV) method (Barnes et al., 1989) does not need a reference spectrum. The SNV method uses the mean data for subtraction from each spectrum and then divides each signal value by the standard deviation of the whole spectrum. Because there is no least squares step in the SNV method, SNV results may be noisier than those of the MSC method. In general, the results of SNV are similar to those of the original MSC, spectrally and regression-wise (Dhanoa et al., 1994).

Orthogonalization is a reference-dependent technique, and the purpose of this method is to remove variability in the spectra that does not relate to the reference value. Wold et al. (1998) introduced this method and several other researchers have improved upon it to reduce model complexity by removing orthogonal compounds from the signal (Sjöblom et al., 1998; Wise, 1998; Andersson, 1999; Fearn, 2000; Trygg and Wold, 2002). Westerhuis et al. (2001) summarized the differences in improved performance of the previous studies and developed another, similar orthogonal signal correction technique called direct orthogonal signal correction (DOSC). DOSC provides a partial least squares (PLS) regression-based solution, and finds an exact solution to the orthogonality constraint in the calculation of the orthogonal components using an approach based solely on least squares steps. Using wavelet analysis and orthogonal signal correction with multivariate calibration, spectra data could be compressed down to 4% of the original matrix size without losing predictive power (Eriksson et al., 2000).

5.2.3.1.2 Resolution Enhancement Methods

Resolution enhancement methods have been used to resolve overlapping bands and explain less distinct bands. Derivative methods, difference spectra, mean centering, and Fourier self-deconvolution have been used as pretreatment methods and resolution enhancement methods (Ozaki et al., 2001; Siesler et al., 2002). Principal correlation analysis (PCA) loading plots (Shimoyama et al., 1998) and two-dimensional correlation spectroscopy (Ozaki and Noda, 2000; Noda and Ozaki, 2004) can be used for resolution enhancement, but they are not pretreatment methods. Derivative methods were explained in the section 5.2.3.1.1 about baseline correction methods, and mean centering will be mentioned in the section 5.2.3.1.3 on centering and normalization.

Difference spectra can be calculated by subtracting all the signal channels of one spectrum from another spectrum (Griffiths, 1987; Griffiths et al., 1987). This method is very useful for spectra that have slight differences and changes, but for reliability requires use of high-accuracy spectral measurements. One example was the use of difference calculations to analyze NIR spectra dependent on temperature, pH, and concentration (Czarnecki et al., 1993).

Fourier self-deconvolution (FSD) has been shown to be more suitable than derivatives for band resolution and band enhancements where spectral features are broader than instrumental resolution (Kauppinen et al., 1981a, 1981b, 1981c). Because FSD has the same data domain as that of the original spectra, it is easy to recognize the relationship of the enhanced bands to the original bands. This method is good for solution-phase infrared spectra exhibiting intermolecular interactions where Lorentzian contours broaden the bands (Lipp, 1986).

5.2.3.1.3 Centering and Normalization Methods

Centering is a spectral scaling method that is applied to each individual variable in all selected samples. Normalization is a scatter correction method (Mark, 1991; Martens and Næs, 1991; Kramer, 1998; Vandenginste, 1998; Næs et al., 2002). Mean centering is a simple method to adjust the spectrum data set to reposition the centroid of the spectrum data to the origin of the coordinate system, and it is calculated by subtracting the average response, calculated from all the spectra in the training set, from each individual response for the variable concerned. After mean centering, all means are zero and variances are spread around zero.

Normalization is the transformation of the spectral points on a unit hypersphere, and the scale of all data is approximately the same. This method is useful for finding similarities between two spectral vectors by measuring the scalar product of these two vectors. However, because normalization changes the geometric configuration of the data points from the original, careful usage is required for exploratory analysis and PCA-related analysis. There are two typical types of normalization procedures for spectral data. One uses vectors normalized to a constant Euclidean norm and the other is mean normalization, where all points of the each spectrum are divided by its mean value.

5.2.3.2 Multivariate Analysis Techniques for Near-Infrared Spectroscopy

The purpose of calibration is to provide the relationship between spectral measurements and analyte information. The mathematical expression of calibration is a calibration model or a calibration equation (Mark and Workman, 2003). Common analytical techniques use a straightforward equation describing a direct relationship between measured signals and values for the samples. This is called univariate calibration and can be used for cases in which the samples are purified and stabilized. However, samples for NIRS generally tend to have a variety of components, and thus there may be interference. For such samples, multivariate calibration is suitable as a mathematical method for removing interference.

Multivariate analysis techniques for NIRS can be categorized into qualitative and quantitative analysis. For these techniques, NIR spectral measurements and wet chemical analysis data are collected for all samples. The number of samples must be adequate to represent both the range of analyte concentrations and the range of variation of interfering materials present in samples likely to be analyzed in the future using the NIR instrument. Because the range of interfering materials may not be fully known, a random collection of samples is advisable to avoid overly similar samples that are not representative of the full variation likely to be encountered. The samples should be treated in the same manner as that which will be used for routine analysis of future samples. Because reference laboratory values obtained from wet chemical analysis may also include some error, it is advisable to analyze multiple aliquots to minimize the possible error. From the full set of samples collected, a calibration set and a validation set are selected. A calibration model or equation is computed using multivariate analysis with only the calibration set data. After creating a calibration model, the calibration model is evaluated by predicting the chemical composition for the "unknown" samples—i.e., samples not included in the calibration set—from the validation data set. If the accuracy of the developed calibration model is not sufficient, the calibration and validation process is repeated with the inclusion of more diverse samples in the calibration data set, and the collection procedure for spectral measurements and the wet chemical analysis are checked again for accuracy, until the validation yields accurate results.

5.2.3.2.1 Qualitative Analysis

Qualitative analysis classifies samples into separate groups based on product identity or quality attribute recognized by chemometric techniques. Chemometric methods for classification and grouping can be categorized as supervised and nonsupervised learning algorithms. If a known training set is used to verify sample identity, supervised learning algorithms can be used. For nonsupervised learning algorithms, no knowledge of the objects to be clustered is required (Walczak and Massart, 1998). Qualitative analysis can use either factor space, which has reduced dimensionality, or wavelength space, which can use whole spectra or data at selected wavelengths. As a compression method to reduce the dimensionality of a data set, principal component analysis (PCA) has often been used (Jolliffe, 1986; Wold et al., 1987). Principal components (PCs) are transformations of the data to a new data set of variables that has some finite number of independent variations occurring in the spectral data and are ordered so that the first few PCs retain most of the variation present in all of the original variables. Principal components have been called eigenvectors, factors, spectral loadings, and loading vectors. They are orthogonal to each another, and thus uncorrelated. There is a relationship between the principal components and concentrations of the constituents, and the principal components can be used to predict unknown samples. The fraction of each principal component, called a score, is the differences between the spectra of samples with different constituent concentrations. The scores are the weighting values for all the calibration spectra. PCA is an exploratory tool and an unsupervised pattern recognition method to uncover trends in the data. Because the goal of PCA is to summarize the data, it can extract features for subsequent classification or regression analysis, but itself is not used as a clustering tool. The goal of cluster analysis is to find the similarities between objects and groups in the data by calculating distances. PCA and cluster analysis do not require the information related to predefined classes of objects, which is necessary for the supervised pattern recognition.

One simple classification method is the Euclidean distance method that calculates the square root of the sum of the squares of the distances between the points along each orthogonal axis of the data space. It is necessary to find the centroid of each cluster, defined as the mean of all points in the cluster, and calculate the distance between the cluster centroids. For a sample located in the region of overlap of two or more clusters, this method might not classify the sample properly. The Euclidean distance describes a circular boundary around the centroid and does not take into account the variability of the values in all dimensions. One of the alternative distance metrics is the Mahalanobis distance method, which weights the differences by the range of variability in the direction of the sample (ASTM, 2005). The Mahalanobis distance can show the variations between the responses at the same wavelengths and the variations among the wavelengths. This method can be extended by selecting more wavelengths and solving for multiple dimensions simultaneously. However, discriminant analysis can determine the aberrations or impurities as matching the group that has the selected wavelength in Mahalanobis distance calculation. If too many wavelengths are used, there can be an overfitting problem. Optimum wavelength selection for all groups is a time-consuming and computationally intensive process. Another distance metric for discriminant analysis is K-nearest neighbor (KNN) classification, which classifies unknown samples according to the classes of the K closest samples (Cover and Hart, 1967). K is the number of neighbors for the decision, and it is typically a positive integer. KNN is a nonlinear, nonparametric, and supervised classification method able to handle irregularly shaped clusters in its multidimensional space. Rather than calculating the distance or using neighbors, another approach for classification is linear discriminant analysis (LDA), which finds an optimum decision boundary or hyperplane between different classes (Massart et al., 1988; Kemsley, 1998). The boundary is located where the variance between the classes is high and the variance within classes is low. LDA finds a linear discriminant function of the two predictors to provide more accurate discrimination than can be achieved by either predictor alone. The discriminant function is calculated to maximize the ratio of between-class variance to within-class variance. For more than two clusters, the problem can be solved piecewise by calculating the boundary of each cluster pair, called piecewise linear discriminant analysis (Shah and Gemperline, 1990; Cho and Gemperline, 1995; Gemperine and Boyer, 1995). One of the variations of LDA is partial least squares discriminant (PLSD) analysis, which uses the partial least squares (PLS) method that was developed for quantitative evaluation. PLSD is based on the PLS2 algorithm, which deals with several dependent variables rather than one dependent variable. Dependent variables have classification information and describe which objects are in the classes of interest. Various discriminant analyses have used a single model to classify all samples in all clusters at the same time. Another approach is to build a model for each cluster separately. Soft independent modeling of class analogy (SIMCA) is based on PCA to find separate spectral sets of each class (Wold, 1976). Because PCA does not have class information, the residual spectral variance and leverage of the new sample's spectrum are used to determine the class. For meaningful results, the model should be built from data that do not have outliers. For irregular clusters exhibiting nonlinearity, artificial neural networks (ANNs) can be useful applied to both qualitative and quantitative analysis (Zupan and Gasteiger, 1993; Zupan, 1994). Generally, a neural network has an input layer (features of the samples), at least one hidden layer, and an output layer (the categories to which the samples are assigned). The layers are composed of nodes, and each node in one layer is linked to every node in the next layer by transfer functions that are trained to produce values at the output nodes corresponding to the classification categories. The common ANN architectures are back-propagation (BP) network (He et al., 2007), radial basis function (RBF) network (Dou et al., 2006), and Kohonen self-organizing map network (Massart et al., 1988).

5.2.3.2.2   Quantitative Analysis

Quantitative analysis for NIRS is based on the measurement of small absorbance changes occurring at multiple wavelengths. A simple approach to quantitative full-spectrum evaluation is the classical least squares (CLS) method, or K-matrix, which assumes that Beer's law holds true and that absorbances are additive. This method is suitable for pure component spectra or samples in which all mixed components are known. Because all components are known, all components can be determined simultaneously, and predictions are independent of the light path length and multiplicative changes. By using known spectra of pure components, qualitative analysis is possible. Because this method uses full spectrum data, the result is more precise than that of other methods using a limited number of variables. However, if there are strong spectral interactions between analytes, performance of the CLS method can be poor. Also, performance depends on the precision of the concentration calculation.

One of the most widely used quantitative analysis methods is inverse least squares (ILS). Whereas the dependent variable in the CLS method is the signal at each wavelength and is calculated as the sum of the properties of the analyte multiplied by the analytical sensitivity, the ILS method uses the property of the analyte as the dependent variable, solved by calculating a solution from multiple independent variables such as the selected wavelengths. It is also known as multiple linear regression (MLR) and P-matrix (Barnett and Bartoli, 1960; Brown et al., 1982; Kisner et al., 1983; Massart et al., 1988; Kramer, 1998). Because the number of selected wavelengths is limited, calculation speed is relatively fast. This method does not need advance knowledge of the composition of the training mixtures, and is thus suitable for complex mixtures that the CLS method cannot handle. This method can determine not only a single component in a complex mixture, but also multiple components at the same time. Because of the dimensionality of the matrix equations, the number of wavelengths used in the model cannot exceed the number of calibration samples. For selecting the wavelengths, collinearity and overfitting should be considered. Principal component regression (PCR) does not require wavelength selection and has been used to improve signals without losing relevant information. PCR is a combination of spectral decomposition by PCA with the ILS regression method using the scores (Bertrand et al., 1984; Cowe and McNicol, 1985; Cowe et al., 1985a, 1985b). Factors are patterns of variability within a group of spectra. The number of factors included in the model is important; the model will be underfit to the data with too few factors, or overfit with too many factors. For selecting the proper number of factors, models can be tested with different numbers of factors through external validation or a cross-validation method. Generally, the number of factors corresponding to the minimum mean square error of prediction from the calibration is chosen from the validation procedure. The PCR method can be used for complex mixtures. However, a large number of samples are required for accurate calibration, and it takes time to calculate. Like PCR, the partial least squares (PLS) method extracts a sequence of factors from the group of calibration spectra, uses a score to indicate a factor's contribution to the spectrum, and regresses on the scores (Wold, 1975; Martens and Jensen, 1987; Geladi and Kowalski, 1986; Martens and Næs, 1988, 1991). The extraction procedure for the factors is different between PCR and PLS. PCR extracts factors based on the covariance between spectra and the information of samples. The first few latent variables have much more information of predictive use. There are two forms of the PLS algorithm. One algorithm is PLS1, which uses a separate set of scores and factors for each constituent of interest. The other algorithm is called PLS2, and it calibrates for all constituents simultaneously. Because it uses a separate set of factors and scores for every constituent of interest, PLS1 requires more time than PLS2. PLS is a single-step decomposition and regression, and factors are directly related to the constituent of interest rather than common spectral variations. General disadvantages of PLS are that it takes time to calculate, a large number of samples are required for calibration, and collinear constituent concentrations for calibration samples should be avoided. ANN can be used in quantitative analysis also. The quantitative values are assigned to nodes in the output layer.

There are always some data that belong to another population, called outliers. It is necessary to determine whether outliers come from actual phenomena or are artifacts arising from some errors while constructing the calibration model (Martens and Næs, 1988; Næs et al., 2002). The accuracy of the calibration is evaluated by the standard error of calibration (SEC), which is calculated from the same data used to create the calibration model. Once the calibration model is developed, the standard error of prediction (SEP) should be used to evaluate how well the model can predict values for unknown samples not included in the calibration data set.

5.3   DEVICE AND APPARATUS

The common requirements for performing NIRS measurements are a radiation source for the desired NIR wavelength, a way of determining the wavelength and the energy content of the radiation at the selected NIR wavelength, and a detector sensitive to the selected NIR wavelength (Figure 5.7).

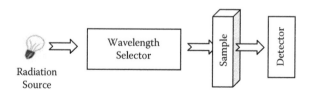

FIGURE 5.7   Common NIR spectroscopy instrument configurations.

The radiation source for spectrometric measurements can be divided into two classes: broadband sources and narrowband sources (Osborne et al., 1993). Broadband sources are thermal sources that produce heat and continuous spectral radiation. The tungsten halogen lamp is a popular and fairly stable thermal source that has a small size and ruggedness, and can provide high energy and an extended lifetime (McClure, 2001). The wavelength range of a tungsten halogen lamp ranges from the visible to infrared regions. Narrowband sources are nonthermal sources and include light-emitting diodes (LEDs), laser diodes (LDs), and lasers (McClure et al., 2002). These are discrete light sources that produce a narrow spectral band. The wavelength region for an LED source depends on the materials for the LED. A gallium arsenide (GaAs) LED produces a peak wavelength of 940 nm and could be used to measure quantities of moisture and fats. An LED made of indium gallium arsenide (InGaAs) spans 1,000 to 1,600 nm, while an LED made of gallium aluminum arsenide (GaAlAs) spans 650 to 900 nm. LD works by sharpening the gain curve of an LED and injecting a large current. While an LED may produce a broad peak 20 to 100 nm in width, an LD can produce a peak of 0.1 to 2 nm in width. The neodymium-doped yttrium aluminum garnet (Nd:YAG) laser is a widely used solid-state laser. Typically, it emits light at a wavelength of 1,064 nm. A titanium-sapphire (Ti:Sapphire) laser is a tunable laser that emits red and NIR light in the range from 650 to 1,100 nm.

One wavelength determination method uses spatial separation of the light by a dispersive system. A prism is a classical dispersing element that requires the use of a narrow slit to select the desired wavelength (Kaye, 1955). However, the dispersion of a prism is low and nonlinear, and the useful wavelength range is limited. More widely used as a dispersing element are diffraction gratings that separate light radiation into the constituent wavelengths of lights. Common diffraction gratings are made by depositing photosensitive material on a flat glass, or metallic or ceramic substrate (Palmer, 2005). Closely spaced and parallel grooves have been etched on the material surface, and aluminum is vacuum deposited on the surface of the grooves to make them reflective.

There are two filter-based wavelength selection methods for obtaining monochromatic light. One is the use of electronically tunable filters, and the other uses optical interference filters. Control methods for electronically tunable filters are voltage, acoustic signal, and other parameters (Gat, 2000). The most common electronically tunable filters are acousto-optical tunable filter (AOTF) and liquid crystal tunable filter (LCTF). AOTFs can behave like longitudinal diffraction gratings when the crystal is excited by sound waves of sufficiently high radio frequency (Figure 5.8). For NIR, tellurium dioxide (TeO_2) is the commonly used crystal (Chang, 1978). The selected wavelength can be adjusted by the radio frequency applied to the crystal. An LCTF consists of a stack of tunable retardation liquid crystal plates and polarizers (Bonaccini et al., 1990; Gat, 2000). The number and thickness of the crystal quartz plates determines the NIR wavelength range. Because of the relaxation time of the crystal, the switching speed of LCTF from one wavelength to another is longer than that of gratings and AOTF. The optical interference filters (Fabry-Perot interferometer) consist of multiple transparent dielectric thin films with different refractive properties (Figure 5.9). The transmission efficiency is up to 70% and bandwidth is

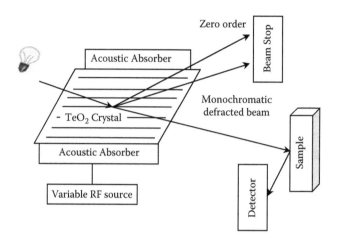

FIGURE 5.8 Schematic of AOTF spectrometer.

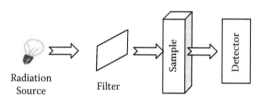

FIGURE 5.9 Instrument schematic using an interference filter.

approximately 10 to 20 nm. Because the filter size can be wide and thin, and cost can be low, optical interference filters are good for on-line and in-field NIR applications. Because a multiple filter system can produce a limited number of wavelength bands, it has been used to calibrate the moisture, protein, and fat in high-volume food applications. Optical interference filters for any wavelength in the NIR region can be made and duplicated easily. It is easy to increase or decrease the bandpass of a filter and the energy falling on the sample.

A Fourier transform–NIR (FT-NIR) system is based on the use of a Michelson interferometer to acquire an interferogram obtained by scanning the difference in the optical path lengths of two beams of a two-arm interferometer (Michelson, 1891). The Fourier transform of the spectrum is produced by constructive and destructive interference at a different rate of each wavelength from two beams (Figure 5.10). Compared to dispersive systems, the FT-NIR system has throughput (Jacquinot, 1954) and multiplex (Fellgett, 1958) advantages. Other FT-NIR advantages include fast scan time, large scan range, high resolution, high sensitivity, and high wavelength accuracy. FT-NIR systems require complex calculation and interpretable procedures; these disadvantages have been improved upon by the use of high-performance computers and mathematical software. Another multiplex method with dispersive spectrometers is Hadamard transform NIR spectrometry (Hammaker et al., 1986; Tilotta et al., 1987). Dispersed energy on a grating from an NIR source

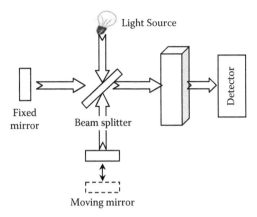

FIGURE 5.10   Diagram of a Michelson interferometer.

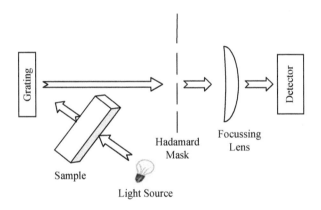

FIGURE 5.11   Schematic diagram of a Hadamard spectrometer.

passes through the multislit mask and is transformed into conventional NIR spectrum with Hadamard mathematics (Figure 5.11). Because of some complex mechanical problems, this method is not currently in commercial use.

The optical devices measuring the intensity of NIR light are detectors that convert the radiation energy into electrical signals. Depending on temperature of operation, and various properties of the detectors, such as quantum efficiency and response speed, each detector has a different performance (McClure, 2001). Detectors are selected depending on the target NIR wavelength range. According to the principles of operation, thermal type and photon type detectors are available. Thermal type detectors have a temperature-sensitive surface that absorbs the NIR energy and has uniform sensitivity at all wavelengths. Examples of thermal detectors are thermistors, thermocouples, Golay detectors, bolometers, and pyroelectric devices (Robinson et al., 2005). But the response times of these detectors are too long to be used for NIR applications.

Widely used photon detectors are phototubes, photomultiplier tubes (PMTs), photoconductive cells, silicon photodiodes, and photodiode arrays. Phototubes and PMTs are based on the photoelectric effect. The wavelength range of PMTs is from the visible to the NIR shorter than 1,000 nm, with low sensitivity for the NIR region. Photoconductive transducers are made by covering a thin film of semiconductor materials on a nonconducting glass surface and sealing it. The semiconductor materials are lead sulfide (PbS), lead selenide (PbSe), indium antimonide (InSb), indium antimonide (InSb), indium gallium arsenic (InGaAs), and mercury cadmium telluride (MCT). The PbS detectors and InGaAs detectors are widely used for the NIR spectral region. The response speed of PbS detectors is slow, but they are easy to make and use. PbS detectors also have a good signal-to-noise ratio, moderate cost, and sensitivity. The coverage range of wavelength for PbS detectors is 1,000 ~ 3,600 nm, and that of InGaAs detectors is 700 ~ 1,700 nm.

The MCT detectors are preferred for the mid-infrared and far-infrared regions, and cooling by liquid nitrogen is necessary to improve the sensitivity and reduce sampling time. Silicon photodiode detectors, which are fast, low noise, and highly sensitive, work most efficiently in the visible and short NIR range of 300 ~ 1,100 nm, peaking at about at 850 nm (McClure, 2001). A device that has aligned one- or two-dimensional infrared photodetection elements is called an NIR multichannel detector, such as photodiode array (PDA) detectors and charge transfer devices (CTDs) (Figure 5.12) (Bilhorn et al., 1987). These devices can measure the entire spectrum in less than one second and have a single-beam design. CTD detectors are more expensive than PDAs, but they have a much better sensitivity, dynamic range, and signal-to-noise ratio than PDAs.

Food samples for NIR analyses are most commonly conducted on bulk materials, raw and untreated, and usually in liquid or solid form. Basic sampling methods are transmission mode and diffuse reflectance mode (Rolfe, 2000). Liquid samples need containers such as cuvettes composed of quartz or fused silica, which must have transparent windows in the spectral region of interest. For the 375 to 2,000 nm spectral region, silicate glass has been used often because of its low cost compared to quartz. Generally, either the transmission mode or diffuse reflectance mode can be used for analyzing liquid samples, while solid samples are analyzed with

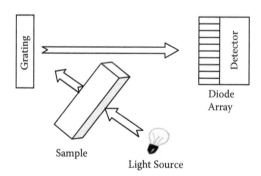

FIGURE 5.12   Diode array in an NIR instrument.

the reflectance mode. Optical fibers of fiber bundles that work based on the total internal reflection are useful to deliver and transfer NIR energy and information. Transmission can be affected by thermal and mechanical stress, and bending the optical fibers can result in loss of transmission. Material selection for the optical fibers depends on the spectral region of interest.

5.4 APPLICATIONS

NIRS has been widely used in quality and safety control for a broad range of raw agricultural products, intermediate products, and final market-ready products. The following discussion topics include NIRS applications for grains, fruits, vegetables, meat, fish, cereal products, dairy products, egg products, beverages, and food safety. Many kinds of sample cells or holders have been developed to accommodate the great variety of food sample types. The most commonly used NIR measurement method for analyzing agricultural products and food materials is the reflectance method, although transmission, transflection, and interaction methods are also used.

5.4.1 GRAINS AND SEEDS

Hart et al. (1962) first applied NIRS to determine moisture in grain products. Since then, NIRS has been applied to the analysis of grains for determining or assessing composition, physical properties, quality, and safety attributes, and for classification between types. The major constituents of grains are moisture, protein, oil, carbohydrates, fiber, and minerals. The quantitative NIR analysis began with measuring the moisture in ground wheat and flour (Norris and Hart, 1965). The quantitative measurement of the protein component in wheat has also been explored (Delwiche et al., 1998). Rubebthaler and Bruinsma (1978) showed that the concentration of amino acids, such as lysine in wheat and barley, can be measured by NIR reflectance spectroscopy. Gill et al. (1979) developed NIR reflectance models for lysine content in barley, while Williams et al. (1984) used NIR reflectance to determine the concentration of four amino acids in wheat (lysine, threonine, tryptophan, and methionine). Glutenin and giladin content in wheat, very important factors for breadmaking, can also be measured by NIR reflectance (Takács et al., 1996; Delwiche et al., 1998). Starch damage can significantly affect water absorption and dough mixing properties of flour; Osborne and Douglas (1981) found that the degree of starch damage in flour could be determined by NIR reflectance at four-second derivative wavelengths (Osborne and Douglas, 1981). Later, Osborne (1998) improved the prediction using precise control of temperature, a monochromator, five-factor PLS regression, and a more precise reference method. NIR models for the volume percentage of wheat flour particles in three particle size ranges were developed as an objective assessment method (Hareland, 1994). Edwards et al. (1996) reported that NIR can be used to measure the degree of yellow pigmentation in durum wheat endosperm, which affects color during the production of breads, rolls, and buns. Wheat kernels are often classified by color, cultivars, and waxy types, but visual classification of samples can sometimes be difficult. Several researchers demonstrated NIR classification of wheat by color in the wheat coat (Ronalds and Blakeney, 1995; Delwiche and Massie, 1996; Dowell, 1997, 1998). Delwiche and Norris (1993) demonstrated that ground hard red winter and ground hard red spring wheat can be differentiated using NIR reflectance spectra, and NIR reflectance also showed a similar performance for differentiating whole grains as well (Delwiche et al., 1995b). For rice, taste is one of the most important factors for quality and depends on the concentrations of protein, moisture, amylose, fatty acids, and minerals. Additionally, the texture quality of cooked rice is affected by amylose. Apparent amylose content (ACC) of ungrounded brown rice or milled rice can be determined by NIR transmittance (Villareal et al., 1994), and ACC of ground milled rice (Delwiche et al., 1995a) and that of intact milled rice (Delwiche et al., 1996) by NIR reflectance. Various constituents in rice were measured by NIR transmittance or reflectance for the Japanese rice (Suzuki et al., 1996; Shimizu et al., 1998; Kawamura et al., 1999).

5.4.2 FRUITS AND VEGETABLES

Among fruits and vegetables, maturity and ripeness are the major quality factors. For assessing maturity, NIR has been used to determine the soluble solids content, sugar content, acidity, dry matter content, and firmness. Other quality factors include defects such as bruises, impurities, internal discoloration, pits, scald, and water core. NIR has been used to detect internal or external defects in produce. The earliest NIR applications for intact fruits or vegetables were for determining soluble solids content in papaya (Birth et al., 1984) and dry matter in onion (Birth et al., 1985); later studies showed that soluble solids in intact cantaloupe could also be measured with the same instrument (Dull et al., 1989). Since then, many studies have been performed using NIR applications to fruits and vegetables, and are well described by Slaughter and Abbott (2004), Saranwong and Kawano (2007), and Schulz and Baranska (2009). Fruit commodities analyzed have included apple, apricot, citrus fruits, banana, cherry, grape, kiwifruit, melon, papaya, paprika, peach, pear, persimmon, pineapple, and strawberry, while vegetable commodities have included carrot, cucumber, onion, potato, sugar beet, and tomato.

A commercial sorter using NIR reflectance techniques to sort fruits for their sweetness was developed in 1988 by the Mitsui Mining and Smelting Co. Ltd. of Japan (Hadfield, 1993). This sorter could handle three fruits per second for apples, peaches, and pears. Another commercial sorter using NIR transmittance for apples and oranges was developed by Fantec Research and Development Co. (Osaka, Japan). Many other commercial NIR-based produce sorters have been developed by companies in Japan, Korea, the Netherlands, the United States, and New Zealand.

5.4.3 MEAT AND MEAT PRODUCTS

NIRS has been used to measure the components and quality attributes of meat and meat products. As a quantitative analysis method, the use of NIR techniques has been investigated in proximal composition analysis of meat and meat products, such as measurement of moisture, fat, protein, and some minerals. As a noninvasive and nondestructive measurement method, the NIR technique has been attempted for determining quality attributes such as sensory tenderness, toughness, hardness, and juiciness. In addition, NIR techniques have been used to address authenticity issues such as the adulteration of high-quality meat by low-quality meat or proteins from other animal or vegetable sources, and also contamination issues such as the detection of fecal matter.

The first investigation of the capability of NIR transmission spectroscopy in quantitative meat analysis was demonstrated by Karl Norris and colleagues (Ben-Gera and Norris, 1968). Kruggel et al. (1981), Martens et al. (1981), and Lanza (1983) later used NIR reflectance to determine the fat, protein, and moisture in meats. Isaksson et al. (1992) compared all root mean square errors of prediction for NIR reflectance and transmittance and found that they were on a similar level. Other NIR applications were reported for studies using raw, homogenized beef and pork (Berg and Kolar, 1991), and intact and minced poultry meat (Cozzolino et al., 1996). Applications of NIR techniques for minor components in meat have also been investigated for fatty acid (Windham and Morrison, 1998), added sodium chloride (Ellekjær et al., 1993), and collagenous substances (Isaksson and Hildrum, 1990; Berg and Kolar, 1991).

The intramuscular fat in intact muscle is a good appearance index for beef grading and was measured by NIR reflectance spectroscopy by Rødbotten et al. (2000). Because tenderness is such an important sensory property, many researchers have sought to use NIRS to predict tenderness (Ben-Gera and Norris, 1968; Hildrum et al., 1994, 1995; Næs and Hildrum, 1997; Byrne et al., 1998; Park et al., 1998; Rødbotten et al., 2000; Liu et al., 2003).

For the detection of adulteration and contamination, the NIR technique was tested for detecting spinal cord contamination in ground beef (Gangidi et al., 2005). Beef adulteration with cheap lamb was investigated with NIRS (McElhinney et al., 1999), and Vis/NIR spectroscopy was determined useful for detecting hamburger adulteration (Ding and Xu, 2000).

5.4.4   Fish and Related Products

Like meat and meat products, NIRS has been applied to measure moisture, fat, and protein contents in fish and fish products and to evaluate texture. The fat content is generally the most important subject for the fish quality assurance.

In early studies, fish samples were homogenized for better results (Gjerde and Martens, 1987; Mathias et al., 1987; Rasco et al., 1991; Sollid and Solberg, 1992). The first NIR analysis of whole intact frozen, then thawed fish was for rainbow trout by Lee et al. (1992), and later whole intact salmon was tested with NIR by Downey (1996) and Solberg et al. (2003).

5.4.5   Cereal Products

NIR technology was adopted early by the flour milling, dough, and cereal products industry. Like in intact wheat grain analysis, NIR was applied to measure protein

(Persson and Sjodin, 2004), moisture (Osborne and Fearn, 1983), ash (Wetzel and Mark, 1977), starch damage (Osborne, 1996), and particle size (Manley et al., 1996) of flour. During the breadmaking process, dough mixing is a critical stage for the quality. The NIR method has been used for quality monitoring during rapid dough mixing. The protein, fat, moisture, starch, and total sugars in bread were analyzed by NIR, and there was no difference between sliced bread samples and samples of dried and ground bread.

5.4.6 MILK AND DAIRY PRODUCTS

The constituents in milk and dairy products that are measured by NIR techniques are moisture, protein, fat, lactose, lactic acid, total solids, and ash. The determination of protein, fat, and lactose in nonhomogenized milk with NIRS has been studied by several research groups (Hall and Chan, 1993), and the potential of NIRS for measuring constituents in raw milk for on-line milk analysis was investigated (Tsenkova et al., 1999). With milk composition analysis, the feasibility of NIRS was evaluated for mastitis diagnosis (Tsenkova et al., 1994; Tsenkova and Atanassova, 2002). NIR analyses of milk products and dairy products other than milk have been performed for composition such as fat, protein, and moisture.

5.4.7 EGGS AND EGG PRODUCTS

Visible/NIR spectroscopy has been used to evaluate the quality of eggs. The most important criterion is freshness, formerly evaluated by air cell height and Haugh unit (HU). As storage time grows longer, the albumen quality worsens such that the thickness of albumen is thinner and the pH of the albumen is increased. Norris (1996) investigated NIRS for egg quality during storage time, but he could not find any relationship between the spectral data and the internal quality changes; too few hours for the storage times used might have been the reason for that result. Another researcher found a high correlation between Vis/NIR transmittance spectra and the air chamber size, weight loss, and pH (Schmilovitch et al., 2002). Kemps et al. (2006) found correlation coefficients of 0.82 and 0.86 for predicted HU and pH, respectively, using Vis/NIR spectroscopy. Vis/NIR spectroscopic techniques have been investigated to detect blood spot in eggs, but small spots and also spots in brown-shelled eggs presented difficulties (De Ketelaere et al., 2004).

5.4.8 FRUIT JUICES

The major constituent of fruit juices is water, while sugars and organic acids are the major soluble solids. Studies by several researchers investigating the feasibility of NIR methods to measure the sugars (Rambla et al., 1997; Rodriguez-Saona et al., 2001; Xie et al., 2009) and acids (Chen et al., 2006; Cen et al., 2007; Shao and He, 2007) in fruit juices found that the performance of transmittance spectra was much better than that of reflectance spectra. NIR methods could detect and determine biological contaminants in fruit juice. Also, NIRS can determine the authenticity of orange juice by discriminant analysis (Evans et al., 1993).

5.4.9   WINE AND BEER

The various constituents in alcoholic beverages have been analyzed by NIRS. The alcohol and sugars are the main components of interest in wine. The concentrations of ethanol, fructose, and tartaric acid in a red and a white wine were determined by NIR measurement (Kaffka and Norris, 1976). Because of matrix variation, accuracy and robustness of calibration of ethanol have been limited. In addition to ethanol and sugar in wines, total phenolics (Medrano et al., 1995), total anthocyanins, and pH (Esler et al., 2002) were also examined by NIRS for assessing the quality of wines.

The feasibility of using NIRS has been tested at each stage of the beer brewing process. Hop is an important raw material for beer, and the hop α-acids, β-acids, and hop storage index were tested with NIR to evaluate the feasibility of usage (Axcell et al., 1981a, 1981b). Chandley (1993) reported better results for hop acids, moisture, and oil. The potential of NIRS for measuring a positive quality parameter, such as hot water extract (HWE), and a negative malting quality parameter, such as β-glucan content, has been evaluated. Ethanol and other compositions in beer have been examined with the feasibility of NIR application.

5.4.10   FOOD SAFETY

Food safety analysis is a part of quality assessment. Aside from insect and mite detection, the detectable concentration level by NIR for toxins, contaminants, or microbes is much higher than the minimum required level at regulatory action. Several researchers have investigated NIR reflectance from wheat kernels as a means to detect contamination by insect or mite (Chambers and Ridgway, 1996; Ridgway and Chambers, 1996, 1998; Ghaedian and Wehling, 1997; Dowell et al., 1998; Wilkins et al., 1998). The mycelia and spores of moldy barley have a high correlation with the concentration of N-acetyl-D-glucosamine in barley, and the NIR model to predict the concentration could be used to quantify the levels of mycelia (Roberts et al., 1991). An NIR probe was more sensitive than visual inspection for measuring the concentration of a fungal toxin and fungal damage in single wheat kernels (Dowell et al., 1999). Delwiche and Kim (2000) showed that single-kernel scab can be detected by hyperspectral NIR image analysis for three wheat varieties. Single wheat kernels containing live or dead insects can be detected by an automated NIR reflectance sensor (Maghirang et al., 2003). For the fruits and vegetables, the feasibility of detection of fecal contamination (Kim et al., 2002, 2004) and defects (Lee et al., 2008) on apples was tested using a Vis/NIR hyperspectral reflectance imaging system, and more recently, the Vis/NIR hyperspectral imaging system was demonstrated for high-speed operation (e.g., three apples per second). The feasibility of detecting fecal and ingesta contamination on poultry carcasses using a Vis/NIR hyperspectral reflectance imaging system was tested by Lawrence et al. (2003) and Park et al. (2006). For food safety concerns in meat and meat products, the NIR technique has been used for detecting microbial spoilage, which poses risks of food poisoning, and spinal cord material, which poses risks for bovine spongiform encephalopathy (BSE).

5.5 PERSPECTIVES

NIRS has been used as a fast, nondestructive, and cost-effective method of food analysis, and it will be used on a variety of food products. It will provide high quality and food safety assurance. Hardware systems will be smaller, faster, and more robust, and software for spectral analyses will be more accurate. The NIR spectrometers will be smaller and the system will become portable for use on a wide range of food products. FT-NIR systems have been developed by many companies to be used for quantitative and qualitative analysis. Hyperspectral imaging systems for the NIR wavelength range will be used widely. The systems have become affordable, and their processing speeds fast enough for many target products. Still, it is necessary to understand the interaction of light and matter on which NIR technology is based. With improved hardware and accurate analysis software, NIRS will provide greater confidence to consumers and quality improvement and cost savings to food-producing companies.

REFERENCES

Andersson, C. A. 1999. Direct orthogonalization. *Chemometr. Intell. Lab.* 47:51–63.

ASTM. 2005. *Standard practices for infrared multivariate quantitative analysis.* Standard E1655-05. Philadelphia: ASTM.

Axcell, B. C., R. Tulej, and J. Murray. 1981a. An ultra-fast system for hop analysis. I. The determination of alpha acids and moisture by near infrared reflectance spectroscopy. *Brew. Dig.* 56:18–19, 41.

Axcell, B. C., R. Tulej, and J. Murray. 1981b. An ultra-fast system for hop analysis. II. The determination of beta acids and prediction of hop storage index by near infrared reflectance spectroscopy. *Brew. Dig.* 56:32–33.

Barnes, R. J., M. S. Dhanoa, and S. J. Lister. 1989. Standard normal variate transformation and de-trending of near-infrared diffuse reflectance spectra. *Appl. Spectrosc.* 43:772–777.

Barnett, H. A., and A. Bartoli. 1960. Least-squares treatment of spectrophotometric data. *Anal. Chem.* 32:1153–56.

Ben-Gera, I., and K. N. Norris. 1968. Direct spectrophotometric determination of fat and moisture in meat products. *J. Food Sci.* 33:64–67.

Berg, H., and K. Kolar. 1991. Evaluation of rapid moisture, fat, protein and hydoxyproline determination in beef and pork using the Infratech food and feed analyzer. *Fleischwirtschaft* 71(7):787–89.

Bertrand, D., P. Robert, and V. Tran. 1984. Traitements mathématiques des spectres NIR de mélanges. *Reports Int. Assoc. Cereal Sci. Technol.* 11:93–97.

Bilhorn, R. B., J. V. Sweedler, P. M. Epperson, and M. B. Denton. 1987. Charge transfer device detectors for analytical optical spectroscopy—Operation and characteristics. *Appl. Spectrosc.* 41(7):1114–25.

Birth, G. S., G. G. Dull, J. B. Magee, H. T. Chan, and C. G. Cavaletto. 1984. An optical method for estimating papaya maturity. *J. Am. Soc. Hortic. Sci.* 109:62–66.

Birth, G. S., G. G. Dull, W. T. Renfroe, and S. J. Kays. 1985. Nondestructive spectrophotometric determination of dry matter in onions. *J. Am. Soc. Hortic. Sci.* 110:297–303.

Bonaccini, D., L. Casini, and P. Stefanini. 1990. Fabry-Perot tunable filter for the visible and near IR using nematic liquid crystals. In *Current developments in optical engineering IV*, ed. R. E. Fischer and W. J. Smith, 221–30. San Diego: SPIE—The International Society for Optical Engineering.

Brown, C. W., P. F. Lynch, R. J. Obremski, and D. S. Lavery. 1982. Matrix representations and criteria for selecting analytical wavelengths for multicomponent spectroscopic analysis. *Anal. Chem.* 54:1472–79.

Byrne, C. E., G. Downey, D. J. Troy, and D. J. Buckley. 1998. Non-destructive prediction of selected quality attributes of beef by near-infrared reflectance spectroscopy between 750 and 1098 nm. *Meat Sci.* 49:399–409.

Cen, H., Y. Bao, Y. He, and D.-W. Sun. 2007. Visible and near infrared spectroscopy for rapid detection of citric and tartaric acids in orange juice. *J. Food Eng.* 82(2):253–60.

Chambers, J., and C. Ridgway. 1996. Rapid detection of contaminants in cereals. In *Near infrared spectroscopy: The future waves*, ed. A. M. C. Davies and P. C. Williams, 484–89. Chichester, UK: NIR Publications.

Chandley, P. 1993. The application of the DESIR technique to the analysis of beer. *J. Near Infrared Spectrosc.* 1:133–139.

Chang, I. C. 1978. Development of an infrared tunable acousto-optic filter. *Proc. SPIE. Practical Infrared Optics* 131:2–10.

Chau, F. T., T. M. Shih, J. B. Gao, and C. K. Chan. 1996. Application of the fast wavelet transform method to compress ultraviolet-visible spectra. *Appl. Spectrosc.* 50:339.

Chen, J. Y., H. Zhang, and R. Matsunaga. 2006. Rapid determination of the main organic acid composition of raw Japanese apricot fruit juices using near-infrared spectroscopy. *J. Agric. Food Chem.* 54(26):9652–57.

Cho, J., and P. J. Gemperline. 1995. Pattern recognition analysis of near-infrared spectra by robust distance method. *J. Chemometr.* 9:169–78.

Colthup, N. B., L. H. Daly, and S. F. Wiberley. 1990. *Introduction to infrared and Raman spectroscopy*. Boston: Academic Press.

Cover, T. M., and P. E. Hart. 1967. Nearest neighbor pattern classification. *IEEE Trans. Inform. Theory* 13:21–27.

Cowe, I. A., and J. W. McNicol. 1985. The use of principal component in the analysis of near infrared spectra. *Appl. Spectrosc.* 39(2):257–66.

Cowe, I. A., J. W. McNicol, and D. C. Cuthbertson. 1985a. A designed experiment for the examination of techniques used in the analysis of near-IR spectra. Part 1. Analysis of spectral structure. *Analyst* 110(10):1227–32.

Cowe, I. A., J. W. McNicol, and D. C. Cuthbertson. 1985b. A designed experiment for the examination of techniques used in the analysis of near-IR spectra. Part 2. Derivation and testing of regression models. *Analyst* 110(10):1233–40.

Cozzolino, D., I. Murray, R. Paterson, and J. R. Scaife. 1996. Visible and near infrared reflectance spectroscopy for the determination of moisture, fat and protein in chicken breast and thigh muscle. *J. Near Infrared Spectrosc.* 4:213–23.

Czarnecki, M. A., Y. Liu, Y. Ozaki, M. Suzuki, and M. Iwahashi. 1993. Potential of Fourier transform near-infrared spectroscopy in studies of the dissociation of fatty acids in the liquid phase. *Appl. Spectrosc.* 47:2162–68.

Davis, R. A., A. J. Charlton, J. Godward, S. A. Jones, M. Harrison, and J. C. Wilson. 2007. Adaptive binning: An improved binning method for metabolomics data using the undecimated wavelet transform. *Chemometr. Intell. Lab.* 85:144–54.

De Ketelaere, B., F. Bamelis, B. Kemps, E. Decuypere, and J. De Baerdemaeker. 2004. Nondestructive measurements of egg quality. *World Poultry Sci. J.* 60:289–302.

Delwiche, S. R., M. M. Bean, R. E. Miller, B. D. Webb, and P. C. Williams. 1995a. Apparent amylose content of milled rice by near-infrared reflectance spectrophotometry. *Cereal Chem.* 72:182–87.

Delwiche, S. R., Y. R. Chen, and W. R. Hruschka. 1995b. Differentiation of hard red wheat by near-infrared analysis of bulk samples. *Cereal Chem.* 72:243–47.

Delwiche, S. R., K. S. McKenzie, and B. D. Webb. 1996. Quality characteristics in rice by near infrared reflectance analysis of whole grain milled samples. *Cereal Chem.* 73:257–263.

Delwiche, S. R., R. A. Graybosch, and C. J. Peterson. 1998. Predicting protein composition, biochemical properties, and dough-handling properties of hard red winter wheat flour by near-infrared reflectance. *Cereal Chem.* 75:412–16.

Delwiche, S. R., and M. S. Kim. 2000. Hyperspectral imaging for detection of scab in wheat. In *Proceedings of SPIE Biological Quality and Precision Agriculture II*, ed. J. A. DeShazer and G. E. Meyer, 13–20. Vol. 4203. Bellingham, WA: SPIE.

Delwiche, S. R., and D. R. Massie. 1996. Classification of wheat by visible and near-infrared reflectance from single kernels. *Cereal Chem.* 73:399–405.

Delwiche, S. R., and K. H. Norris. 1993. Classification of hard red wheat by near-infrared diffuse reflectance spectroscopy. *Cereal Chem.* 70:29–35.

Dhanoa, M. S., S. J. Lister, R. Sanderson, and R. J. Barnes. 1994. The link between multiplicative scatter correction (MSC) and standard normal variate (SNV) transformations of NIR spectra. *J. Near Infrared Spectrosc.* 2:43–47.

Ding, H. B., and R. J. Xu. 2000. Near-infrared spectroscopic technique for detection of beef hamburger adulteration. *J. Agric. Food Chem.* 48:2193–98.

Dou, Y., H. Mi, L. Zhao, Y. Ren, and Y. Ren. 2006. Radial basis function neural networks in non-destructive determination of compound aspirin tablets on NIR spectroscopy. *Spectrochim. Acta A* 65(1):79–83.

Dowell, F. E. 1997. Effect of NaOH on visible wavelength spectra of single wheat kernels and color classification efficiency. *Cereal Chem.* 74:617–20.

Dowell, F. E. 1998. Automated color classification of single wheat kernels using visible and near-infrared reflectance. *Cereal Chem.* 75:142–44.

Dowell, F. E., M. S. Ram, and L. M. Seitz. 1999. Predicting scab, vomitoxin, and ergosterol in single wheat kernels using near-infrared spectroscopy. *Cereal Chem.* 76:573–76.

Dowell, F. E., J. E. Thorne, and J. E. Baker. 1998. Automated nondestructive detection of internal insect infestation of wheat kernels by using near-infrared reflectance spectroscopy. *J. Econ. Entomol.* 91:899–904.

Downey, G. 1996. Non-invasive and non-destructive percutaneous analysis of farmed salmon flesh by near infra-red spectroscopy. *Food Chem.* 55:305–11.

Dull, G. G., G. S. Birth, D. A. Smittle, and R. G. Leffler. 1989. Near infrared analysis of soluble solids in intact cantaloupe. *J. Food Sci.* 54(2):393–95.

Edwards, N. M., J. E. Dexter, D. C. Sobering, and P. C. Williams. 1996. Whole grain prediction of durum wheat yellow pigment by visible/near infrared reflectance spectroscopy. In *Near infrared spectroscopy: The future waves*, ed. A. M. C. Davies and P. C. Williams, 462–65. Chichester, UK: NIR Publications.

Ellekjaer, M. R., K. I. Hildrum, T. Næs, and T. Isaksson. 1993. Determination of the sodium chloride content of sausages by near infrared spectroscopy. *J. Near Infrared Spectrosc.* 1:65–75.

Eriksson, L., J. Trygg, E. Johansson, R. Bro, and S. Wold. 2000. Orthogonal signal correction, wavelet analysis, and multivariate calibration of complicated process fluorescence data. *Anal. Chim. Acta* 420:181–95.

Esler, M. B., M. Gishen, I. L. Francis, R. G. Dambergs, A. Kamouris, W. U. Cynkar, and D. R. Boehm. 2002. Effects of variety and region on near infrared reflectance spectroscopic analysis of quality parameters in red wine grapes. In *Near infrared spectroscopy: Proceedings of the 10th International Conference*, ed. A. M. C. Davies and R. K. Cho, 249–53. Chichester, UK: NIR Publications.

Evans, D. G., C. N. G. Scotter, L. Z. Day, and M. N. Hall. 1993. Determination of the authenticity of orange juice by discriminant analysis of near infrared spectra. A study of pretreatment and transformation of spectral data. *J. Near Infrared Spectrosc.* 1:33–44.

Fearn, T. 2000. On orthogonal signal correction. *Chemometr. Intell. Lab.* 50:47–52.

Fellgett, P. 1958. A contribution to the theory of the multiplex spectrometer. *J. Phys. Rad.* 19:187–91.

Gangidi, R. R., A. Proctor, F. W. Pohlman, and J. F. Meullenet. 2005. Rapid determination of spinal cord content in ground beef by near-infrared spectroscopy. *J. Food Sci.* 70:397–400.

Gat, N. 2000. Imaging spectroscopy using tunable filters: A review. *SPIE* 4056:50–64.

Geladi, P., D. MacDougall, and H. Martens. 1985. Linearization and scatter correction for near-infrared reflectance spectra of meat. *Appl. Spectrosc.* 39:491–500.

Geladi, P., and B. R. Kowalski. 1986. Partial least-squares regression: A tutorial. *Anal. Chim. Acta* 185:1–17.

Gemperline, P. J., and N. R. Boyer. 1995. Classification of near-infrared spectra using wavelength distances: Comparison of Mahalanobis distance and residual variance. *Anal. Chem.* 67:160–66.

Ghaedian, A. R., and R. L. Wehling. 1997. Discrimination of sound and granary-weevil-larva-infested wheat kernels by near-infrared diffuse reflectance spectroscopy. *J. AOAC Int.* 80:997–1005.

Gill, A. A., C. Starr, and D. B. Smith. 1979. Lysine and nitrogen measurement by infra-red reflectance analysis as an aid to barley breeding. *J. Agric. Sci.* 93:727–33.

Gjerde, B., and H. Martens. 1987. Predicting carcass composition of rainbow trout by near-infrared reflectance spectroscopy. *J. Anim. Breed Genet.* 104:137–48.

Goicoechea, H. C., and A. C. Olivieri. 2001. A comparison of orthogonal signal correction and net analyte preprocessing methods. Theoretical and experiment study. *Chemometr. Intell. Lab.* 56(2):73–81.

Griffiths, P. R. 1987. Mid infrared Fourier transform spectrometry. In *Laboratory methods in vibrational spectroscopy*, ed. H. A. Willis, J. H. van der Maas, and R. G. J. Miller, 121–44. Chichester, UK: John Wiley & Sons.

Griffiths, P. R., J. A. Pierce, and G. Hongjin. 1987. Curve-fitting and Fourier self-deconvolution for the quantitative representation of complex spectra. In *Computer-enhanced analytical spectroscopy*, ed. H. L. C. Meuzelaar and T. L. Isenhour, 29–54. New York: Marcel Dekker.

Hadfield, P. 1993. Technology: A sweet frequency for oranges. *New Scientist* 1883:20.

Hall, J. W., and K. Chan. 1993. Near-infrared spectroscopic analysis of bovine milk for fat, protein and lactose. In *Proceedings of Cheese Yield and Factors Affecting Its Control, IDF Seminar*, 230–39. Bruxelles, Belgium: International Dairy Federation Publishing.

Hammaker, R. M., J. A. Graham, D. C. Tilotta, and W. G. Fateley. 1986. What is Hadamard transform spectroscopy? In *Vibrational spectra and structure*, ed. J. R. Durig, 401–85. Amsterdam: Elsevier.

Hareland, G. A. 1994. Evaluation of flour particle size distribution by laser diffraction, sieve analysis and near-infrared reflectance spectroscopy. *J. Cereal Sci.* 21:183–90.

Hart, J. H., K. H. Norris and C. Golumbic. 1962. Determination of the moisture content of seeds by near-infrared spectrophotometry of their methanol extracts. *Cereal Chem.* 39:94–99.

He, Y., X. Li, and X. Deng. 2007. Discrimination of varieties of tea using near infrared spectroscopy by principal component analysis and BP model. *J. Food Eng.* 79(4):1238–42.

Hildrum, K. I., T. Isaksson, T. Næs, B. N. Nilsen, M. Rødbotten, and P. Lea. 1995. Near infrared reflectance spectroscopy in the prediction of sensory properties of beef. *J. Near Infrared Spectrosc.* 3:81–87.

Hildrum, K. I., B. N. Nilsen, M. Mielnik, and T. Næs. 1994. Prediction of sensory characteristics of beef by near-infrared spectroscopy. *Meat Sci.* 38:67–80.

Hruschka, W. R. 2001. Data analysis: Wavelength selection methods. In *Near-infrared technology in the agricultural and food industries*, ed. P. C. Williams and K. H. Norris, 39–58. St. Paul, MN: AACC.

Isaksson, T., and K. I. Hildrum. 1990. Near infrared transmittance (NIT) analysis of meat products. In *Proceedings of the 3rd International Conference of Near Infrared Spectroscopy*, ed. R. Biston and N. B. Thill, 202–6. Brussels.

Isaksson, T., C. E. Miller, and T. Næs. 1992. Nondestructive NIR and NIT determination of protein, fat, and water in plastic-wrapped, homogenized meat. *Appl. Spectrosc.* 46:1685–94.

Jacquinot, P. 1954. The luminosity of spectrometers with prisms, gratings, or Fabry-Perot etalons. *J. Opt. Soc. Am.* 44:761–65.

Jolliffe, I. T. 1986. *Principal component analysis.* New York: Springer-Verlag.

Kaffka, K. J., and K. H. Norris. 1976. Rapid instrumental analysis of composition of wine. *Acta Alimenta* 5:267–79.

Karstang, T. V., and R. Manne. 1992. Optimized scaling—A novel approach to linear calibration with closed data sets. *Chemometr. Intell. Lab.* 14(1–3):165–73.

Kauppinen, J. K., D. J. Moffatt, D. G. Cameron, and H. H. Mantsch. 1981a. Noise in Fourier self-deconvolution. *Appl. Opt.* 20(10):1866–79.

Kauppinen, J. K., D. J. Moffatt, H. H. Mantsch, and D. G. Cameron. 1981b. Fourier self-deconvolution: A method for resolving intrinsically overlapped bands. *Appl. Spectrosc.* 35(3):271–76.

Kauppinen, J. K., D. J. Moffatt, H. H. Mantsch, and D. G. Cameron. 1981c. Fourier transforms in the computation of self-deconvoluted and first-order derivative spectra of overlapped band contours. *Anal. Chem.* 53:1454–57.

Kawamura, S., M. Natsuga, and K. Itoh. 1999. Determination of undried rough rice constituent content using near-infrared transmission spectroscopy. *Trans. ASAE* 42(3):813–18.

Kaye, W. 1955. Near-infrared spectroscopy—Part 2. Instrumentation and techniques. A review. *Spectrochim. Acta* 7:181–204.

Kemps, B., F. Bamelis, B. De Ketelaere, K. Mertens, B. Kamers, K. Tona, E. Decuypere, and J. De Baerdemaeker. 2006. Visible transmission spectroscopy for the assessment egg freshness. *J. Sci. Food Agric.* 86:1399–406.

Kemsley, E. K. 1998. *Discriminant analysis and class modeling of spectroscopic data.* New York: John Wiley & Sons.

Kim, M. S., A. M. Lefcourt, K. Chao, Y. R. Chen, I. Kim, and D. E. Chan. 2002. Multispectral detection of fecal contamination on apples based on hyperspectral imagery. Part I. Application of visible and near-infrared reflectance imaging. *Trans. ASAE* 45(6):2027–37.

Kim, M. S., A. M. Lefcourt, Y. R. Chen, and S. Kang. 2004. Uses of hyperspectral and multispectral laser induced fluorescence imaging techniques for food safety inspection. *Key Eng. Mater.* 270–273:1055–63.

Kisner, H. J., C. W. Brown, and G. J. Kavarnos. 1983. Multiple analytical frequencies and standards for the least-squares spectrometric analysis of serum lipids. *Anal. Chem.* 55:1703–7.

Kramer, R. 1998. *Chemometric techniques for quantitative analysis.* New York: Marcel Dekker.

Kruggel, W. G., R. A. Field, M. L. Riley, H. D. Radloff, and K. M. Horton. 1981. Near infrared reflectance determination of fat, protein and moisture in fresh meat. *J. Assoc. Off. Anal. Chem.* 64(3):692–96.

Lanza, E. 1983. Determination of moisture, protein, fat, and calories in raw pork and beef by near infrared spectroscopy. *J. Food Sci.* 48:471–74.

Lawrence, K. C., W. R. Windham, B. Park, and R. J. Buhr. 2003. A hyperspectral imaging system for identification of fecal and ingesta contamination on poultry carcasses. *J. Near Infrared Spectrosc.* 11(4):269–81.

Lee, K., S. Kang, S. R. Delwiche, M. S. Kim, and S. Noh. 2008. Correlation analysis of hyperspectral imagery for multispectral wavelength selection for detection of defects on apples. *Sens. Instrum. Food Qual. Saf.* 2(2):90–96.

Lee, M. H., A. G. Cavinato, D. M. Mayes, and B. A. Rasco. 1992. Noninvasive short-wavelength near-infrared spectroscopic method to estimate the crude lipid content in muscle of intact rainbow trout. *J. Agric. Food Chem.* 40:2176–81.

Lipp, E. D. 1986. Application of Fourier self-deconvolution to the FT-IR spectra of polydimethylsiloxane oligomers for determining chain length. *Appl. Spectrosc.* 40(7):1009–11.

Liu, Y., B. G. Lyon, W. R. Windham, C. E. Realini, T. Dean, S. Pringle, and S. Duckett. 2003. Prediction of color, texture, and sensory characteristics of beef steaks by visible and near infrared reflectance spectroscopy. A feasibility study. *Meat Sci.* 65:1107–15.

Maghirang, E. B., F. E. Dowell, J. E. Baker, and J. E. Throne. 2003. Automated detection of single wheat kernels containing live or dead insects using near-infrared reflectance spectroscopy. *Trans. ASAE* 46(4):1277–82.

Manley, M., A. E. J. McGill, and B. G. Osborne. 1996. Whole wheat grain hardness measurement by near infrared spectroscopy. In *Near infrared spectroscopy: The future waves*, ed. A. M. C. Davies and P. C. Williams, 466–70. Chichester, UK: NIR Publications.

Mark, H. 1991. *Principle and practice of spectroscopic calibration.* New York: John Wiley & Sons.

Mark, H., and J. Workman. 2003. *Statistics in spectroscopy.* London: Elsevier.

Martens, H., E. A. Bakker, and K. I. Hildrum. 1981. Application of near infrared reflectance spectrometry in the analysis of meat products. In *The 27th European meeting of meat researchers workers*, 561–564. Vienna, Austria.

Martens, H., and S. Å. Jensen. 1983. Two-stage partial least squares regression on latent factors: A new NIR calibration method. In *Proceedings of the 7th World Cereal and Bread Congress, Progress in cereal chemistry and technology*, ed. J. Holas and J. Kratochvil, 607–47. Amsterdam.

Martens, H., and T. Næs. 1987. Multivariate calibration by data compression. In *Near-infrared technology in agricultural and food industries*, ed. P. C. Williams and K. Norris, 57–87. St. Paul, MN: American Association of Cereal Chemists.

Martens, H., and T. Næs. 1991. *Multivariate calibration.* Chichester, UK: John Wiley & Sons.

Martens, H., and E. Stark. 1991. Extended multiplicative signal correction and spectral interference subtraction: New preprocessing methods for near infrared spectroscopy. *J. Pharm. Biomed. Anal.* 9:625–35.

Massart, D. L., B. G. M. Vandeginste, S. N. Deming, Y. Michotte, and L. Kaufman. 1988. *Chemometrics: A textbook.* New York: Elsevier.

Mathias, J. A., P. C. Williams, and D. C. Sobering. 1987. The determination of lipid and protein in freshwater fish using near-infrared reflectance spectroscopy. *Aquaculture* 61:303–11.

McClure, W. F. 2001. Near-infrared instrumentation. In *Near-infrared technology in the agricultural and food industries*, ed. P. C. Williams and K. H. Norris, 109–28. St. Paul, MN: AACC.

McClure, W. F., and A. M. C. Davies. 1988. Fast Fourier transforms in the analysis of near-infrared spectra. In *Analytical applications of spectroscopy*, ed. C. S. Creaser and A. M. C. Davies, 414–36. London: Royal Society of Chemistry.

McClure, W. F., D. Moody, D. L. Stanfield, and O. Kinoshita. 2002. Hand-held NIR spectrometry. Part II. An economical no-moving parts spectrometer for measuring chlorophyll and moisture. *Appl. Spectrosc.* 56:720–24.

McElhinney, J., G. Downey, and C. O'Donnell. 1999. Quantitation of lamb content in mixtures with raw minced beef using visible, near and mid-infrared spectroscopy. *J. Food Sci.* 64:587–591.

Medrano, R., S. H. Yan, M. Maudoux, V. Baeten, and M. Meurens. 1995. Wine analysis by NIR. In *Leaping ahead with near infrared spectroscopy: Proceedings of the Sixth International Conference on Near Infrared Spectroscopy*, ed. G. D. Batten, P. C. Flinn, L. A. Welsh, and A. B. Blakeney, 303–6. Victoria: Royal Australian Chemistry Institute.

Michelson, A. A. 1891. Visibility of interference—Fringes in the focus of a telescope. *Phil. Mag.* 31(190):256–59.

Misiti, M., Y. Misiti, G. Oppenheim, and J. M. Poggi. 2008. *Wavelet toolbox for user's guide.* Natick, MA: MATLAB, The Mathwork.

Næs, T., and K. I. Hildrum. 1997. Comparison of multivariate calibration and discriminant analysis in evaluating NIR spectroscopy for determination of meat tenderness. *Appl. Spectrosc.* 51:350–57.

Næs, T., T. Isaksson, T. Fearn, and T. Davies. 2002. *A user-friendly guide to multivariate calibration and classification*. Chichester, UK: NIR Publications.

Noda, I., and Y. Ozaki. 2004. *Two-dimensional correlation spectroscopy*. Chichester, UK: John Wiley & Sons.

Norris, K. H. 1996. History of NIR. *J. Near Infrared Spectrosc.* 4:31–37.

Norris, K. H., and J. R. Hart. 1965. Direct spectrophotometric determination of moisture content of grain and seeds. In *Principles and methods of measuring moisture in liquids and solids*, ed. A. Waxler, 19–25. New York: Reinhold.

Osborne, B. G. 1996. Near infrared spectroscopic studies of starch and water in some processed cereal foods. *J. Near Infrared Spectrosc.* 4:195–200.

Osborne, B. G., and S. Douglas. 1981. Measurement of the degree of starch damage in flour by near infrared reflectance analysis. *J. Sci. Food Agric.* 32:328–332.

Osborne, B. G., and T. Fearn. 1983. Collaborative evaluation of universal calibrations for the measurement of protein and moisture in flour by near infrared reflectance. *J. Food Technol.* 18:453–60.

Osborne, B. G., T. Fearn, and P. H. Hindle. 1993. *Practical near infrared spectroscopy with applications in food and beverage analysis*. Harlow, UK: Longman Scientific and Technical.

Osborne, B. G. 1998. Improved NIR prediction of flour starch damage. In *Proceedings of the 48th Australian Cereal Chemistry Conference*, ed. L. O'Brien, A. B. Blakeney, A. S. Ross, C. W. Wrigley. 434–438. Melbourne, Australia: Royal Australian Chemistry Institute.

Ozaki, Y., Y. Katsumoto, J. H. Jiang, and Y. Liang. 2003. Spectral analysis in the NIR region. In *Useful and advanced information in the field of near infrared spectroscopy*, ed. S. Tsuchikawa, 1–16. Trivandrum, India: Plenum Press.

Ozaki, Y., S. Morita, and Y. Du. 2007. Spectral analysis. In *Near-infrared spectroscopy in food science and technology*, ed. Y. Ozaki, W. F. McClure, and A. A. Christy, 43–72. Hoboken, NJ: Wiley-Interscience.

Ozaki, Y., and I. Noda. 2000. *Two-dimensional correlation spectroscopy*. AIP Conference Series. New York: American Institute of Physics.

Ozaki, Y., S. Šašič, and J. H. Jiang. 2001. How can we unravel complicated near infrared spectra? Recent progress in spectral analysis methods for resolution enhancement and band assignments in the near infrared region. *J. Near Infrared Spectrosc.* 9:63–95.

Palmer, C. 2005. Diffraction Grating Handbook, 6th ed. Rochester, NY: Newport Corporation.

Park, B., Y. R. Chen, W. R. Hruschka, S. D. Shackelford, and M. Koohmaraie. 1998. Near-infrared reflectance analysis for predicting beef longissimus tenderness. *J. Anim. Sci.* 7:2115–20.

Park, B., K. C. Lawrence, W. R. Windham, and D. P. Smith. 2006. Performance of hyperspectral imaging system for poultry surface contaminant detection. *J. Food Eng.* 75(3):340–48.

Persson, J.-A., and R. Sjodin. 2004. Efficiency all the way from grain to flour. *In Focus* 28(1):4–5.

Rambla, F. J., S. Garrigues, and M. de la Guardia. 1997. PLS-NIR determination of total sugar, glucose, fructose and sucrose in aqueous solutions of fruit juices. *Anal. Chim. Acta* 344(1–2):41–53.

Rasco, B. A., C. E. Miller, and T. L. King. 1991. Utilization of NIR spectroscopy to estimate the proximate composition of trout muscle with minimal sample pretreatment. *J. Agric. Food Chem.* 39:67–72.

Ridgway, C., and J. Chambers. 1996. Detection of external and internal insect infestation in wheat by near-infrared reflectance spectroscopy. *J. Sci. Food Agric.* 71:251–64.

Ridgway, C., and J. Chambers. 1998. Detection of insects inside wheat kernels by NIR imaging. *J. Near Infrared Spectrosc.* 6:115–19.

Roberts, C. A., R. R. Marquardt, A. A. Frohlich, R. L. McGraw, R. G. Rotter, and J. C. Henning. 1991. Chemical and spectral quantification of mold in contaminated barley. *Cereal. Chem.* 68:272–75.

Robinson, J. W., E. M. S. Frame, and G. M. Frame II. 2005. *Undergraduate instrumental analysis.* New York: Marcel Dekker.

Rødbotten, R., B. N. Nilsen, and K. I. Hildrum. 2000. Prediction of beef quality attributes from early post mortem near infrared reflectance spectra. *Food Chem.* 69:427–36.

Rodriguez-Saona, L. E., F. S. Fry, M. A. McLaughlin, and E. M. Calvey. 2001. Rapid analysis of sugars in fruit juices by FT-NIR spectrosopy. *Carbohydr. Res.* 336(1):63–74.

Rolfe, P. 2000. *In vivo* near-infrared spectroscopy. *Ann. Rev. Biomed. Eng.* 2:715–54.

Ronalds, J. A., and A. B. Blakeney. 1995. Determination of grain colour by near infrared reflectance and near infrared transmittance spectroscopy. In *Leaping ahead with near infrared spectroscopy,* ed. G. D. Battern, P. C. Flinn, L. A. Welsh, and A. B. Blakeney, 148–53. Victoria: Royal Australian Chemistry Institute.

Rubenthaler, G. L. and B. L. Bruinsma. 1978. Lysine estimation in cereals by near-infrared reflectance. *Crop Sci.* 18:1039–1042.

Saranwong, S., and S. Kawano. 2007. Fruits and vegetables. In *Near-infrared spectroscopy in food science and technology,* ed. Y. Ozaki, W. F. McClure, and A. A. Christy, 219–45. Hoboken, NJ: John Wiley & Sons.

Savitzky, A., and M. J. E. Golay. 1964. Smoothing and differentiation by simplified least squares procedures. *Anal. Chem.* 36(8):1672–39.

Schmilovitch, Z., A. Hoffman, H. Egozi, and E. Klein. 2002. Determination of egg freshness by NNIRS (near-near infrared spectroscopy). In: *Proceedings of Agricultural Engineering Conference* (Paper number 02-AP-023). Budapest, Hungary.

Schulz, H., and M. Baranska. 2009. Fruits and vegetables. In *Infrared spectroscopy for food quality analysis and control,* ed. D.-W. Sun, 321–53. London: Elsevier.

Schulze, G., A. Jirasek, M. M. L. Yu, A. Lim, R. F. B. Turner, and M. W. Blades. 2005. Investigation of selected baseline removal techniques as candidates for automated implementation. *Appl. Spectrosc.* 59:545–74.

Shah, N. K., and P. J. Gemperline. 1990. Combination of the Mahalanobis distance and residual variance pattern recognition techniques for classification of near-infrared reflectance spectra. *Anal. Chem.* 62:465–70.

Shao, Y. N., and Y. He. 2007. Nondestructive measurement of the internal quality of bayberry juice using Vis/NIR spectroscopy. *J. Food Eng.* 79(3):1015–19.

Shimizu, N., J. Katsura, T. Yanagisawa, B. Tezuka, Y. Maruyama, S. Inoue, C. G. Eddison, I. A. Cowe, R. P. Withey, A. B. Blakeney, T. Kimura, S. Yoshizaki, H. Okadome, H. Toyoshima, and K. Ohtsubo. 1998. Evaluating techniques for rice grain quality using near infrared transmission spectroscopy. *J. Near Infrared Spectrosc.* 6:A111–16.

Shimoyama, M., T. Ninomiya, K. Sano, Y. Ozaki, H. Higashiyama, M. Watari, and M. Tomo. 1998. Near infrared spectroscopy and chemometrics analysis of linear low-density polyethylene. *J. Near Infrared Spectrosc.* 6(1):317–24.

Siesler, H. W., Y. Ozaki, S. Kawata, and H. M. Heise. 2002. *Near infrared spectroscopy— Principles, instruments, applications.* Weinheim, Germany: Wiley-VCH.

Sjöblom, J., O. Svensson, M. Josefson, H. Kullberg, and S. Wold. 1998. An evaluation of orthogonal signal correction applied to calibration transfer of near infrared spectra. *Chemometr. Intell. Lab.* 44(1–2):229–44.

Slaughter, D. C., and J. A. Abbott. 2004. Analysis of fruits and vegetables. In *Near-infrared spectroscopy in agriculture,* ed. C. A. Roberts, J. Workman Jr., and J. B. Reeves III, 377–98. Madison, WI: American Society of Agronomy, Crop Science Society of America, Soil Science Society of America.

Solberg, C., E. Saugen, L. P. Swensen, T. Bruun, and T. Isaksson. 2003. Determination of fat in live farmed Atlantic salmon using non-invasive NIR techniques. *J. Sci. Food Agric.* 83:692–96.

Sollid, H., and C. Solberg. 1992. Salmon fat content estimation by near infrared transmission spectroscopy. *J. Food Sci.* 57:792–93.

Steiner, J., Y. Termonia, and J. Deltour. 1972. Comments on smoothing and differentiation of data by simplified least square procedure. *Anal. Chem.* 44(11):1906–1909.

Stork, C. L., D. J. Veltkamp, and B. R. Kowalski. 1998. Detecting and identifying spectral anomalies using wavelet processing. *Appl. Spectrosc.* 52(10):1348–52.

Suzuki, Y., S. Takahashi, M. Takebe, and K. Komae. 1996. Factors that affect protein contents in brown rice estimated by a near-infrared instrument (in Japanese with English abstract). *Nippon Shokuhin Kogyo Gakkaishi* 43(2):203–10.

Szymanski, H. A. 1964. *IR theory and practice of infrared spectroscopy.* New York: Plenum Press.

Takács, G., T. Pokorny, and I. Bányász. 1996. Application of a portable near infrared analyser for wheat measurement. In *Near infrared spectroscopy: The future waves,* ed. A. M. C. Davies and P. C. Williams, 92–96. Chichester, UK: NIR Publications.

Tilotta, D. C., R. M. Hammaker, and W. G. Fateley. 1987. A visible-near-infrared Hadamard transform spectrometer based on a liquid crystal spatial light modulator array: A new approach in spectrometry. *Appl. Spectrosc.* 41(6):727–34.

Trygg, J., and S. Wold. 2002. Orthogonal projections to latent structures (O-PLS). *J. Chemometr.* 16(3):119–28.

Tsenkova, R., and S. Atanassova. 2002. Mastitis diagnostics by near infrared spectra of cow's milk, blood and urine using soft independent modelling of class analogy classification. In *Near infrared spectroscopy: Proceedings of the 10th International Conference,* ed. A. M. C. Davies and R. K. Cho, 123–28. Chichester, UK: NIR Publications.

Tsenkova, R., S. Atanassova, K. Toyoda, Y. Ozaki, K. Itoh, and T. Fearn. 1999. Near-infrared spectroscopy for dairy management: Measurement of unhomogenized milk composition. *J. Dairy Sci.* 82:2344–51.

Tsenkova, R., K. I. Yordanov, K. Itoh, Y. Shinde, and J. Nishibu. 1994. Near infrared spectroscopy of individual cow milk as a means for automated monitoring of udder health and milk quality. In *Dairy systems for the 21st century: Proceedings of the Third International Dairy Housing Conference,* ed. R. Bucklin, 82–91. St. Joseph, MI: ASAE.

Vandenginste, B. G. M., D. L. Massart, L. M. C. Buydens, S. de Jorg, P. J. Lew, and J. Smeyers-Verbeke. 1998. *Handbook of chemometrics and qualimetrics.* Amsterdam: Elsevier.

Villareal., C. P., N. M. De La Cruz, and B. O. Juliano. 1994. Rice amylose analysis by near-infrared transmittance spectroscopy. Cereal Chem. 71:292–263.

Walczak, B., and D. L. Massart. 1998. Multiple outlier detection revisited. *Chemometr. Intell. Lab. Syst.* 41(1):1–15.

Westerhuis, J. A., S. de Jong, and A. K. Smilde. 2001. Direct orthogonal signal correction. *Chemometr. Intell. Lab.* 56:13–25.

Wetzel, D., and H. Mark. 1977. Spectroscopic cellulose determination as a criterion of flour purity with respect to bran. *Cereal Foods World* 22:481.

Wilkins, D. R., I. A. Cowe, B. B. Thind, J. W. McNicol, and D. C. Cuthbertson. 1986. The detection and measurement of mite infestation in animal feed using near infra-red reflectance. *J. Agric. Sci.* 107:439–48.

Williams, P. C., K. R. Preston, K. H. Norris, and P. M. Starkey. 1984. Determination of amino acids in wheat and barley by near-infrared reflectance spectroscopy. *J. Food Sci.* 49:17–20.

Windham, W. R., and W. H. Morrison. 1998. Prediction of fatty acid content in beef neck lean by near infrared reflectance analysis. *J. Near Infrared Spectrosc.* 6:229–34.

Wise, B. 1998. Orthogonal signal correction. In *Functions in the MATLAB user area: Eigenvector research*, Wenatchee, WA. http://www.eigenvector.com/MATLAB/OSC. html (accessed January 12, 2010).

Wold, H. 1975. Soft modeling by latent variables: The nonlinear iterative partial least squares approach. In *Perspectives in probability and statistics*, ed. J. Gani, 520–40. London: Academic Press.

Wold, S. 1976. Pattern recognition by means of disjoint principal components models. *Pattern Recogn.* 8:127–39.

Wold, S., K. Esbensen, and P. Geladi. 1987. Principal component analysis. *Chemometr. Intell. Lab. Syst.* 2:37.

Wold, S., H. Antti, F. Lindgren, and J. Öhman. 1998. Orthogonal signal correction of near infrared spectra. *Chemometr. Intell. Lab.* 44:175–85.

Xie, L., X. Ye, D. Liu, and Y. Ying. 2009. Quantification of glucose, fructose and sucrose in bayberry juice by NIR and PLS. *Food Chem.* 114(3):1135–40.

Zupan, J. 1994. Introduction to artificial neural network (ANN) methods: What they are and how to use them. *Acta Chim. Slov.* 41:327–52.

Zupan, J., and J. Gasteiger. 1993. *Neural networks for chemists: An introduction.* Weinheim, Germany: VCH-Verlag.

6 Quality Measurements Using Nuclear Magnetic Resonance and Magnetic Resonance Imaging

Michael J. McCarthy, Sandra P. Garcia, Seongmin Kim, and Rebecca R. Milczarek

CONTENTS

6.1 INTRODUCTION

Quality evaluation of agricultural products has historically been primarily based on the surface appearance. In the case of fruit products the surface should be the appropriate color and free of defects, such as scars, bruises, and insect damage. However, the nutritional and sensory qualities of a fruit product are often based on the internal chemical and physical properties of the product. Good quality fruit should have the appropriate maturity, texture, chemical composition, structure, and the absence of defects (e.g., bruises, browning, mold, insect damage). Measurement of the internal quality is currently performed on a subsample of the product that is destructively analyzed. This subsample approach is suboptimal because the range of individual variations within the batch yield many fruit that differ from the desired quality specifications. This has naturally given rise to the need to develop nondestructive methods to quantify fruit quality.

There are a number of potential technologies that have been investigated for the measurement of internal fruit quality. These technologies include near-infrared spectroscopy, mid-infrared spectroscopy, Raman spectroscopy, ultrasound spectroscopy, dielectric imaging, x-ray imaging, and nuclear magnetic resonance spectroscopy. Each spectroscopy has advantages and disadvantages in application to fruit quality measurement (Irudayaraj and Reh, 2008). A sensor system for any fruit would most likely be constructed from two or more of these technologies combined to yield measurements of the most important quality parameters. Magnetic resonance imaging measurements are sensitive to a wide range of product attributes, including internal structure, defects (browning, rot, mold, insect damage), major component concentrations (oil, water, and sugar), and maturity, and the measurement is rapid enough for in-line application (McCarthy, 1994).

6.2 PRINCIPLES OF MAGNETIC RESONANCE

6.2.1 PHENOMENA OF MAGNETIC RESONANCE

Magnetic resonance is a phenomenon that occurs between atomic particles and an external magnetic field. The atomic particles responsible for this interaction are the electrons and the nucleus. The interaction between the atomic particles and an external magnetic field is similar to what happens when iron filings are placed near a bar magnet. The filings become oriented and a magnetic field is induced in the metal. However, unlike the filings, the physical orientation of the atomic particles is not altered. At most common magnetic field strengths only the magnetic moment of the atomic particles is induced. The phenomenon of resonance is observed in these systems because they absorb and emit energy at specific frequencies. The specific frequency depends on the individual atomic particle and the strength of the applied magnetic fields.

Work discussed in this chapter will focus on using the nucleus as the atomic particle, and in this case the phenomenon is referred to as *nuclear magnetic resonance* (NMR). Common nuclei with magnetic moments include ^1H, ^{31}P, ^{15}N, and ^{23}Na. The most commonly studied nucleus in food systems is the ^1H. In both medical and food applications of NMR, the technique is referred to simply as magnetic resonance (MR). The term *nuclear* is omitted so that patients/consumers will not confuse this technique with nuclear procedures that use radioactive materials. MR is a safe experimental procedure and does not harm or alter the sample, operator, or environment (McCarthy, 1994).

MR is a very useful spectroscopy because the signal emitted from a sample is sensitive to the number of nuclei, to the chemical and electronic surroundings of the nuclei at the molecular level, and to the diffusion or flow of the nuclei. This sensitivity to a range of sample characteristics is responsible for the wide use of MR in chemistry, biochemistry, biotechnology, petrophysics, plastics, engineering, building materials, consumer products, medicine, and food technology. When MR is used to make internal images of objects it is commonly called *magnetic resonance imaging*. Magnetic resonance imaging (MRI) is a spectroscopic technique based on the magnetic properties of nuclei. MRI is the extension of MR spectroscopy to obtain the signal as a function of spatial coordinates within the sample.

6.2.2 BLOCH MODEL

MR is based on the interaction between an external magnetic field and atomic particles, electrons, and nuclei that possess spin, more precisely spin angular momentum. NMR focuses only on the nuclear interactions. Atomic nuclei have a spin property that is quantized to discrete values. Nuclear spin or, more precisely, nuclear spin angular momentum depends on the precise atomic composition. Only nuclei that have a nonzero spin will have a magnetic moment that interacts with an external magnetic field. Almost every element in the periodic table has an isotope with a nonzero nuclear spin. However, protons, ^1H (spin ½ nuclei), are our main interest because they are the most abundant isotope for hydrogen and have a strong signal.

For an accurate mathematical description of a nucleus with spin and its interaction, quantum mechanical or statistical mechanical principles are required. However, when performing NMR, exceedingly large numbers of nuclei act largely independently so that at the macroscopic level the collection of particles appears continuous, the so-called ensemble behavior. For these, all states of the ensemble can be characterized by a simple vector quantity, which is referred to as the nuclear magnetization. Thus, we can discuss the NMR phenomenon using classical mechanics (Callaghan, 1991).

In a macroscopic sample, the net magnetization moment, $M = (M_x, M_y, M_z)$, can be derived from the sum of all the nuclear magnetic moments. In such a model the macroscopic angular momentum vector is simply M/γ, where M is the magnetization and γ is the gyromagnetic ratio. Then, the motion for the classical magnetization vector is

$$\frac{dM}{dt} = \gamma M \times B \qquad (6.1)$$

The solution to Equation 6.1 when B is a magnetic field of amplitude B_0 corresponds to a precession of the magnetization about the field at the Larmor frequency, $\omega_0 = \gamma B_0$, which is one of the fundamental relationships in NMR.

The resonance phenomenon occurs after the application of a transverse magnetic field oscillating at the Larmor frequency, ω_0. This perturbing radio frequency (RF) pulse, applied perpendicular to B_0, is given by

$$B_1(t) = B_1[\cos(\omega_0 t)i - \sin(\omega_0 t)j] \qquad (6.2)$$

where i, j, and k are unit vectors along the x, y, and z axes, respectively. Then the applied RF pulse yields

$$\frac{dM_x}{dt} = \gamma\left[M_y B_0 + M_z B_1 \sin(\omega_0 t)\right]$$

$$\frac{dM_y}{dt} = \gamma\left[M_z B_1 \cos(\omega_0 t) - M_x B_0\right] \tag{6.3}$$

$$\frac{dM_z}{dt} = \gamma\left[-M_x B_1 \sin(\omega_0 t) - M_y B_1 \cos(\omega_0 t)\right]$$

Under the initial condition of $M(t = 0) = M_0 k$, the solution of Equation 6.3 is

$$M_x = M_0 \sin(\omega_1 t)\sin(\omega_0 t)$$

$$M_y = M_0 \sin(\omega_1 t)\cos(\omega_0 t) \tag{6.4}$$

$$M_z = M_0 \cos(\omega_1 t)$$

where $\omega_1 = \gamma B_1$. Equation 6.4 implies that upon application of a rotating magnetic field of frequency ω_0, the magnetization simultaneously precesses about the static field B_0 at ω_0 and about the RF field B_1 at ω_1. In the laboratory frame, the transverse magnetization precessing at the Larmor frequency will induce an oscillatory electromagnetic field (e.m.f.) at ω_0 in the NMR probe. This alternating current, or signal, is heterodyned with a reference frequency, ω_r. This signal detection method is inherently phase sensitive, such that we obtain separate quadrature phase output signals that are each respectively proportional to orthogonal phases of the magnetization, in effect detecting M_x and M_y. The signal may oscillate at an offset frequency $\Delta\omega = \omega_0 - \omega_r$. By applying a 90° RF pulse ($\omega_1 = \pi/2$), the laboratory magnetization at time t is given by

$$M(t) = M_0\left[\cos(\omega_0 t)i + \sin(\omega_0 t)j\right] \tag{6.5}$$

which can be expressed in complex notation as

$$M(t) = M_0 \exp(i\omega_0 t) \tag{6.6}$$

and the resulting complex NMR signal at the offset $\Delta\omega$ is out of phase by φ, described as

$$S(t) = S_0 \exp(i\phi + i\Delta\omega t) \tag{6.7}$$

where S_0 is the arbitrary initial signal intensity that depends upon the electronics of the detection system and is proportional to the total transverse magnetization of the sample M_0. In the time domain, this primary NMR signal is measured as an oscillating and decaying e.m.f. induced by the magnetization in free precession and is known as the free induction decay (FID).

Relaxation occurs when the spins dissipate the absorbed RF energy and return to their equilibrium state. As the relaxation occurs, the magnetization recovers in the longitudinal plane. Longitudinal and transverse relaxation processes occur simultaneously. T_1 relaxation, referred to as longitudinal relaxation or spin-lattice relaxation, is the return of longitudinal magnetization to thermal equilibrium (i.e., from a higher energy state to a lower energy state) and is associated with the exchange of energy between the spin system and surrounding thermal reservoir or lattice. This process can be described as

$$\frac{dM_z}{dt} = \frac{-(M_z - M_0)}{T_1} \tag{6.8}$$

And the magnetization at certain time t is given by

$$M_z(t) = M_z(0)\exp\left(-\frac{t}{T_1}\right) + M_0\left[1 - \exp\left(-\frac{t}{T_1}\right)\right] \tag{6.9}$$

where T_1 is the time required for M_z to return to 63% of its original value following an excitation pulse. At room temperature, T_1 is a few seconds for protons in water and shorter for most other liquids. A typical T_1 relaxation curve is shown in Figure 6.1, as well as data from a saturation recovery experiment that is used to measure T_1.

The transverse relaxation, also called T_2 relaxation or spin-spin relaxation, is the return of transverse magnetization to equilibrium. T_2 relaxation involves intermolecular and intramolecular interactions such as vibrations or rotations among spins. As the excited spins release their energy among themselves, the coherence of transverse magnetization disappears gradually. T_2 relaxation can be described by a first-order relaxation process as

$$\frac{dM_{x,y}}{dt} = \frac{-M_{x,y}}{T_2} \tag{6.10}$$

with solution

$$M_{x,y}(t) = M_{x,y}(0)\exp\left(-\frac{t}{T_2}\right) \tag{6.11}$$

where T_2 is spin-spin relaxation time or transverse relaxation time, which is the time required for $M_{x,y}$ to decay to 37% of its initial value. T_2 is generally less than or equal to T_1. Figure 6.2 shows a typical T_2 relaxation curve and the data from a Carr, Purcell, Meiboom, Gill (CPMG) experiment (Callaghan, 1991) used to measure the T_2 of a sample.

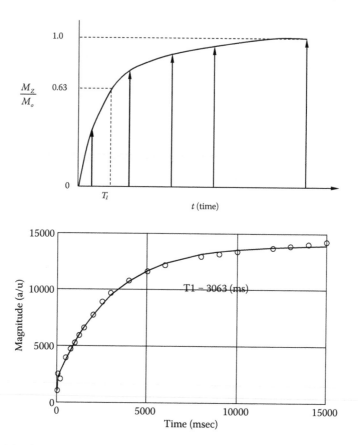

FIGURE 6.1 Shown in the top plot is a typical T_1 relaxation curve. The arrows are the size of the z-component of the net magnetization. Shown in the lower plot is a saturation recovery experiment used to measure the spin-lattice relaxation time; the sample is tap water at 43.85 MHz proton frequency.

The Bloch equations are a combination of Equations 6.3, 6.8, and 6.10, which describe the behavior of a magnetization vector in the laboratory frame and are shown below:

$$\frac{dM_x}{dt} = \gamma[M_y B_0 + M_z B_1 \sin(\omega t)] - \frac{M_x}{T_2}$$

$$\frac{dM_y}{dt} = \gamma[M_z B_1 \cos(\omega t) - M_x B_0] - \frac{M_y}{T_2} \qquad (6.12)$$

$$\frac{dM_z}{dt} = \gamma[-M_x B_1 \sin(\omega t) - M_y B_1 \cos(\omega t)] - \frac{(M_z - M_0)}{T_1}$$

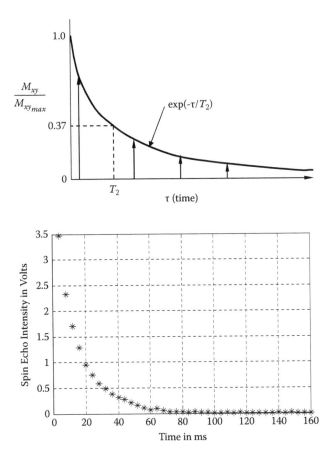

FIGURE 6.2　Shown in the top plot is a typical T_2 relaxation curve. The arrows are the size of the z-component of the net magnetization. In the bottom plot is the CPMG decay curve from a dehydrated plum acquired at 10 MHz proton frequency.

6.2.3　MAGNETIC RESONANCE IMAGING

In conventional NMR spectroscopy, the spectrum of nuclear precession frequencies provides information on the chemical and electronic environment of the spins. The polarizing magnetic field, B_0, is uniform across the sample. The static field inhomogeneities can be removed by careful adjustment of the currents in magnet shim coils. The frequency of energy is determined by the exact magnetic field strength that the spins experience. MRI techniques use this field dependence to localize proton frequencies to different spatial positions. This localization is accomplished by applying linear magnetic field gradients to disturb the main magnetic field homogeneity such that the local Larmor precession frequency becomes spatially dependent. Although these magnetic field gradients perturb the static magnetic field, B_0, they are much smaller, so that the Larmor frequency is affected only by components parallel to

B_0 and orthogonal components will be averaged to zero. Hence, the local Larmor frequency can be defined as

$$\omega(r) = \gamma B_o + \gamma G \cdot r \tag{6.13}$$

where G is the field gradient vector and r is the position vector.

For a small element of volume dV at position r with local spin density $\rho(r)$ in the presence of the field gradient, the NMR signal is given by

$$dS(G,t) = \rho(r)dV \exp[i\gamma(B_0 + G \cdot r)t] \tag{6.14}$$

When the reference frequency is chosen to be γB_0, so called on resonance, neglecting the effect of the main magnetic field, the signal oscillates at $\gamma G \times r$. The integrated signal over the entire set of volume elements is given by

$$S(t) = \iiint \rho(r) \exp(i\gamma G \cdot rt)dr \tag{6.15}$$

where dr represents a volume integration.

For simplification, and to further display that Equation 6.15 is a sum of oscillating terms that has the form of a Fourier transformation, Mansfield (Callaghan, 1991) introduced the concept of a reciprocal space vector k given by

$$k = \frac{\gamma G t}{2\pi} \tag{6.16}$$

The k-vector has a magnitude expressed in units of reciprocal space, m^{-1}, and k-space may be traversed by moving either in time or in gradient magnitude. The factor of 2π appears since γ is expressed in units of radians per second per Gauss and G, the units of Gauss per meter. However, the direction that k-space is traversed is determined by the direction of the gradient G. Using the k-space construct, and the concept of Fourier transform and its inverse,

$$S(k) = \iiint \rho(r) \exp(i2\pi\gamma k \cdot r) \, dr \tag{6.17}$$

$$\rho(r) = \iiint S(k) \exp(-i2\pi\gamma k \cdot r)c \tag{6.18}$$

Equations 6.17 and 6.18 state that the signal $S(k)$ and the nuclei density $\rho(r)$ are mutually conjugate. This is the fundamental relationship in NMR imaging. For the actual NMR experiment, data are attained over a limited range in k and at discrete intervals. Hence, Equations 6.17 and 6.18 should be interpreted as discrete Fourier transforms for which there exists a simple relation between the range and resolution in both spaces.

For two-dimensional imaging, k-space is acquired in two dimensions. The images presented in this chapter are two-dimensional even though the sample is three-dimensional in nature. When the magnetic resonance signal is sampled in the presence of a gradient, points are obtained along a single line in k-space. This line is oriented along one of the Cartesian axes, and the associated gradient is known as the read gradient. The intercept of this line along the orthogonal axis can be changed by imposing the G_y gradient for a fixed period before sampling begins. Then G_y is named the phase gradient since it imparts a phase modulation to the signal, dependent on the position of volume elements along the y-axis. The third dimension will be called the slice selection and is a plane in the sample that is excited using a frequency-selective RF pulse in the presence of a static magnetic field gradient, where the selective pulse is composed of an on-resonance RF field modulated by a lower-frequency envelope of finite duration (Blumich, 2000). The resulting signal will represent only the magnetization from the selected plane or slice. This approach avoids having to acquire the entire three-dimensional time-domain data set.

A spin-echo pulse sequence is one of the most commonly used pulse sequences in MRI (Figure 6.3). The Gaussian-shaped RF pulse along with the applied G_z gradient confines the region of interest to a slice in the x-y plane. During the interval τ, the preread gradient pulse, G_x, allows the acquisition of an echo by encoding a constant amount of phase, and simultaneously, a variable amount of phase from the applied G_y pulse is encoded along the y-direction. The amount of phase encoded is dependent on the magnitude of G_y. In order to acquire the signal to generate the x-y image of the slice, a series of experiments (or excitations) are performed where G_y is incremented in steps δG_y from $(N_y/2)\delta G_y$ to $-(N_y/2)\delta G_y$. During the signal acquisition, the phase along x is constantly evolving and the signal is composed of the sum over the frequency contributions from all the sample volume elements. For the pulse sequence of Figure 6.3, k_x-space is attained by sampling the echo at specific time intervals, designated as t, and k_y-space is traversed by changing the G_y gradient amplitude in steps denoted by δG_y. Two-dimensional Fourier transformation of the k-space data set yields an image of the magnetization density in the sample. Then the NMR signal is given by

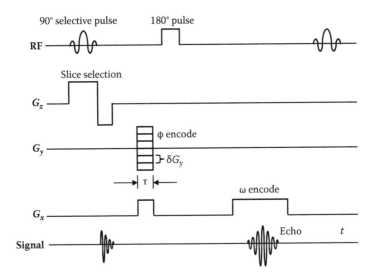

FIGURE 6.3   Spin-echo pulse sequence.

$$S(k_x,k_y) = \int_{-b/2}^{b/2} \left[\int_{-\infty}^{\infty} \int_{-\infty}^{\infty} \rho(x,y,z) \exp\{i2\pi(k_x x + k_y y)\} \, dx \, dy \right] dz \qquad (6.19)$$

where b is the slice thickness. For convenience, the outer integral is ignored since it represents the process of averaging across the slice, giving

$$S(k_x,k_y) = \int_{-\infty}^{\infty} \int_{-\infty}^{\infty} \rho(x,y) \exp\{i2\pi(k_x x + k_y y)\} \, dx \, dy \qquad (6.20)$$

Reconstruction of $\rho(x, y)$ from $S(k_x, k_y)$ simply requires that we calculate the inverse Fourier transform:

$$\rho(x,y) = \int_{-\infty}^{\infty} \int_{-\infty}^{\infty} S(k_x,k_y) \exp\{-i2\pi(k_x x + k_y y)\} \, dk_x \, dk_y \qquad (6.21)$$

where

$$k_x = \frac{\gamma G_x t}{2\pi}$$

and

$$k_y = \frac{\gamma G_y \tau}{2\pi}$$

Taking a viewpoint of the imaging experiment that a certain region of k-space should be sampled with some specific resolution in order to obtain an image with the appropriate range and resolution, an imaging scheme will work as long as k-space is adequately sampled. For example, k_y-space could be acquired by starting with negative G_y amplitude and proceeding to positive G_y, or vice versa. Second, we could have used an x-gradient of negative amplitude during τ and omit the 180° RF pulse. Furthermore, the sampling of k-space need not be confined to a Cartesian grid (Callaghan and Eccles, 1987; Callaghan, 1991), but whatever the sampling strategy, a means of transforming the acquired data from the time-domain to the spatial-domain is required. Additional details of MR theory and pulse sequence development are given in the books by Blumich (2000), McCarthy (1994), and Callaghan (1991).

6.3 EQUIPMENT

All NMR equipment includes a spectrometer, magnet, and RF coil. The spectrometer is used to apply energy to the sample and record the decay of energy from the sample. The magnet is used to polarize the spins in the sample, and the RF probe is used to couple the sample to the spectrometer. MRI spectrometers additionally include magnetic field gradient coils, which are used to induce linear gradients in the applied field for spatially localizing the signal. A diagram of the basic components is shown in Figure 6.4.

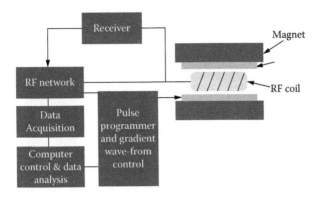

FIGURE 6.4   Basic hardware components of an MRI spectrometer.

   Permanent magnets exist in a large number of different configurations. These con-figurations include single-sided, where the field and signal are acquired at a distance from the magnet surface; cylindrical cavities, like a Halbach design with the sample in the center of the cylinder; and as two plates, where the sample is located between the plates. Permanent magnets are made of an assembly of individual magnetized bricks. Materials that are used to fabricate the bricks include alnico alloys or rare earth alloys. The bricks are often coupled to an iron yoke or face plate to increase the magnetic field strength/uniformity. Field strength for these types of magnets ranges from 0.05 to 2.35 T (proton resonance frequencies from 2 to 100 MHz). Almost all past and current applications of MR for process control have utilized permanent magnets. Figures 6.5 and 6.6 show two permanent magnets, one used for imaging of fruit quality and the other a single-sided system used to measure relaxation times of fruit tissue.

6.4   APPLICATIONS

Properties of the sample that influence the signal include nuclei density, nuclei motion, sample temperature, sample viscosity, sample structure, and composition. There exist a variety of MR experiments that are used individually or in combina-tion to characterize a sample. Several examples will be used to describe how MR and MRI can be used to quantify the quality of fruits based on the internal structure, and the response of the signal to changes in the internal structure of the fruit. The internal structure of a product significantly influences the MR signal and permits noninvasive and noncontact detection of quality (Chen et al., 1991).

   The MR techniques that are used fall into the following general categories: qual-ity assessment based on either the rate of signal decay or the structure of the sample, or a combination of the two (McCarthy, 1994). The signal will decay at a specific rate for healthy tissue and at a different rate for damaged tissue. The difference between these decay rates ranges from 25% to several orders of magnitude. These different decay rates yield variations in image intensity. The second method is based on the structure of the sample. Defects that are related to structure will be seen as a change in the structure, for example, the freeze damage in an orange or kiwi (Kerr et al.,

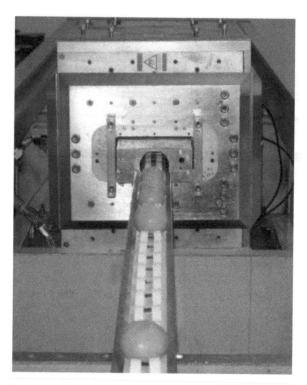

FIGURE 6.5 Permanent-magnet–based MRI system adapted to a fruit-sorting conveyor. (Photo courtesy of Uri Rapoport, ASPECT Magnet Technologies Ltd.)

FIGURE 6.6 Permanent-magnet–based relaxation time measurement system for application as a single-sided sensor (magnet built by ABQMR, Inc.).

1997; Gambhir et al., 2005). These techniques are referred to as relaxation-weighted imaging experiments.

MR image data are a two-dimensional array of the signal intensity value from a small volume inside the sample. Each element of the two-dimensional data set is called a volume element (or voxel) and represents the value of signal intensity in each region of the sample. The signal intensity of each volume element in a spin-echo image is a function of the intrinsic properties of the sample: proton density (ρ), proton spin-lattice, and spin-spin relaxation times. A volume element has three spatial dimensions: two define in-plane resolution, and the third defines the slice thickness. This volume contains protons from which the MR signal is recorded. Equations 6.22 and 6.23 defines the signal intensity (S) of each pixel for a spin-echo-based imaging protocol.

$$S_{\text{spin echo}} = K\rho[1 - \exp(-TR/T_1)]\exp(-TE/T_2) \qquad (6.22)$$

$$S_{\text{gradient echo}} = K\rho\sin(\theta)(1 - \exp(-TR/T_1)(\exp(-TE/T_2^*)/(1 - \cos(\theta)\exp(-TR/T_1)) \quad (6.23)$$

where TR is repetition time, the time interval from the beginning of pulse sequence until the next pulse sequence starts; TE is the time from the middle of the first RF pulse to the middle of the echo signal; and θ is the flip angle of the RF pulse.

A spin-echo image that has a value of TE such that the intensity is influenced by differences in T_2 as a function of position is shown in Figure 6.7. This is an image of a Fuji apple with internal browning where the signal decay rate in the section with browning is slower than in normal tissue. The image is displayed with a grayscale map that shows the internal browning as dark and the healthy tissue as lighter in color. This image contrast is common for internal browning, water core, bruises, and rot in apples or pears.

Shown in Figure 6.8 are two images of the central plane in a tomato; one is a gradient recalled echo image, where the signal decay enhances the regions in the pericarp with greater amounts of air, and the other is a spin-echo image, where this influence is minimized. Figures 6.7 and 6.8 demonstrate just a small portion of the contrast mechanisms available when utilizing MRI.

The challenge is in optimizing the data acquisition parameters for MRI to detect quality factors. This can be compounded when the goal is to measure multiple quality factors, for example, seeds, hollow center, and mold damage in citrus. Utilization of statistical approaches can greatly reduce the effort in optimizing the data acquisition. An approach that has been used on multivarite data for optimization of other sensor technologies is chemometrics. Chemometrics is the discipline that uses multivariate statistical methods to (1) design or select optimal chemical measurement procedures and (2) provide maximum chemical information from a measurement

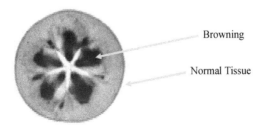

FIGURE 6.7 Spin-echo–based image of a Fuji apple with internal browning.

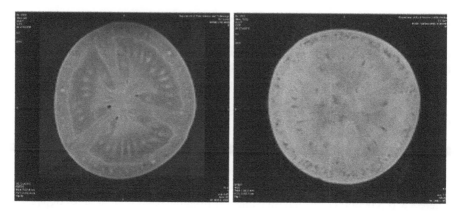

FIGURE 6.8   Proton images of a red tomato. The left has a spin-echo image showing intensity differences based on T_2. The right has a gradient recalled echo image that shows contrast based on proton density.

in an efficient way (Otto, 1999). Chemometric techniques have traditionally been used with multidimensional measurement techniques such as gas chromatography and infrared spectroscopy. However, chemometric techniques can also be used for data reduction of both single images and sets of images. For example, in chemical shift imaging, an MR spectrum is measured at every pixel of the image. Using the chemometric technique of principal components analysis (PCA), one can determine which parts of the MR spectrum are most important in differentiating among pixels of different intensities in the image. This technique can be applied to groups of pixels (usually regions of interest encompassing different structures in the object being imaged) to provide information on how areas in the interior of the object differ in terms of their MR properties. Using the chemometric regression technique of partial least squares (PLS), one can also use a "learning" set of images of objects in different classes to classify future "test" images. The image learning set is regressed onto a set of logical indices (e.g., an image's index for a certain class would be 1 if the image belongs to the class or 0 if the image does not belong to the class), and the resulting multivariate regression model is used to predict the class membership of new images (Wise et al., 2004). This method thus would enable sorting of objects based on their MR images. The PLS model can be continually updated by adding images to the learning set, increasing the accuracy of the sorting procedure. We will describe an example using chemometrics applied to MRI data of processing tomatoes to predict peelability and citrus to predict seed count.

6.4.1   Processing Tomato Peelability

The state of California produces more than 90% of the processing tomatoes in the United States and approximately 35% of the world's supply (Hartz et al., 2008). Considering that Americans consume about 72 kg of processed tomatoes per person per year (USDA, 2008), it is important that California tomatoes are processed efficiently and in a way that ensures high nutritional and flavor quality for the consumer.

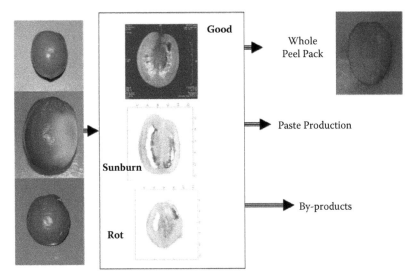

FIGURE 6.9 Diagram of using MRI to sort processing tomatoes based on internal properties.

Peeling is the first step in producing value-added tomato products, such as diced and whole-peel-packed tomatoes. Insufficient or improper peeling deteriorates tomato quality and results in product loss. Thus, tomato processors seek a method for predicting the peelability of individual fruit and directing each fruit to the appropriate end product, as shown in Figure 6.9.

Processing tomatoes were collected from the flume in a commercial processing facility and imaged. Each tomato was then placed in an MRI and twenty-eight different image types acquired. After imaging the fruit were tagged with a radio frequency identification (RFID) chip and sent through a pilot peeling process. The possible peeling outcomes were whole peeled, some skin, broken and some skin, broken, unpeeled, fell out somewhere in the process (three locations), or other (generally lost RFID tag). "Whole peeled" is the only desired outcome from the process. The MRI image data sets were used to predict the peeling outcomes using a partial least squares–discriminant analysis (PLS-DA). The X-block data for PLS-DA were the average intensities of the pericarp region of interest (ROI) pixels in the MR images of each tomato sample. The Y-block consisted of class assignment of each tomato sample. Figure 6.10 is a sample of the data from this study. Each point represents an individual tomato fruit. The shape and color of the data point indicate the actual peeling outcome for a single fruit. The fruit samples are arranged along the X-axis in time order. The months of July and August 2008 are covered in this data set. The scores generated by the PLS-DA model are measured along the Y-axis. The red dashed line is the cutoff threshold for classifying a fruit in the "whole peeled" class. Consider an aqua circle point above the dashed line; one of these points represents a correct prediction: the fruit was "whole peeled," and the model predicts it to be "whole peeled." Consider a red triangle point above the dashed red line; this shows an incorrect prediction: the fruit was

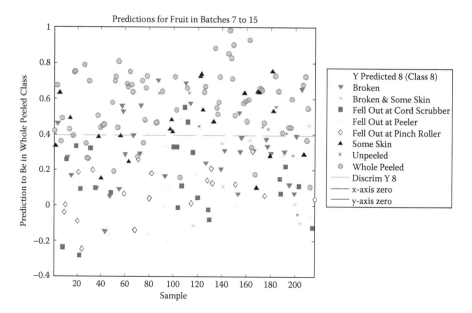

FIGURE 6.10 PLS-DA model predictions for processing tomatoes to be in the "whole peeled" class.

actually in the "broken" class, but the model predicted it to be "whole peeled." In summary the model correctly classified 83% of the whole peeled fruit into the "whole peeled" class and misclassified 17% of the fruit into the "whole peeled" class (Milczarek, 2009).

The MRI images provided useful information concerning the peeling outcomes of the tomatoes. The model presented above was optimized in a statistical sense; however, actual implementation in a production facility would involve optimizing the model from a business perspective. The use of chemometrics also permits us to determine which image types, and hence which MR parameters, are most important in predicting peelability. This approach facilitates the next step in implementation by significantly reducing the data taken to be compatible with the tomato peeling process. Combined with other MRI experiments to predict tomato maturity and identify defects (e.g., mold damage, insect damage), a comprehensive approach to measuring multiple quality factors can be achieved.

6.4.2 Mandarin Seed Count

Seedless mandarin oranges are considered to be of higher quality than seeded fruit, so citrus packers seek methods for identifying fruit that contain seeds. There is no external evidence that an orange contains seeds, so an internal, nondestructive technology is needed for this purpose. Due to the structural and physical similarities between seeds and pits, development of MR-based seed detection in citrus fruit has benefited from some lessons learned during research on nondestructive detection of pits in processed cherries, plums, and other stone fruit. Processors of these commodities are interested in nondestructive detection of pits for reasons of safety and quality grading. Han et al. (1992) determined that x-ray imaging could detect peaches with split pits. Zion et al. (1994) and Law (1973) found that NMR projections and near-infrared scattering, respectively, could detect cherry pits. However, in all three of the previously mentioned cases, orientation of the fruit was critical to the success rate of the measurement. For the near-infrared measurement, variability due to orientation can be eliminated when a hyperspectral approach is used (Qin and Lu, 2005). Timm et al. (1991) found that point measurement of near-infrared light transmission could detect pits, but that the signal was unreliable due to variations in cherry size, orientation, pitter scar, and external defects. However, using light transmission imaging, the same researchers were able to achieve 95% detection of cherries with a pit, not over 5% false rejection of pitted cherries (excluding defective cherries), and 100% detection of loose pits for cherries moving at up to twenty cherries per second in a single lane. Thus, pit detection research indicates that one can increase the success rates for identifying hard, solid inclusions in soft-fleshed fruit by using (1) multivariate approaches and (2) two-dimensional images (as opposed to projections or spectra averaged over a whole fruit).

These results have been extended to the task of detecting seeds in citrus fruit using MR imaging. Kim et al. (2008) used the ASPECT Magnet Technologies Ltd. system described in Section 6.5.1 to obtain fast spin-echo MR images of "W. Murcott" mandarin oranges. These images were used to detect both seeds and freeze damage. In the image analysis, the authors used two different masks: one near the center axis of the fruit (for seed detection) and one between the rind and the first mask (for freeze damage detection). Hernandez-Sanchez et al. (2006) and Barreiro et al. (2008) used fast (1 s or less for a 128 × 128 pixel image) MRI pulse sequences to identify seeds in "Nardocott" mandarin oranges. In these two studies, seeds were hypointense and locules (juicy flesh of the fruit) were hyperintense. This pixel intensity pattern lent itself to combining the seed pixels with the central air column pixels to make one image object. Various measurements (perimeter, aspect ratio, etc.) of this object were then compared for seeded vs. seedless fruit.

A different approach is to use MRI pulse sequences that make seeds hyperintense and both locules and air hypointense. Using MR images of this type, Lau (2008) designed a genetic algorithm that segmented the images into wedges and then determined the area of interest in each image where seeds were likely to occur. The genetic algorithm approach identified 92% of seeded fruit and 91% of seedless fruit, vs. 87% of seeded fruit and 89% of seedless fruit in the study by Hernandez-Sanchez et al. (2006).

One can also develop images in which seeds are hyperintense (and in high contrast with surrounding tissue) using the technique of multivariate MRI. In the 2007 citrus harvest season, 140 "W. Murcott" mandarin oranges were imaged on the ASPECT Magnet Technologies Ltd. system, as described in Milczarek and McCarthy (2009). Three congruent MR images were obtained for each fruit. The MR image types were fast spin-echo (FSE), turbo fast low-angle shot (turbo FLASH), and gradient recalled echo (GRE). These three images comprised the X-block data for a PLS regression analysis. The Y-block data were binary images in which pixels corresponding to seeds were coded with a value of 1 and all other pixels were coded 0.

The PLS regression of the Y-block onto the X-block resulted in a 1-latent variable model that was then applied to the X-block to create a prediction image. In the prediction image, seeds were hyperintense. This prediction image was then thresholded, creating a binary image that contained both seeds (desired) and smaller, spurious objects (undesired). To eliminate the undesired objects, a morphological classification routine was applied to each thresholded image. An object was identified as a seed if its size, eccentricity (ellipsoidal shape parameter), and solidity were within certain bounds. After morphological screening, 97% of seedless fruit and 56% of seeded fruit were ientified by the computer algorithm. That is, the combination of multivariate MRI, prediction image thresholding, and morphological selection (all of which were automated) resulted in a high success rate for identifying seedless fruit, but an unacceptably low success rate for identifying seeded fruit.

To improve the success rate for automated identification of seeded fruit, a follow-up study was performed two years later with ninety new fruit of the same cultivar. The fruit came from three batches that were collected at different times during the growing season, thus creating a diverse data set. Twelve MR images, as opposed to three, were obtained for each fruit. The image types included the FSE, turbo FLASH, and GRE sequences from the first study as well as diffusion-weighted and spin-echo images. Settings within some of the pulse programs (inversion time for the turbo FLASH sequence, for example) were varied to create a total of twelve images. The PLS regression, thresholding, and morphological feature screening were the same as in the first study. In this case, the algorithm correctly identified 97% of the seedless fruit and 84% of the seeded fruit—a marked improvement over the results from the first study. Analysis of the relative contribution of the twelve image types to the predictive model showed that the FSE, turbo FLASH, and diffusion-weighted images had the most influence on the model. This indicates that these three pulse sequences should be pursued further for industrial implementation of the MR sorting technology. Thus, MRI, along with various postprocessing techniques, is effective for identifying seeds in intact citrus fruit. A summary of the success rates for identifying seeded and seedless fruit from various studies is shown in Table 6.1.

6.4.3 Olive Accession Evaluation Using Relaxation Time Analysis

Quality parameters of fruit are also often assessed using only relaxation time data. Relaxation times can be related to moisture content, bruise damage, texture, maturity, and other quality parameters (McCarthy, 1994). An example of this is the application of relaxation time analysis to characterize different olive accessions.

In 2004, the United States grew olives on 13,000 Ha, producing 94,000 metric tons of fruit, of which only 8% was crushed (USDA, 2005), to produce 383,000 gallons of oil (Olive Oil Source, 2005). Globally, oil production dominates, with olives grown on 7,455,049 ha producing 14 million metric tons in 2005 and providing 460 million gallons of oil (Olive Oil Source, 2005). Due to the health benefits of olive oil, both U.S. and global demand are on the rise (CRB, 2005). Characterization of different olive accessions would greatly facilitate stakeholder utilization of collections. Since oil content varies widely in olive accessions (Tous et al., 2002) and oil production is the most important use of olives worldwide, characterization of oil content

TABLE 6.1

Seedless and Seeded Fruit Identification Rates for MRI Screening of Mandarin Oranges

Study	Percentage of Seedless Fruit Correctly Identified	Percentage of Seeded Fruit Correctly Identified	Total Sample Size	MR Sequence(s) Used	Special Features
Kim et al. (2008)	90	92	56	FSE[a]	Annular region of interest near center of fruit
Hernandez-Sanchez et al. (2006)	89	87	33	FLASH[b]	
Barreiro et al. (2008)	100	96	78	FLASH[b]	Classification function based on 3 morphological features
Barreiro et al. (2008)	100	100	15	COMSPIRA[c]	Classification function based on 3 morphological features
Lau (2008)	91	92	100	Unspecified	Genetic algorithm used for region of interest selection
This work—2007 study	97	59	140	Multivariate—3 image types	
This work—2009 study	97	84	90	Multivariate—12 image types	

[a] Fast spin echo.
[b] Fast low-angle shot.
[c] Combined spiral and radial acquisition.

is a critical factor for selection of an accession. A total of 201 olive accession sets harvested in 2007–2008 from the National Clonal Germplasm Repository (NCGR) in Davis, California, were examined.

Olive samples were collected at harvest date for each olive accession. Harvest dates were based on 2005–2006 date at which 90% of fruit on each tree achieved mature color. Greater than thirty olives samples for each harvest from each tree (olive accession) were delivered, and thirty samples were hand selected and divided into three subsets for further NMR testing. Then each subset including ten olive samples was weighted before NMR measurements. After measuring the weight of each subset, the bulk NMR measurement on the subset was performed using a portable MR system (QM, Quantum Magnetics, Inc., San Diego CA., United States). A subset out of the three subsets was selected, and the weight of each olive sample in the subset was measured and numbered for another individual NMR measurement using an MRI system (ASPECT, ASPECT Magnetic Technologies Ltd., Israel). Five samples were selected for T_1 and T_2 measurements, and only T_2 measurements were performed on the other five samples out of ten olive samples. Finally, five olive samples used for T_1 and T_2 measurements were used to measure moisture in a vacuum oven with twenty-four 0.25 In•Hg at 80°C for 24 hours.

The Quantum Magnetics (QM) system was used to calculate the rate of decay of three subsets of each accession. The frequency of the QM system varied according to the temperature of the room; therefore, it was necessary to calibrate with each day of use. The frequency was calibrated using a phantom sample. The calibrated frequencies ranged from 1,705.5 to 1,711.0 kHz. CPMG pulse sequence was used to measure the spin-spin relaxation time constant. The number of echoes was 2,048 with echo time 2.69 ms. The number of data points per echo was fifty-eight. The number of scans was 4, and the number of excitations was 8,192.

The NMR measurements of each intact olive sample of a selected subset were performed on a horizontal-bore 1 T magnet. A solenoid type NMR coil with inner diameter of 60 mm was used to acquire NMR data. A sample holder was used to place each olive in the middle of the NMR coil. Saturation recovery pulse sequence was used to measure spin-lattice time constant (T_1). The saturation recovery times were set to 5, 100, 200, 400, 500, 800, 1,000, 1,200, 2,000, 2,800, 3,500, 5,000, and 7,000 ms. A CPMG pulse sequence was used to measure spin-spin relaxation time constant (T_2). The number of echoes was 2,048 with echo time 3.2 ms. The number of data points per echo was 128. The number of scans was 1, and the number of excitations was 256.

The acquired NMR signals were processed with commercial programming software (MATLAB R2006b, Mathworks, Natick, MA., United States) to calculate NMR relaxation time constants and related factors. The calculated factors from the CPMG signals acquired by the QM system were mono- and biexponential curve fitting results related to relaxation time constants (M20q, T2q, M210q, M220q, T21q, and T22q). The calculated factors out of CPMG signals acquired by the ASPECT system are mono- and biexponential curve fitting results related to relaxation time constants (M20, T2, M210, M220, T21, and T22). The calculated factors out of saturation recovery signals are relaxation time constants (T_1) and signal magnitudes (M10). Multivariate data analysis software (PLS Toolbox 4.0, Eigenvector, United States) was used to analyze the relationships between the acquired NMR signals and the calculated NMR relaxation time constant–related factors.

The relationship between the measured physical properties and the estimated NMR-related values was examined. Table 6.2 shows a summary of the linear regression relationship between measured physical properties and estimated NMR properties. The linear regression relationship between the initial weight (Wi) of olives and the magnitude of the spin-lattice relaxation time constant (M10) acquired by the ASPECT system shows a reasonably high R^2 value (0.792), and between the initial weight of olives and the sum of magnitude of spin-spin relaxation time constants (M20q_s) obtained from biexponential curve fitting acquired by QM system a high R^2 value (0.831). They show positive linear relationships.

The linear regression relationship between dry-based moisture content (MCd) of olives and spin-lattice relaxation time constant (T1) acquired by the ASPECT system shows a reasonably high R^2 value (0.669), and between dry-based moisture content (MCd) of olives and the sum of spin-spin relaxation time constants

TABLE 6.2

Linear Relationship between Measured Physical Properties and Estimated NMR Properties

X	Y	Regression Equation	R^2	SSE	RMSE
M10	Wi	Y = 0.000763X + 1.566	0.792	240.9	1.1
M20q_s	Wi	Y = 0.0000495X + 1.054	0.831	194.7	0.9891
T1	MCd	Y = 0.1581X + 31.41	0.669	37910	22.11
T2q_s	MCd	Y = 0.1992X + 45.64	0.531	137900	26.32
T1	DM	Y = −0.0287X + 62.59	0.623	3927	4.442
T2q_s	DM	Y = −0.0371X + 60.38	0.520	4996	5.011
1/T1	DM	Y = 9285X + 28.23	0.614	4023	4.496
1/T2q_s	DM	Y = 688.7X + 29.01	0.404	6163	5.608

Note: Wi = initial weight of olives, MCd = dry-based moisture content, DM = percent dry matter, M10 = magnitude obtained from monoexponential spin-lattice relaxation time constant curve fitting using data acquired by ASPECT system, M20q_s = magnitude sum obtained from biexponential spin-spin relaxation time constant curve fitting using data acquired by QM system, T1 = spin-lattice relaxation time constant acquired by ASPECT system, T2q_s = sum of spin-spin relaxation time constants obtained from biexponential curve fitting acquired by QM system, 1/T1 = inverse of the spin-lattice relaxation time constant acquired by ASPECT system, 1/T2q_s = inverse of the sum of spin-spin relaxation time constants obtained from biexponential curve fitting acquired by QM system.

(T2q_s = T2w + T2o) obtained from biexponential curve fitting acquired by the QM system a reasonable R^2 value (0.531). They show positive linear relationships.

The linear regression relationship between percent dry matter (%DM) of olives and spin-lattice relaxation time constant (T1) acquired by the ASPECT system shows a reasonably high R^2 value (0.623), and between percent dry matter (%DM) of olives and the sum of spin-spin relaxation time constants (T2q_s = T2w + T2o) obtained from biexponential curve fitting acquired by the QM system a reasonable R^2 value (0.520). They show negative linear relationships.

The linear regression relationship between percent dry matter (%DM) of olives and the inverse of the spin-lattice relaxation time constant (1/T1) acquired by the ASPECT system shows a reasonably high R^2 value (0.614), and between percent dry matter (%DM) of olives and the inverse of sum of spin-spin relaxation time constants (1/T2q_s = 1/T2w + 1/T2o) obtained from biexponential curve fitting acquired by the QM system a reasonable R^2 value (0.404). They show positive linear relationships.

Nine olive accessions were selected based on average percent dry matter (%DM) for further analysis. The average percent dry matter ranges from a minimum value of 25.8% to a maximum value of 61.3%, with about a 5% gap. Figure 6.11 shows the relationship between average percent dry matter (%DM) and average T_2 measured by the QM system (T2q) of nine selected olive accessions. The linear regression relationship between percent dry matter (%DM) of olives and the spin-spin relaxation time constant acquired by the QM system (T2q) shows a fairly high R^2 value (0.872).

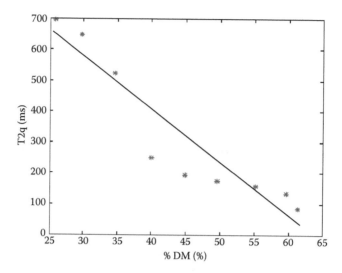

FIGURE 6.11 Relationship between average percent dry matter (%DM) and average T_2 measured by the QM system (T2q) of nine selected olive accessions.

TABLE 6.3
Measured Physical and NMR-Related Mean Values of Selected Olive Accessions (units in ms)

| | | | | ASPECT | | | | QM | | |
| | | | | Mono | | Bi | | Mono | Bi | |
Access	Harvest Date	W (g)	%DM (%)	T_1	T_2	T_2w	T_2o	T_2	T_2w	T_2o
A3:6	11/15/07	9.36	25.8	1421	27.7	42.2	7.8	696.4	736.5	75.5
B13:2	11/15/07	9.71	29.7	1232	28.9	43.2	8.3	647.2	697.1	66.2
B10:9	11/01/07	6.41	34.6	852.2	39.5	54.6	6.1	523.6	632.8	70.3
A10:4	11/15/07	5.11	40.0	628	23.8	33.1	5.9	249.9	319.9	54.2
B7:4	01/17/08	6.55	44.9	797.8	37.7	51.8	6.4	192.9	218.3	30.1
A3:4	12/27/07	5.17	49.6	446.4	31.9	40.9	7.8	175.0	210.8	37.8
B9:3	01/17/08	4.62	55.2	443.0	26.7	36.1	7.7	158.9	199.7	41.1
B5:1	12/07/07	3.05	59.6	277.2	17.2	23.1	6.7	134.2	187.5	35.2
A11:11	12/27/07	4.04	61.3	319.9	21.8	29.7	4.6	83.8	114.9	39.9

Note: W = average weight of an olive, %DM = average % dry matter of olive samples, T_1 = spin-lattice relaxation time constant, T_2 = spin-spin relaxation time constant, $T_2w = T_2$ of water component, $T_2o = T_2$ of oil component.

Table 6.3 shows measured percent dry matter and average NMR-related values of the selected nine olive accessions.

A total of 201 olive accessions harvested during the 2007–2008 season were examined. The relationships between measured physical properties and NMR-related factors were studied. A good correlation between percent dry matter and NMR relaxation time parameters was obtained, permitting the evaluation of each accession and comparison to others. The key factor is that NMR analysis of the olives is rapid and nondestructive, permitting the analysis of hundreds of individual olives from each accession. This type of analysis would have been cost- and time-prohibitive if destructive oil and moisture content measurements had been done on the entire sample set.

6.5 PERSPECTIVES ON MAGNETIC RESONANCE SENSORS

A recent review of the applications and status of using NMR/MRI to measure the quality of horticultural products provides an excellent summary of the potential and current state of the art to measure internal quality factors (Hills and Clark, 2003). Over thirty different fruits and vegetables have been examined for internal quality factors. NMR and MRI have been shown be effective in quantifying maturity/ripeness, bruises, voids, tissue breakdown, heat/chill injury, infection, and insect damage. A summary to show the application of NMR/MRI to several products is shown in Table 6.4. The checkmarks indicate that NMR/MRI has been used to quantify attributes listed in the first column for the products in the subsequent columns.

The results to date for applying NMR and MRI to detect the internal quality of fruits and vegetables are extensive and impressive. These applications include coupling a conveyor with research-grade MR systems (Zion et al., 1994, 1997; Chen et al., 1996; Kim et al., 1999; Hernandez-Sanchez et al., 2004; Hernandez et al., 2005; Kim and McCarthy, 2006). What then is limiting the utilization of MR to quality sorting of agricultural products? Hills and Clark (2003) have identified several areas that need to be addressed. These areas requiring research include:

- Can a system be built that is cost-effective? This includes the issues of a magnet that is operable in an industrial environment.

TABLE 6.4
Quality Factors That Can Be Detected Using NMR/MRI are Shown as Checked in the Table

	Apples	Citrus	Fresh Tomatoes	Potatoes	Fresh-Cut Fruits	Stone Fruits
Hollow	√	√		√		√
Pests and insects	√	√	√	√	√	√
Bruising and cracks	√	√		√		√
Mold and rot	√	√	√	√	√	
Mold under skin		√		√	√	√
Browning	√				√	
Seed size/number		√			√	
Maturity	√		√		√	
Sugar level Acidity ratio	√	√	√			√
Mealiness	√		√		√	√

- Research is needed to determine if low-field magnet-based sensors can adequately assess quality at a rate of ten to twelve fruits per second for application in a packinghouse.
- Given the large number of quality factors and the large number of possible pulse sequences for generating images, the time to optimize detection of each factor is extensive.

Over the last ten years these limitations have been addressed and practical solutions implemented for incorporating MR-based sensors in fruit packing operations and other industrial production systems.

6.5.1 COST-EFFECTIVENESS

MR systems have historically been extremely expensive, resulting from both equipment costs and location requirements. However, recent advances in permanent magnet design and construction have dramatically reduced the cost of MR-based systems. These newer magnets for NMR/MRI perform well in industrial environments. Several current applications in the polymer processing industry and oil well characterization use permanent-magnet–based NMR sensors (McCarthy and Bobroff, 2000). The keys are identifying needs and designing an appropriate magnet. An appropriate magnet system has not been available for fruit and or food systems until recently. As Hills and Clark (2003) noted in their review, to make high field magnets of sufficient homogeneity there needed to essentially be a breakthrough in magnet technology. This breakthrough has been accomplished by ASPECT Magnet Technologies Ltd. The magnet design offers a large volume of very homogeneous fields, high magnetic field strengths (1.0 to 1.5 T), and no external magnetic fields outside the magnet. The magnet itself is designed for operation in an industrial environment. Shown in Figure 6.5 is the magnet adapted to a fruit-sorting conveyor and the images are shown in Figure 6.12. The imaging volume in the magnet for sorting of mandarin fruit is $90 \times 110 \times 110$ mm. The magnet design has been scaled from 30 mm to up to 250 mm imaging volume (and the range of the volume can be extended).

In addition to the dramatic improvement in magnet technology, the advances in the electronics industry have resulted in cost reduction and performance improvements for the spectrometer that is coupled to the magnet. There are now a number of vendors that produce low-cost spectrometers, and some that sell entire low-cost systems. Some of the vendors of low-cost and portable spectrometers include TecMag (www.tecmag.com), Spincore (www.spincore.com), Magritek (www.magritek.com), and SpinTrak (www.process-nmr.com). Combined with novel magnets, these permit the construction of sensors that are cost-competitive in the fruit packing and food processing industries.

6.5.2 PERFORMANCE

MRI systems need to be able to operate at a rate compatible with sorting eight to twelve fruits per second, which is the current standard commercial packinghouse line speed. Each fruit is normally conveyed in an individual cup and is referred to as singulated.

(a) Turbo FLASH   (b) Fast Spin-Echo

FIGURE 6.12 MR images from a clementine citrus fruit with seeds. Images are 64 mm field of view with a resolution of 128 × 128. (Courtesy of Uri Rapoport, ASPECT Magnet Technologies Ltd., www.aspect-mr.com.)

The successful integration of MR-based quality sensors will require high throughput while maintaining the singulated identity of the fruit. Achieving this rate of sorting can be accomplished by having the fruit move through the sensor at line speed (1 to 2 m/s), having the fruit slow down through the sensor (<1 m/s), or having the fruit be conveyed into the sensor, stop, and then be conveyed out of the sensor. All of these approaches fundamentally require the sensor to make a rapid measurement. Permanent-magnet-based MRI systems have historically been slower than superconducting-based systems as a result of limitations imposed by the formation of eddy currents in the magnet pole face. These limitations have been overcome with purposely designed pole faces, and now permanent-magnet–based MRI systems have very rapid data acquisition times. Shown in Figure 6.11 are two images from a 1 T permanent magnet imaging system (Figure 6.5). Both of these images have been acquired in less than 1 s. This demonstrates the feasibility to achieve MRI data at packinghouse speeds, although this will most likely require several fruits to be imaged simultaneously.

6.5.3 PARAMETER OPTIMIZATION

The time required to optimize MR data acquisition to achieve selection of fruit based on specific quality standards is a significant hurdle to the application of this sensor technology. However, this hurdle is the same one encountered for all sensor technologies. Each type of fresh fruit has a natural range of quality parameters, and these will vary to some degree for each variety, season, growing location, and perhaps cultivation practice. Application of chemometrics as discussed in the example of predicting peelability of tomato fruit is an efficient method to screen the wide range of potential contrast mechanisms available in NRM/MRI. A sample of the type of X-block data needed in the tomato example is shown in Figure 6.13. Four types of images were chosen to encode differences in the tomato pericarp that were expected to impact the ease of peeling. A balance needs to be chosen on this type of experiment since a large

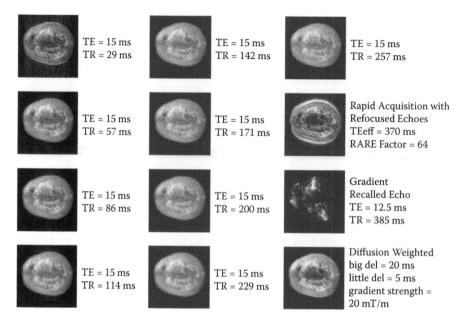

FIGURE 6.13   A range of image types, spin echo, gradient recalled echo, rapid acquisition with refocused echoes, and a diffusion-weighted spin echo used to build the X-block for the prediction of tomato peelability.

number of images and a large number of individual tomatoes are needed to achieve sufficient data to make sound conclusions. Additionally, there is a need to have quantitative measurements to use in developing the model from a subset of the X-block, which is subsequently tested on the remaining X- and Y-blocks.

The definition of quantitative measurements for quality factors is often difficult to achieve. Consider the current approach to detect freeze damage in citrus fruit. A sample is taken from a lot of citrus fruit. The sample fruits are then destructively investigated for determination of freeze damage by using a segment cut. The segment cut proceeds by removing both the stem and blossom end of the fruit such that a center section of fruit remains that is 1 to 1.5 inches in width. This center segment of the fruit is carefully opened to show the segments (opened from one cut through only the peel). The segments are then inspected, and an orange must show damage to the entire length of both sides of two segments to be classified as freeze damaged. Damage is a water-soaked appearance or evidence of previous water soaking or the presence of crystals. The tolerance in a lot of fruit for freeze damage is 15%. This definition of freeze damage is useful; however, it is difficult to translate to a quantitative scale. Researchers have proposed using the definition of damage to be the number of damaged cells compared to the total number of cells. The comparison between number of cells damaged and the segment cut analysis has yet to be performed. Many quality factors lack easily measured or easily translated quantitative scales. This step in sensor development/implementation is often as time-consuming as the optimization of the procedure, and in many cases the optimization and development of a quantitative scale are done iteratively.

REFERENCES

Barreiro, P., C. Zheng, D.-W. Sun, N. Hernandez-Sanchez, J. M. Perez-Sanchez, and J. Ruiz-Cabello. 2008. Non-destructive seed detection in mandarins: Comparison of automatic threshold methods in FLASH and COMSPIRA MRIs. *Postharvest Biology and Technology* 47:189–98.

Blumich, B. 2000. *NMR imaging of materials.* Oxford: Clarendon Press.

Callaghan, P. T. 1991. *Principles of nuclear magnetic resonance microscopy.* Oxford: Clarendon Press.

Callaghan, P. T., and C. D. Eccles. 1987. Sensitivity and resolution in NMR imaging. *Journal of Magnetic Resonance* 71:426–45.

Chen, P., M. J. McCarthy, and R. Kauten. 1989. NMR for internal quality evaluation of fruits and vegetables. *Transactions of the ASAE* 32(5):1747–53.

Chen, P., M. J. McCarthy, S.-M. Kim, and B. Zion. 1996. Development of a high-speed NMR technique for sensing maturity of avocados. *Transactions of the ASAE* 39(6):2205–9.

Gambhir, P. N., Y. J. Choi, D. C. Slaughter, J. F. Thompson, and M. J. McCarthy. 2005. Proton spin-spin relaxation time of peel and flesh of navel orange varieties exposed to freezing temperature. *Journal of the Science of Food and Agriculture* 85(14):2482–86.

Han, Y. J., S. V. Bowers III, and R. B. Dodd. 1992. Nondestructive detection of split-pit peaches. *Transactions of the ASAE* 35(6):2063–67.

Hartz, T., G. Miyao, J. Mickler, M. Lestrange, S. Stoddard, J. Nunez, and B. Aegerter. 2008. *Processing tomato production in California.* University of California Agricultural and Natural Resources Publication 7228. Richmond, CA: University of California.

Hernandez, N., P. Barreiro, M. Ruiz-Altisent, J. Ruiz-Cabello, M. Encarnacion, and M. E. Fernandez-Valle. 2005. Detection of seeds in citrus using MRI under motion conditions and improvement with motion correction. *Concepts in Magnetic Resonance Part B* 26B:81–92.

Hernandez-Sanchez, N., P. Barreiro, M. Ruiz-Altisent, J. Ruiz-Cabello, and M. E. Fernandez-Valle. 2004. Detection of freeze injury in oranges by magnetic resonance imaging of moving apples. *Applied Magnetic Resonance* 26:431–45.

Hernandez-Sanchez, N., P. Barreiro, and J. Ruiz-Cabello. 2006. On-line identification of seeds in mandarins with magnetic resonance imaging. *Biosystems Engineering* 95(4):529–36.

Hills, B. P., and C. J. Clark. 2003. Quality assessment of horticultural products by NMR. *Annual Reports on NMR Spectroscopy* 50:75–120.

Irudayaraj, J., and C. Reh, eds. 2008. *Nondestructive testing of food quality.* Ames, IA: Blackwell Publishing.

Kerr, W. L., C. J. Clark, M. J. McCarthy, and J. S. de Ropp. 1997. Freezing effects in fruit tissue of kiwifruit observed by magnetic resonance imaging. *Scientia Horticulturae* 69(3–4):169–79.

Kim, S.-M., P. Chen, M. J. McCarthy, and B. Zion. 1999. Fruit internal quality evaluation using on-line nuclear magnetic resonance sensors. *Journal of Agricultural Engineering Research* 74(3):293–301.

Kim, S.-M., and M. J. McCarthy. 2006. Analysis of characteristics of in-line magnetic resonance sensor. *Key Engineering Materials* 321–23:1221–24.

Kim, S.-M., R. Milczarek, and M. McCarthy. 2008. Fast detection of seeds and freeze damage of mandarins using magnetic resonance imaging. *Modern Physics Letters B* 22(11):941–46.

Lau, B. T. 2008. Intelligent citrus seed identification. In *3rd IEEE Conference on Industrial Electronics and Applications*, Singapore, 79–83.

Law, S. E. 1973. Scatter of near-infrared radiation by cherries as a means of pit detection. *Journal of Food Science* 38:102–7.

McCarthy, M. J. 1994. *Magnetic resonance imaging in foods*. New York: Chapman & Hall.

McCarthy, M. J., and S. Bobroff. 2000. Nuclear magnetic resonance and magnetic resonance imaging for process analysis. In *Encyclopedia of analytical chemistry*, ed. R. A. Meyers, 8264–81. Chichester, England: John Wiley & Sons Ltd.

McCarthy, M. J., P. Chen, S. Kim, and B. Zion. 2000. Applications of magnetic resonance imaging and spectroscopy in sorting of avocados and other fruits. In *Instrumentacao Agropecuaria 2000*, ed. P. E. Cruvinel, L. A. Colnago, and A. T. Neto, 3–10. São Carlos, Brazil: Embrapa-Cnpdia.

Milczarek, R. R. 2009. Multivariate image analysis: An optimization tool for characterizing damage-related attributes in magnetic resonance images of processing tomatoes. PhD thesis, University of Californa, Davis.

Milczarek, R. R., and M. J. McCarthy. 2009. Low-field MR sensors for fruit inspection. In *Magnetic resonance microscopy*, ed. S. L. Codd and J. D. Seymour, 289–99. Weinheim, Germany: Wiley-VCH Verlag GmbH & Co.

Olive Oil Source. 2005. http://www.oliveoilsource.com/statistics.htm (accessed September 21, 2009).

Otto, M. 1999. *Chemometrics: Statistics and computer application in analytical chemistry*. New York: Wiley-VCH.

Qin, J., and R. Lu. 2005. Detection of pits in tart cherries by hyperspectral transmission imaging. *Transactions of the ASAE* 48(5):1963–70.

Timm, E. J., P. V. Gilliland, G. K. Brown, and H. A. Affeldt Jr. 1991. Potential methods for detecting pits in tart cherries. *Transactions of the ASAE* 7(1):103–9.

Tous, J., A. Romero, J. Plana, and J. F. Hermoso. 2002. Behavior of ten Mediterranean olive cultivars in the northeast of Spain. *Acta-Horticulturae* 586:113–16.

U.S. Department of Agriculture, Economic Research Service. 2005. http://usda.mannlib. cornell.edu/usea/usd (accessed September 21, 2009).

U.S. Department of Agriculture, Economic Research Service. 2008. *Vegetables and melons yearbook—89011*. http://usda.mannlib.cornell.edu/usda/ers/89011/Table085.xls (accessed April 14, 2009).

Wise, B. M., N. B. Gallagher, J. M. Shaver, and W. Winding. 2004. Introduction to multivariate image analysis. Center for Process Analytical Chemistry presentation and notes. Eigenvector Research, Inc.

Zion, B., S. M. Kim, M. J. McCarthy, and P. Chen. 1997. Detection of pits in olives under motion by nuclear magnetic resonance. *Journal of the Science of Food and Agriculture* 75(4):496–502.

Zion, B., M. J. McCarthy, and P. Chen. 1994. Real-time detection of pits in processed cherries by magnetic resonance projections. *Lebensmittel-Wissenschaft und Technologie* 27(5):457–62.

7 Ultrasound Systems for Food Quality Evaluation

Ki-Bok Kim and Byoung-Kwan Cho

CONTENTS

7.1 INTRODUCTION

Ultrasonic measurement is a promising, nondestructive sensing method of characterizing materials by transmitting mechanical waves at frequencies above 20 kHz through a material, and investigating the characteristics of the transmitted or reflected ultrasound waves. Offering advantages including nondestructiveness, rapidity, relatively low cost, and potential for automatic application, ultrasonic techniques provide a very useful and versatile method for investigating the physical properties of materials. Ultrasonic waves may be reflected and transmitted when they pass from one medium to another. The amount of energy reflected and transmitted through materials depends on their relative acoustic impedances, as defined by the density and velocity of the penetrating wave within and between material interfaces. In addition to ultrasound energy attenuation, the time of flight and velocity could also serve as good indicators of material properties or variation of material characteristics, since ultrasound velocity is dependent on the density and elastic properties of the medium. Hence, ultrasound properties are closely related to the physical properties, such as structure, texture, and physical state, of the components of the media. In addition, due to foreign objects such as bone, glass, or metal fragments in food products, strong reflection and refraction at the interfaces of the host tissue and foreign object interface are anticipated. Such objects have different physical properties than the medium, and consequently, changes in ultrasound time of flight and velocity are expected.

At the early stage of application of ultrasound, after development of a depth finder by Langevine in 1921 in France, ultrasound technology was studied for military purposes, such as sonar to detect battleships and submarines, and to measure the position of moving underwater objects. Ultrasound technology has found industrial applications, such as nondestructive testing, measurement of the thickness of thin films, cleaners, metal welding, etching, and measurement of flow rate and water level, as well as medical applications, such as diagnosis and treatment of disease. Recently, there has been extensive research on the application of ultrasound technology to biological engineering, including cleaning foodstuffs and containers, extraction of proteins and fats, maturation of food, food process monitoring, and nondestructive quality evaluation of agricultural, marine, and livestock food products.

The application of ultrasound in the field of food process can be divided into two branches. One is the application of high energy level ultrasound for surface cleaning, dehydration, promotion of drying and filtration, activity control of microorganisms and enzymes, cell division, removal of gas from liquid materials, promotion of heat transfer, acceleration of extraction, and so on. The other branch is the application of low energy level ultrasound for controlling the food process and evaluating the quality of food. Since low energy level ultrasound does not cause physical or chemical alteration to the properties of the material, nondestructive internal measurement of materials is possible. In this chapter, basic principles of ultrasound and its applications to agricultural and food products are explored.

7.2   PRINCIPLES

7.2.1   PROPAGATION OF ULTRASOUND

If stress is applied to an elastic material, it produces a progression of waves, which are referred to as elastic waves. Two kinds of elastic waves can arise in an isotropic material: longitudinal and shear waves. The direction of particle displacement in a longitudinal wave is parallel to the direction of wave propagation, while the direction of particles in a shear wave is perpendicular to the direction of propagation. There are two types of shear waves perpendicular to one another, horizontal shear waves and vertical shear waves, which propagate horizontally and vertically with the shear planes. In a fluid, the stress state, which is a pure pressure, only has a longitudinal component; however, in a solid, the stress state includes a shear component as well. The wave propagation in an unbounded isotropic elastic solid medium is expressed as follows:

$$\rho\frac{\partial^2 u}{\partial t^2} = (\lambda + G)\nabla\nabla \cdot u - G\nabla \times \nabla \times u \tag{7.1}$$

where ρ is the density, u is the displacement of a particle in the medium, λ is the Lame constants, and G is the shear modulus.

If there is only longitudinal deformation without transverse deformation ($\nabla \times u = 0$), Equation 7.1 can be written as follows:

$$\nabla^2 u = \frac{1}{c_p^2}\frac{\partial^2 u}{\partial t^2} \tag{7.2}$$

where c_p is the velocity of a longitudinal wave and is defined as follows:

$$c_p^2 = \frac{\lambda + 2G}{\rho} = \frac{E}{\rho}\frac{(1-\mu)}{(1+\mu)(1-2\mu)} \tag{7.3}$$

where E is the Young's modulus of the medium and μ is the Poisson's ratio.

If only transverse deformation exists ($\nabla \times u = 0$) in the medium, Equation 7.1 can be written as follows:

$$\nabla^2 u = \frac{1}{c_s^2}\frac{\partial^2 u}{\partial t^2} \tag{7.4}$$

where c_s is the velocity of a shear wave and is defined as follows:

$$c_s^2 = \frac{G}{\rho} \tag{7.5}$$

The relationship between E and G is as follows:

$$E = 2(1+\mu)G \tag{7.6}$$

Using the above equations, the mechanical properties of materials, such as Young's modulus (E), shear modulus (G), and Poisson's ratio (μ), can be calculated with the measured parameters of the density (ρ), and the velocity of the longitudinal and shear waves (c_p, c_s) of the materials.

7.2.2 ACOUSTIC IMPEDANCE

When an incident ultrasound wave propagates on an interface between two media of different properties, reflection, refraction, and scattering occur. These phenomena could be good indicators of the media properties. Ultrasound reflection and refraction can be well explained using acoustic impedance. Acoustic impedance (Z) is defined as the ratio of the sound pressure (P) to the particle velocity (v) in a sound wave.

$$Z = P/v \tag{7.7}$$

It is equal to the product of the density (ρ_0) of the medium and the velocity of sound (c) in the medium. The unit is kg/m²-s.

$$Z = \rho_0 c \tag{7.8}$$

Since the acoustic impedance of Equation 7.7 includes the characteristics of the wave and the medium simultaneously, it is called the specific acoustic impedance. The acoustic impedance of Equation 7.8 is defined as the characteristic acoustic impedance because it contains the information of the physical properties of the medium.

7.2.3 REFLECTION AND TRANSMISSION

A right-propagating wave normally incident on an interface between two materials of different properties is shown in Figure 7.1. The wave creates a reflection from and a transmission across the interface. If the acoustic impedances of the materials are Z_1 and Z_2, respectively, the coefficients of wave transmission (T) and reflection (R) are respectively expressed as follows:

$$T = \frac{4Z_1 Z_2}{\left(Z_1 + Z_2\right)^2} \tag{7.9}$$

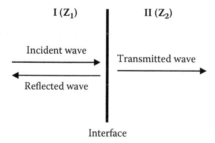

FIGURE 7.1 Normal reflection and transmission at an interface.

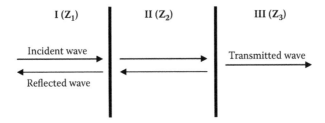

FIGURE 7.2 Impedance matching layer.

$$R = \frac{(Z_1 - Z_2)^2}{(Z_1 + Z_2)^2} \qquad (7.10)$$

In the case where a layer of intermediate impedance is imposed between the two materials, as shown in Figure 7.2, the coefficient of transmission is defined as follows:

$$T = \frac{4}{2 + \left(\dfrac{Z_3}{Z_1} + \dfrac{Z_1}{Z_3}\right)\cos^2 k_2 L + \left(\dfrac{Z_2^2}{Z_1 Z_3} + \dfrac{Z_1 Z_3}{Z_2^2}\right)\sin^2 k_2 L} \qquad (7.11)$$

where L is the thickness and k_2 is the wave number of medium II.

For optimum transmission, T should be 1, and hence the impedance of the intermediary layer is found to be as follows:

$$Z_2 = \sqrt{Z_1 Z_3} \qquad (7.12)$$

For maximum transmission, the thickness of the intermediary layer should be a quarter wavelength.

7.3 DEVICE AND APPARATUS

7.3.1 ULTRASONIC TRANSDUCER

Ultrasonic transducers are the devices that generate and receive ultrasound and convert electric energy into ultrasound energy, and vice versa. Although numerous methods can be used to generate ultrasound waves, piezoelectric transducers are the most common types. The piezoelectric effect describes the relationship between the mechanical stress and the electrical voltage in a piezoelectric material, and it is reversible: an applied mechanical stress generates a voltage and an applied voltage changes the shape of the piezoelectric material. The piezoelectric effect occurs when an electric charge develops on the faces of a piezoelectric element that is mechanically deformed. Conversely, an electric signal or voltage applied across the faces will cause deformation. Thus, deformation of a piezoelectric material at high frequencies generates ultrasound vibrations that propagate as waves in the material through a suitable couplant.

Figure 7.3 presents a simple description of the piezoelectric effect of a piezoelectric material with electrodes on the top and bottom. When voltage (V) is applied to both electrodes, because a piezoelectric material is dielectric, it has a clamp capacitance that is given as

$$C_0 = \frac{\varepsilon^S A}{d}$$

(7.13)

where ε^S is a clamped dielectric constant under a condition of zero strain, A is the cross-sectional area, and d is the thickness. Because of piezoelectricity, from Hooke's law, the stress T can be described as

$$T = C^D S + hD$$

(7.14)

where h is a piezoelectric constant. C^D is an elastic stiffness constant and is obtained under a constant dielectric displacement, D, and an electric field, E.

Conversely, if the stress is applied to a piezoelectric material, the electric field E can be described as

$$E = hS + \beta^S D$$

(7.15)

where β^S is dielectric impermeability under constant or zero strain.

With Equations 7.14 and 7.15, the generation and reception of ultrasound can be described.

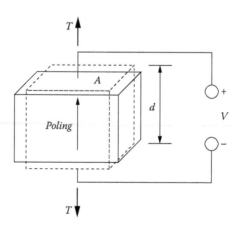

FIGURE 7.3 Simple description of piezoelectric effect of piezoelectric material with electrodes on the top and bottom.

Several parameters, such as central frequency, bandwidth, diameter, resolution, near-field length, acoustical focal length, front matching layer, backing material, and so on, should be considered when designing and fabricating an ultrasound transducer. Among various design parameters, the characteristics of ultrasound are mainly dependent on the kind of piezoelectric material. Conventional materials include lead zirconate titanate (PZT), lead metaniobate (PMN), and barium titanate (BaTi). Piezoelectric elements from most synthetic materials are available in either extensional or transverse modes for direct excitation of longitudinal or shear waves. Lead zirconate titanate is a good electromechanical converter, but it has limited resolution since it is not superior in the reciprocal action, converting mechanical motion to an electrical signal. It maintains its piezoelectric properties up to 300°C.

Two types of PZT material (PZT4 and PZT5) are used in an ultrasonic transducer. Both types are effective transmitters. However, PZT4 is superior in terms of generating high-level ultrasound energy, whereas PZT5 is better for pulse type applications. Barium titanate elements are chemically stable and maintain their usefulness at temperatures up to 100°C. This material is very effective for the generation of ultrasound energy, but poor for its reception. It is often used in ultrasonic cleaning equipment. A disadvantage of piezoelectric material is its property variation during aging. Lead metaniobate, meanwhile, may be used at temperatures of 500°C and has high thermal stability. It is also sensitive for receiving ultrasound energy.

Recently, a new class of single-crystal piezoelectric materials, including lead magnesium doped with lead titanate (PMN-PT), has been synthesized, and these materials show enhanced piezoelectric properties compared to PZT and PMN. Polymer and polymer-ceramic composite piezoelectric transducers are also used. Polymers such as polyvinylidene fluoride (PVDF) and copolymers like vinylidene fluoride–trifluoroethylene (PVDF-TrFE) have some advantages over piezoelectric ceramic transducers, although their energy conversion efficiency is lower (Kawai, 1969; Ohigashi et al., 1985). Advantages of these materials are their conformability and low acoustic impedance. Polymer transducers have acoustic impedance near that of the human body and, consequently, are used in the medical profession to effectively couple ultrasound into humans. The relatively low acoustic impedance of these transducers should also make them useful for coupling ultrasound energy into fruits (Shmulevich et al., 1996).

There are several types of piezoelectric ultrasonic transducers: immersion, contact, and noncontact transducers. Immersion transducers must be sealed to protect the interior from water. Contact transducers can be grouped into normal-beam and angle-beam types. The element of the normal-beam transducer can be selected to generate either longitudinal or shear waves. Angle-beam transducers can be configured with piezoelectric elements mounted at several different angles so as to facilitate longitudinal-wave, shear-wave, and surface-wave testing. Transducers can be selected to generate an ultrasound beam at a specific angle for a specific type of material. Figure 7.4 shows the layout of a focusing contact type ultrasound transducer for fruit (Kim et al., 2004).

The major components of an ultrasound transducer are a wear plate (front matching layer) for focusing the ultrasound energy and direct contact with the object's surface, a piezoelectric material, and a backing material. The operating frequency,

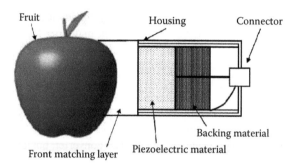

FIGURE 7.4   Layout of contact type ultrasonic transducer for fruit.

f_r, of the ultrasonic transducer is determined by the velocity of the wave propagation (*c*) and the thickness (*t*) of the piezoelectric materials at the first thickness vibration mode, as delineated in the following:

$$f_r = \frac{c}{2t} \tag{7.16}$$

The basic purpose of the transducer wear plate is not only to protect the piezoelectric element from the testing environment, but also to serve as an acoustic transformer between the piezoelectric material and the load material. The reflection and transmission coefficients are determined by the acoustic impedances of adjacent media. In an ideal situation 100% transmission of the ultrasound energy into the load would be achieved. A front matching layer with appropriate acoustic impedance improves the transmission and also provides protection for the piezoelectric material from wear with the load material. The optimal acoustic impedance Z_2 of the front matching layer is determined by the quarter-wavelength propagation condition (Kino, 1987):

$$Z_2 = \sqrt{Z_1 \cdot Z_3} \tag{7.17}$$

where Z_1 and Z_3 are the acoustic impedances of the piezoelectric materials and the load material, respectively.

The thickness *d* of the matching layer is given by

$$d = \frac{\lambda}{4} \tag{7.18}$$

where λ is the wavelength of the ultrasound wave in the front matching layer.

The backing material plays an important role in a piezoelectric ultrasonic transducer. It reduces ringing resulting from reverberation of pulses and scatters, and thus the bandwidth of the transducer is widened. The backing material should be highly attenuative and is used to control the vibration of the transducer by absorbing the energy radiated from the back face of the piezoelectric material. Moreover, the backing material supporting the active element influences the damping characteristics. When the acoustic impedance of the backing material matches that of the active element, a heavily damped transducer that displays good range resolution can be realized; however, lower signal amplitude may also result. As the acoustic impedance mismatch between the piezoelectric material and the backing material increases, the transducer resolution decreases due to a longer waveform; however, it may have higher signal amplitude.

In order to find the appropriate acoustic impedance of the backing material, the response signals of ultrasonic transducers were analyzed using a KLM model simulation under various conditions for the acoustic impedance of the backing material. Figure 7.5 shows the KLM model for a piezoelectric transducer. In the KLM model, presented by Krimholtz, Leedom, and Matthaei in 1970), an electrical port is connected to the center node of two acoustic ports, representing the front and back faces of the transducer. In the electrical port, L_s and C_0 represent the inductance of the series inductor and the clamped capacitance, respectively. The transformer is an ideal electromechanical transformer that conserves power during transformation. The series inductance, L_s, the electrical transformer ratio, ϕ, and the capacitance, C', are given by the following equations (Krimholtz et al., 1970):

$$L_s = \frac{1}{(2\pi f_0)^2 C_0} \tag{7.19}$$

$$\phi = k_t \sqrt{\frac{1}{2 f_0 C_0 Z_c A}} \sin c\left(\frac{f}{2 f_0}\right) \tag{7.20}$$

$$C' = \frac{C_0}{k_t^2 \sin c(f/f_0)} \tag{7.21}$$

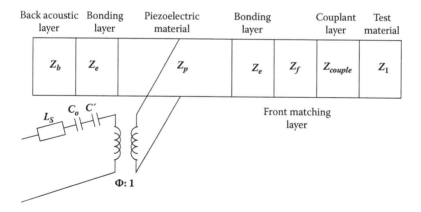

FIGURE 7.5 Description of transmission line for piezoelectric transducer in thickness mode of vibration based on the KLM model.

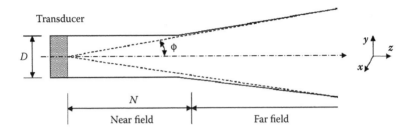

FIGURE 7.6 The near and far fields of the piezoelectric ultrasonic transducer.

where k_t, f_o, Z_c, and A are the electromechanical coupling factor, the resonance frequency, the characteristic acoustic impedance, and the cross-sectional area of the piezoelectric disk, respectively.

In Figure 7.5, Z_b, Z_e, Z_p, Z_{couple}, and Z_l are the characteristic acoustic impedances of the back acoustic material, bonding layer, piezoelectric material, coupling material, and load material, respectively. The KLM model includes all impedances, both acoustic and electrical, as well as signal amplitudes in both the forward and backward directions as a function of frequency. The round-trip transfer function of the transducer can be represented by a series of simple transfer matrices cascaded together (Kervel and Thijssen, 1983; Castillo et al., 2003) and a complex spectrum can be generated, from which a waveform in the time domain can be calculated by an inverse Fourier transform. The sensitivity and the bandwidth of the transducer were characterized by measuring the peak amplitude of the waveform and its frequency spectrum.

Major ultrasound beam parameters include intensity and spreading of the beam as it propagates. Intensity is the strength of the beam per unit of cross-sectional area. As the wave front first moves away from the transducer, it passes through a formative region near the element face. This formative region, called the near field, is explained by considering each small area on the surface of the transducer as a point source emitting spherical waves. According to Huygen's principle, interference between waves emitted by numerous sources causes variations in pressure as the beam propagates in the near field (Figure 7.6). Beyond the near field, in a region called the far field, such numerous wave fronts support rather than interfere with each other, and a wave front of more nearly uniform intensity is formed. For flat, disk-shaped transducer elements with $\lambda \ll D$, the intensity of the far field diminishes uniformly as the wave propagates. For full confidence in an ultrasound inspection, interrogation areas should be limited to the far field. The length of the near field is approximately equal to

$$N = \frac{D^2}{4\lambda} \tag{7.22}$$

where N, D, and λ are the length of the near field (mm), the diameter of the transducer (mm), and the ultrasound wavelength (mm).

The beam pattern in the near field has a very irregular interior, with many peaks and valleys, especially near the transducer face. The lateral extent of the near field is roughly confined to the size of the transducer, although it should be noted that it is difficult to precisely define the edge of such an irregular field. As the beam progresses into the far field, its topology becomes much smoother, eventually evolving into a well-defined single main lobe with low-intensity side lobes. The edges of this beam now spread linearly with distance, and the width of the main lobe, as given by the half-angle ϕ to the first zero on each side, asymptotically diverges at a constant angle inversely proportional to the transducer diameter, and therefore the near-field beam diameter. In the near field, the beam is nearly collimated until it approaches the transition distance, N. At this point the beam starts to spread slightly. As it travels into the far field, the beam widens further and eventually approaches a constant angle of divergence ϕ_d.

To quantitatively define the transition distance between the near-field and far-field regions, and to more precisely determine the amount of beam spreading in the far field, the radiation pattern from an ultrasonic transducer should be solved mathematically. The geometry of the problem for a circular transducer of radius a and diameter D is given in Figure 7.7; a circular coordinate system is initially assumed. The transducer face was assumed to be vibrating with a sinusoidal velocity of $u = u_0 cos(\omega t)$ perpendicular to the ρ-θ plane. Assuming all portions of the transducer face are oscillating with the same velocity and are in phase with each other, as would be the case for a rigid, piston-like transducer, the total pressure at the observation point is the integral of the incremental pressure:

$$p(r,\phi,t) = \frac{kZu_0}{2\pi} \int \frac{\cos(\omega t - kr')}{r'} dS$$

$$= Zu_0 \left[\cos(\omega t - kr') - \cos(\omega t - k\sqrt{a^2 + z^2}) \right]$$

(7.23)

where dS (= $\rho d\rho d\rho$) is a point source of incremental size, k is a propagation constant of the wave (= $2\pi/\lambda$), Z is the acoustical impedance of the intervening medium, and r' is the distance from the source to observation point.

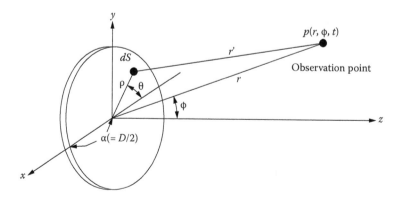

FIGURE 7.7 Geometry for calculating the near-field on-axis pressure field from a circular transducer.

In Equation 7.23, the first term, $Zu_0 \cos(\omega t - kr')$, is the familiar form for a pressure wave that appears to be coming from the center of the transducer, whereas the second term, $Zu_0 \cos(\omega t - k\sqrt{a^2 + z^2})$, appears to be a wave coming from a point at the edge (radius $= a$) of the transducer. A combination of these two waves, with phases that change at different rates as z varies, provides a destructive and constructive interference pattern, which produces the irregularities found in the near field. Figure 7.8 is a plot of the magnitude of Equation 7.23, and it reveals rapid variation of on-axis pressure in the near field of a circular transducer. There is a multitude of points in the near field where the pressure actually diminishes to zero. In addition, the rapidity of the spatial oscillation of the pattern decreases as one moves farther away from the face of the transducer. At large distances from the face of the transducer, the resultant pressure amplitude is no longer oscillatory but behaves as a slowly decreasing field.

When the beam is observed at a large distance from the transducer, the pressure at the far field may be approximated by

$$p(r, \phi, t) = \pi a^2 K \cos(\omega t - kr) \left[\frac{2J_1(ka\sin\phi t)}{ka\sin\phi} \right] \qquad (7.24)$$

where J_1 is a Bessel function of the first kind with order 1.

The term in the brackets of Equation 7.24 is called a directional factor, $H_c(\phi)$, and is given as

$$H_c(\phi) = \left[\frac{2J_1(ka\sin\phi)}{ka\sin\phi} \right] \qquad (7.25)$$

The directional factor $H_c(\phi)$ shows that the pattern may be considered to be a circularly symmetric function of the angle ϕ via the term $\sin\phi$; this equation is valid even for large ϕ. An angular plot of the logarithm of the square of Equation 7.25 in terms of decibels is shown in Figure 7.9 in polar coordinates. To obtain such a specific angular plot, a value must be given for the transducer radius a. For Figure 7.9,

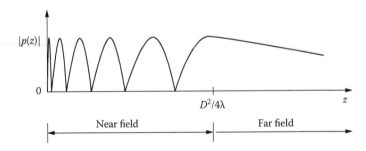

FIGURE 7.8 Variation of the magnitude of on-axis pressure field from a circular transducer of diameter D at a particular time, $t = 0$.

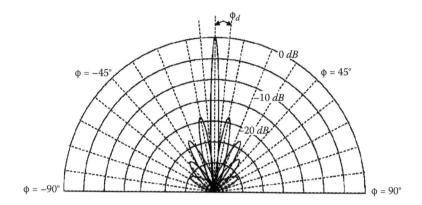

FIGURE 7.9 An angular plot of the beam pattern of the circular disk transducer with $ka = 10\pi$.

the transducer radius is assumed to be five wavelengths wide, and hence $a = 5\lambda$ or $ka = 10\pi$. From Figure 7.9, the ultrasound beam from the transducer consists of a main lobe and several side lobes, and the main lobe can be obtained from the first zero conditions of J_1 as

$$J_1(\pm 3.83) = 0 \qquad (7.26)$$

The angular position of the first zero defines the amount of divergence ϕ_d (half angle) of the main lobe as it propagates from the source; from Equation 7.26,

$$\phi_d = \sin^{-1}\left(\frac{3.83}{ka}\right) \qquad (7.27)$$

Equation 7.27 is only valid for the circular disk transducer.

Figure 7.10 is a photo of the developed ultrasonic transducers for fruit with 100 kHz central frequencies; the diameter and thickness of the piezoelectric element are 40 and 20 mm, respectively (Kim et al., 2004). The front matching layer and the backing material of the fabricated transducer are Teflon and a mixture of epoxy and tungsten powder backing material, respectively. The front matching layer has an acoustic lens with a curvature of 60 mm and thickness at the center of 4.3 mm to allow direct contact with the surface of the fruit.

7.3.2 ULTRASOUND TESTING TECHNIQUE

7.3.2.1 Pulse-Echo Method

A pulse echo describes a technique where a pulsed ultrasound beam is transmitted through a couplant into the material to be tested, travels to another surface of the material, and is then reflected back to a transducer that may or may not be a transmitting transducer. The echo wave originating from a flaw in the testing material

FIGURE 7.10 Developed ultrasonic transducer for fruit. (Reprinted with permission from Kim, K. B., M. S. Kim, S. Lee, and M. Y. Choi, *Review of Progress in Quantitative Nondestructive Evaluation*, 24A:1047–53, 2004.)

is indicated according to the transit time from the transmitter to the flaw and to the receiving transducer. Figure 7.11 shows a typical pulse-echo screen view. The main band occurs at the start of the screen display, and the reflecting surface is parallel to the probe contact surface; note that the same probe may be used as the transmitter and receiver. Flaws in the testing material will reflect or scatter part of the incident energy, and hence a flaw echo will appear on the screen display as a horizontal line corresponding to the flaw depth in the material. If a back-wall echo appears in addition to the echo from the flaw, its transit time will correspond to the thickness of the specimen. By measuring the transit times accurately, it is possible to determine where flaws are located. As shown in Figure 7.11b, the time t from the main bang to the time of arrival of echo from the flaw can be calculated as

$$t = \frac{2}{c}l \qquad (7.28)$$

where c is the phase velocity of the testing material.

7.3.2.2   Through-Transmission Method

Through transmission is used in several cases, particularly for highly attenuative materials such as fruit, where a pulse-echo trip results in a significant loss in signal

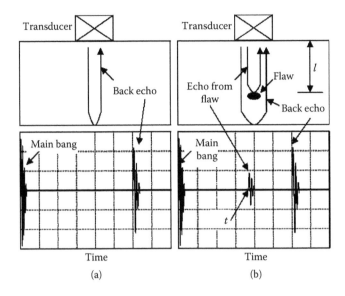

FIGURE 7.11 Typical screen displays obtained by the pulse-echo method with (a) no flaw and (b) flaw.

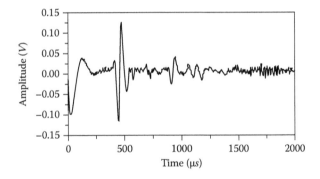

FIGURE 7.12 The transmitted ultrasonic signal through an apple by 100 kHz ultrasonic transducer. (Reprinted with permission from Kim, K. B., M. S. Kim, S. Lee, and M. Y. Choi, *Review of Progress in Quantitative Nondestructive Evaluation*, 24A:1047–53, 2004.)

strength. Since only one trip through the material is required, the oscilloscope indication will be only one half the distance from time zero to the location of the signal pulse. Figure 7.12 shows a typical ultrasound transmitted signal through an apple by the developed ultrasound transducer for fruit. Advantages of the through-transmission technique include the possibility to test highly attenuative materials, detection of near-surface flaws, and shielding of the receiving transducer from front-surface reflections and the effects of beam spread. Disadvantages are transducer alignment difficulties and the requirement of highly uniform coupling.

7.3.2.3 Doppler Method

In the previous section pulse-echo and through-transmission methods were described briefly. Another important application of ultrasound is measurement of the velocity of moving fluids in food process monitoring, such as particle velocity in a pipe or channel. The Doppler principle is by far the most popular means of acoustically measuring the velocity of a fluid or particle. Waves reflected from moving objects are shifted in frequency by an amount proportional to the velocity of the scattering objects. The change in frequency embodies the Doppler effect, which states that the observed frequency of a wave depends on the motional states of the wave sources and the observer. Figure 7.13 is an example of a Doppler shift describing a moving object scattering a source wave of frequency f_T into a reflected wave having a different frequency, f_R. In Figure 7.13, the Doppler shift frequency, f_D, is defined as the difference between the source frequency (f_T) and the received frequency (f_R):

$$f_D = f_R - f_T = -\frac{v}{c}(\cos\theta_T + \cos\theta_R)f_T \tag{7.29}$$

where v is the velocity of the moving object, c is the velocity of the wave, and θ_T and θ_R are the incident and reflection angles, respectively.

7.3.2.4 Ultrasound Data Displays

The amplitudes of the reflected signal in the pulse-echo technique and the transmitted signal in the through-transmission technique, respectively, are related to flaw geometry and material conditions of the ultrasound test material. These signals are usually displayed as voltages on a cathode ray tube (CRT); this technique is called A-mode or A-scanning (Figure 7.14b). Reflections from the test material due to changes in acoustical impedance are received back at the receiving transducer, and the total transit time from the initial pulse transmission, called the main bang, to reception of the echo is proportional to the depth of the flaw or bottom. The signal amplitude can be related to flaw size, and hence the amplitude is a good indicator of

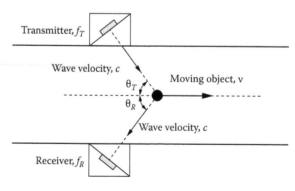

FIGURE 7.13 Description of the Doppler shift frequency.

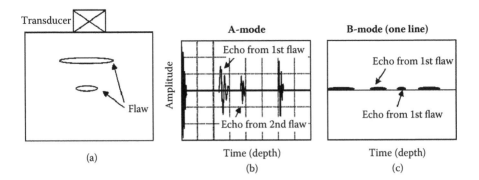

FIGURE 7.14 Comparison of the screen displays between A-mode and B-mode. In the B-mode the brightness of the beam is modulated by the amplitude obtained by A-mode.

flaw size when the flaw is small compared with the beam diameter, and the location of the pulse on the baseline of the display is proportional to flaw depth. The advantage of the A-mode is that it gives positional information quickly with simple equipment. Its weakness is that this depth information is one-dimensional only, along the line of beam propagation.

An improved method of presenting flaw information is called B-mode or B-scanning. When the echo distances from an A-mode signal are synchronized with the transducer location on the surface of the test material, a cross-sectional view of the test material can be displayed on the CRT, as indicated in Figure 7.14. B-mode stands for "brightness" modulation of the displayed beam. The position of the echo is determined by its acoustic transit time and beam direction in the plane. Alternatively, an imaging plane contains the propagation or depth axis.

C-scan ultrasound examination is a useful tool for the nondestructive evaluation of the item to be inspected, since it can provide high-resolution imaging of subsurface regions, which would otherwise be inaccessible with the conventional techniques. It is a hybrid mode incorporating characteristics of A-mode and B-mode. As in the B-mode, the brightness of the display line is modulated according to the amplitude of the received echoes. It is similar to the A-mode, however, in that echoes are collected only in one dimension, along the steadily held direction of the beam. These signals are presented on the horizontal axis of the display. A C-scan image provides a two-dimensional view of a test material, as shown in Figure 7.15b, and is obtained by synchronizing flaw or discontinuity signals with the motion of the transducer as it is moved across the entire surface of the specimen. Data in the form of the return ultrasound signal amplitude or time of flight are collected along a two-dimensional grid, as the transducer is moved over the surface of the material to be inspected, in immersion or contact. The C-scan system shows the locations of defects, but in the simplest form it does not show their depth. However, the depth can be determined with varying gray scale or color-coded displays, where the distance of the flaw echo from the screen origin is given as a gray shade or a color on the C-scan. C-scans can be made by using either pulse-echo or through-transmission inspection techniques.

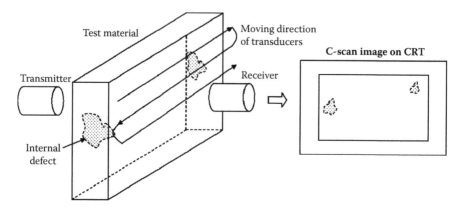

FIGURE 7.15 Basic configuration for C-scan and C-scan image obtained by through-transmission scanning.

7.3.2.5 Noncontact Ultrasonic Measurement

Conventional ultrasound techniques depend on contact measurement using a coupling medium between the transducer and the test material to overcome the acoustic impedance mismatch between air and the material. Use of a coupling medium, such as oils or gels, might change or destroy sample materials by absorption and interaction of the couplants. Further, the coupling medium could be a source of contamination if the test specimens are food materials. Hence, extensive research has been performed to develop an air-coupled ultrasonic transducer to overcome the limitations of the conventional contact ultrasonic measurement technology.

Considerable advances have been achieved in air-coupled ultrasonic transducer design. The most important element of an air-coupled transducer is the matching layer, which determines the efficiency of ultrasound transmission from the piezoelectric material to the medium. Fox et al. (1983) developed a matching layer for air-coupled transducers with silicon rubber. Haller et al. (1992) designed a special matching layer using tiny glass spheres in a matrix of silicone rubber. Bhardwaj (2000) developed an efficient acoustic matching layer with soft polymers to optimize air-coupled ultrasound transducers of 100 kHz to 5 MHz frequency.

In addition to the development of air-coupled transducers, effective signal processing techniques are necessary for successful noncontact ultrasound measurement. Recent development of a real-time noncontact ultrasound analysis method based on a synthetic aperture imaging technique appears to show good performance (Bhardwaj, 2000). The method greatly improves upon the signal-to-noise ratio and time-of-flight accuracy by borrowing from the well-established field of radar signal processing. In this mode of operation, the instrument employs a frequency-swept chirp excitation at the source transducer, followed by a particular form of digital signal processing at the receiver known as pulse compression.

Two air-coupled ultrasonic transducers were used, as shown in Figure 7.16. The transducers operate as both a transmitter and a receiver. Since the system has two

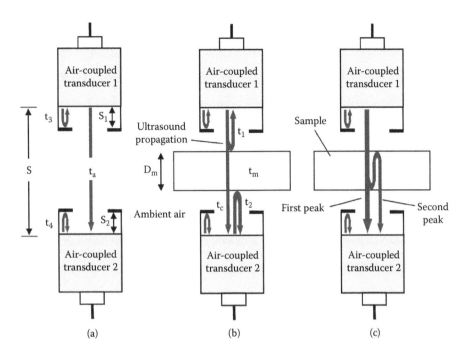

(a) (b) (c)

FIGURE 7.16 Schematic of noncontact ultrasonic measurement (a) without and (b) with a sample material and (c) attenuation coefficient measurement.

channels for data acquisition, it can provide four operation modes: two reflection (one for each of the two transducers) and two transmission (one used as a transmitter and the other as a receiver, and vice versa) modes. Using the four operation modes, the sample thickness and velocity of the sample can be estimated in real time by the following equations (Bhardwaj, 2000):

$$D_m = V_a \times [t_a - (t_1 + t_2)/2] = S - [(V_a \times t_1/2) + (V_a \times t_2/2)] \tag{7.30}$$

$$V_m = D_m/t_m = D_m/[t_c - (t_1 + t_2)/2] \tag{7.31}$$

where D_m is the sample thickness, V_a and V_m are the respective velocities of ultrasound in air and through the sample, S is the distance between transducer 1 and transducer 2, t_m is the time of flight in the test material, t_a is the time of flight between transducer 1 and transducer 2 in air, t_c is the time of flight between transducer 1 and transducer 2 in the sample, t_1 is the round-trip time of flight between transducer 1 and the sample, and t_2 is the round-trip time of flight between the sample and transducer 2.

Another parameter that should be determined is the attenuation coefficient, an indicator of the attenuated ultrasound energy through the sample. The attenuation coefficient can be calculated using multiple peaks observed in the through-

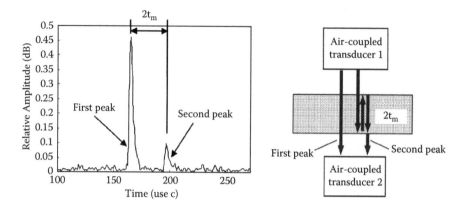

FIGURE 7.17 A typical noncontact ultrasonic signal transmitted through cheese. (Reprinted with permission from Cho, B., and J. M. K. Irudayaraj, *Journal of Food Science*, 68(7):2243–47, 2003.)

transmission mode of measurement (Figure 7.16c). The area underneath the peaks above −6 dB depicting the ultrasound energy is termed the integrated response (IR) (Bhardwaj, 2000). The IR provides information on the decrease in energy of the transmitted ultrasound signal through a sample in the time domain. A typical noncontact ultrasound through-transmission signal of a food material is shown in Figure 7.17. The first peak is a directly transmitted signal through air and the sample material, while the second peak denotes a transmitted signal with a material internal reflection. Other periodic peaks are multireflected by the sample material, which are rarely observed in food materials. The attenuation coefficient can be estimated as the difference between the integrated response of the first peak and the second peak divided by twofold of the sample thickness.

7.4 APPLICATIONS

Nondestructive ultrasonic testing is a versatile and successful technique that can be applied to a wide variety of material analysis applications. The state-of-the-art technology of sensors, microprocessors, and methods for signal analysis affords promise of new possibilities for the application of ultrasound techniques to nondestructive and nonhazardous testing. Ultrasonic material analysis is based on a simple principle of physics. That is, the motion of a wave is affected by the medium through which it travels. Thus, changes in one or more of the parameters associated with the passage of an elastic wave through a material—time of flight, attenuation, scattering, and frequency—are correlated with changes in physical properties such as hardness, elastic modulus, density, homogeneity, and particle size and structure. Ultrasound techniques can be applied to various biosystems, such as food, agricultural products, and biological materials, owing to their nondestructive nature. In the following section, two representative applications are reviewed.

7.4.1 Determination of Fruit Firmness by Nondestructive Ultrasonic Measurement

Firmness is one of the major quality indicators for fruit and provides a useful guide for producers, quality inspectors, and consumers. In particular, firmness is used as an indication of the handling characteristics of many types of fruit, and picking and grading of fruit may be based on firmness measurement. Traditional destructive methods such as application of a penetrometer and compression tests have been used to estimate the firmness of fruit. Firmness tests using destructive methods have the advantage of high accuracy, but the fruit is destroyed and wasted. As a nondestructive method, ultrasound techniques have been used for evaluating the quality of agricultural products (Sarkar et al., 1983; Mizrach et al., 1989, 1999, 2000; Kim et al., 2006). Most ultrasonic transducers used in previous research for fruit were not suitable for fruit because they were made for industrial usage. Hence, the fruits should be sliced uniformly so as to allow contact with the surface of the ultrasonic transducers. Recently, an ultrasonic transducer for fruit was successfully developed (Kim et al., 2004). The developed transducer has a curved wear plate (front matching layer) to allow direct contact with the surface of the fruit. In this study, the relationship between ultrasound parameters obtained by the developed ultrasonic transducer for fruit and the firmness of apples was evaluated, and estimation models for the firmness of the apple were developed and validated.

As fruit samples, Korean apples (*Malus pumila*, 'Sansa' cultivar) were experimented with. Tests to measure the ultrasound velocity and attenuation for the whole fruit were carried out using the through-transmission mode of the ultrasound measurement setup. To generate an ultrasonic wave, a high-power and low-frequency ultrasound pulser was connected to the ultrasonic transmitter. Figure 7.18 shows the overall experimental setup consisting of ultrasonic transducers, an apple sample, an ultrasound pulser, an oscilloscope, and a personal computer. As ultrasonic parameters, the transmission velocity and attenuation were calculated. After ultrasonic measurement,

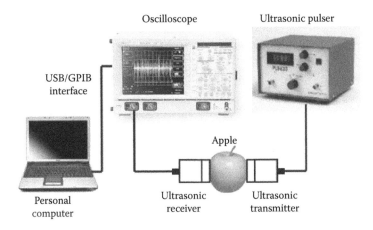

FIGURE 7.18 Ultrasonic measurement setup. (Reprinted with permission from Kim, K. B., *Postharvest Biology and Technology*, 52:44–48, 2009, Elsevier.)

compression tests were conducted using a universal testing machine (UTM). Parallel plates were used to compress the apple samples; one plate was fixed and the other plate was moved to the apple sample situated between the two plates. As firmness parameters, apparent elastic modulus and rupture point were obtained by compressing the equatorial region of the apple using a crosshead controlled by UTM.

The relationships between ultrasonic parameters (velocity and attenuation) and firmness factors (apparent elastic modulus and rupture point) were analyzed, as shown in Figures 7.19 and 7.20. The correlation coefficients (R) between the apparent elastic modulus and ultrasonic velocity and attenuation were 0.884 and −0.867, respectively. The correlation coefficients between the rupture point and ultrasonic velocity and

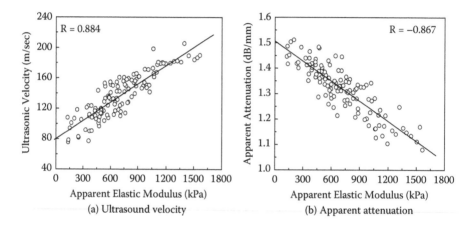

FIGURE 7.19 Variation of ultrasonic velocity and apparent attenuation with apparent elastic modulus for apple. (Reprinted with permission from Kim, K. B., *Postharvest Biology and Technology*: 52:44–48, 2009, Elsevier.)

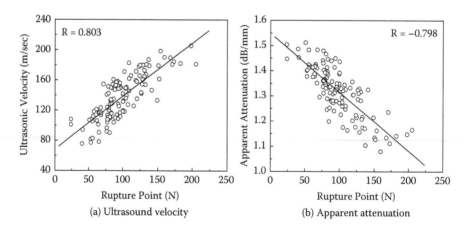

FIGURE 7.20 Variation of ultrasonic velocity and apparent attenuation with rupture point for apple. (Reprinted with permission from Kim, K. B., *Postharvest Biology and Technology*, 52:44–48, 2009, Elsevier.)

attenuation were meanwhile lower than that of the apparent elastic modulus. From the results of the correlation analysis, the correlation between ultrasonic parameters and firmness suggests that it is possible to predict the shelf life of a batch of apples by measuring their ultrasonic parameters. In order to estimate the capability of determining the apparent elastic modulus and rupture with ultrasonic velocity and attenuation, the following two multiple linear regression models were proposed:

$$E = a_0 + a_1 V + a_2 \alpha \tag{7.32}$$

$$R = a_0 + a_1 V + a_2 \alpha \tag{7.33}$$

where E is the apparent elastic modulus (kPa), R is the rupture point (N), V is the ultrasonic velocity (m/s), α is the apparent attenuation (dB/mm), and a_0, a_1, and a_2 are the regression coefficients.

The results of the regression analysis indicate that the firmness of the apple is predictable using ultrasonic measurement. The calibration equations for firmness were developed and validated based on the results of the regression analysis using Equations 7.32 and 7.33. The performance of the developed calibration equation was tested by evaluating the coefficient of determination (R^2), standard error of calibration (SEC), standard error of prediction (SEP), and bias. The SEC measures the scatter of the actual measured value (firmness) about the values calculated by the calibration equation. The developed calibration equation was then used to predict the firmness using measured ultrasonic parameters that were not used in developing the calibration equation. The calibration equations were developed with 120 different apple samples with a coefficient of determination of 0.833 and an SEC of 132.1 kPa for the apparent elastic modulus, and with a coefficient of determination of 0.696 and an SEC of 18.65 kPa for rupture, respectively. The calibration equations, Equations 7.32 and 7.33, were validated with one hundred different apple samples. The coefficient of determination, SEP, and bias were 0.800, 113.04 kPa, and −35.57 kPa, respectively, for the apparent elastic modulus and 0.684, 17.24 kPa, and 0.64 kPa for the rupture point. From Figures 7.21 and 7.22, the measured values are in good agreement with the predicted values, and this study demonstrates the feasibility of using ultrasound to estimate the firmness of fruit. As a conclusion, the proposed technique for firmness measurement based on ultrasonic velocity and attenuation holds promise for the development of practical instruments to measure the firmness of fruit.

7.4.2 NONDESTRUCTIVE MEASUREMENT OF CHEESE TEXTURE USING ULTRASOUND

For cheese, texture is considered one of the most important sensory attributes for consumer acceptability, since it determines the identity and quality of specific varieties. The texture of cheese is related to its structure formed with a complex interaction of components, such as casein, fat, water, and other minor elements. Although instrumental analyses of texture properties of cheese are important for objective

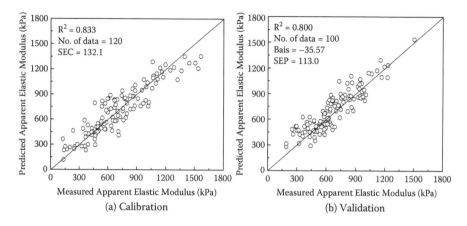

FIGURE 7.21 Relationship between apparent elastic modulus of apple by UTM and that predicted by Equation 7.1. (a) The calibration data of 120 Korean '*Sansa*' apples and (b) the validation data of 100 Korean '*Sansa*' apples. (Reprinted with permission from Kim, K. B., *Postharvest Biology and Technology*, 52:44–48, 2009, Elsevier.)

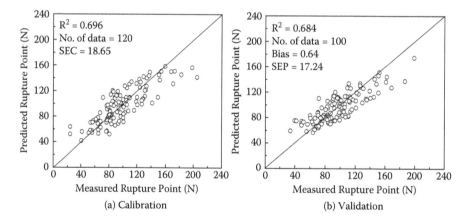

FIGURE 7.22 Relationship between rupture point of apple by UTM and that predicted by Equation 7.2. (a) The calibration data of 120 Korean '*Sansa*' apples and (b) the validation data of 100 Korean '*Sansa*' apples. (Reprinted with permission from Kim, K. B., *Postharvest Biology and Technology*, 52:44–48, 2009, Elsevier.)

quality measurement, most methods are either partially or totally destructive. In addition, these methods are time consuming, labor intensive, and are performed off-line. To determine the texture of cheese nondestructively, an ultrasound measurement technique was used. The ultrasonic wave propagation velocity and spectra of cheese samples using pulse-echo and through-transmission techniques were explored. The ultrasonic system consists of two 2.25 MHz contact transducers, a pulser-receiver, a position control system, and an A/D conversion card linked to a personal computer, as shown in Figure 7.23.

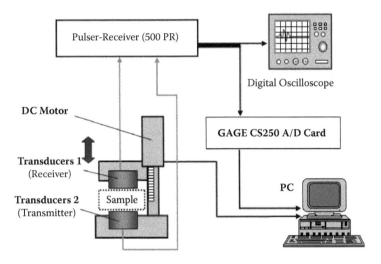

FIGURE 7.23 Generalized schematic diagram of ultrasonic system. (Reprinted with permission from Cho, B., J. M. K. Irudayaraj, and S. Omato, *Applied Engineering in Agriculture*, 17(6):827–32, 2001.)

7.4.2.1 Through-Transmission Measurement

A cheese sample was placed between two ultrasonic transducers, and the upper transducer, which receives the ultrasound signal, was lowered until contact was made with the sample at a preset contact distance. The measured ultrasonic time domain signal was transformed to the frequency domain spectrum using a fast Fourier transform (FFT) algorithm. From the ultrasonic spectrum, the peak frequency of each spectrum was determined. A representative spectrum of an ultrasound transmission signal is shown in Figure 7.24.

7.4.2.2 Pulse-Echo Measurement

The cheese sample was placed on a transducer and the wave propagation velocity through the sample was measured. An ultrasound coupling gel was utilized to minimize the acoustic impedance mismatch between air and the sample. The velocity was computed by dividing the total distance traveled by the time of flight. Time of flight is the time interval between the peak of the first signal through the sample and the peak of the second signal. Figure 7.25 shows a representative ultrasound A-mode signal for measuring the ultrasound propagation velocity.

The ultrasonic propagation velocity tends to increase with an increase in the Young's modulus and hardness values (Figure 7.26). The Young's modulus and hardness of the cheese sample could be estimated by the ultrasound velocity using a linear regression to an R^2 of 0.827 (SEE = 10.47 kPa) and 0.85 (SEE = 2.347 kPa), respectively.

The ultrasonic transmission signal in the time domain was transformed to a frequency domain spectrum using an FFT algorithm, and the peak frequency was estimated. The peak frequency increases with increasing Young's modulus and hardness (Figure 7.27); however, the results show high variability ($R^2 = 0.710$, SEE = 13.37

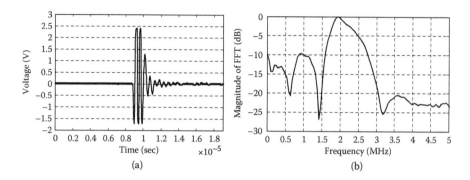

FIGURE 7.24 Ultrasonic spectrum of cheese sample. (a) Time domain signal. (b) FFT of time domain signal. (Reprinted with permission from Cho, B., J. M. K. Irudayaraj, and S. Omato, *Applied Engineering in Agriculture*, 17(6):827–32, 2001.)

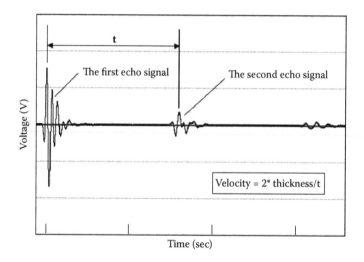

FIGURE 7.25 Ultrasonic velocity measurement through cheese sample. (Reprinted with permission from Cho, B., J. M. K. Irudayaraj, and S. Omato, *Applied Engineering in Agriculture*, 17(6):827–32, 2001.)

kPa and $R^2 = 0.436$, SEE = 4.355 kPa, respectively). It is indicated that the peak frequency of the ultrasound spectrum is related to the physical properties of the biological material (Park and Whittaker, 1990). The shift of the peak frequency on the ultrasonic spectrum may reflect differences in cheese structure. High-frequency components are normally attenuated by a relatively lower concentration of structural matrix, tiny cracks, and defects that might reduce the elasticity of the cheese sample and cause the spectrum to shift toward lower frequencies. Of the cheese samples, reduced-fat cheddar, which contains a higher concentration of structural matrix per unit volume than full-fat cheddar, is firmer and has a higher Young's modulus and hardness; hence, it has a higher peak frequency on the spectrum.

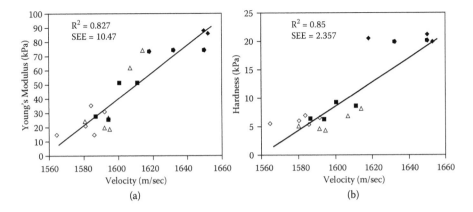

FIGURE 7.26 Relationship between contact ultrasonic velocity and (a) Young's modulus and (b) hardness of cheddar cheese (2% fat milk sharp) (•), sharp (o), extra sharp (■), and New York aged (△). (Reprinted with permission from Cho, B., J. M. K. Irudayaraj, and S. Omato, *Applied Engineering in Agriculture*, 17(6):827–32, 2001.)

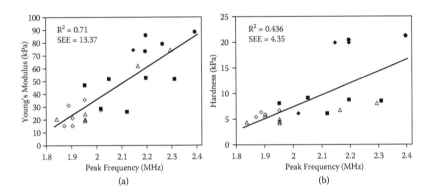

FIGURE 7.27 Relationship between contact ultrasonic peak frequency and (a) Young's modulus and (b) hardness of cheddar cheese (2% fat milk sharp) (•), sharp (o), extra sharp (■), and New York aged (△). (Reprinted with permission from Cho, B., J. M. K. Irudayaraj, and S. Omato, *Applied Engineering in Agriculture*, 17(6):827–32, 2001.)

7.4.2.3 Noncontact Measurement

The textures of five varieties of hard cheese (sharp cheddar, reduced-fat sharp cheddar, Asiago, Romano, and Parmesan) were measured using a noncontact ultrasound system. The cheese sample was prepared as a 25 × 25 × 25 mm size cube consistently using a specially designed wire cutter. The prepared cheese sample was placed between the two air-coupled transducers. The ultrasound parameters, such as velocity and attenuation coefficient, were measured as shown in Figure 7.28. Six texture parameters of cheese—fracturability, hardness, springiness, cohesiveness, chewiness, and gumminess—were measured using an Instron testing machine with a crosshead speed of 50 mm/min, and the results were compared with ultrasound parameters.

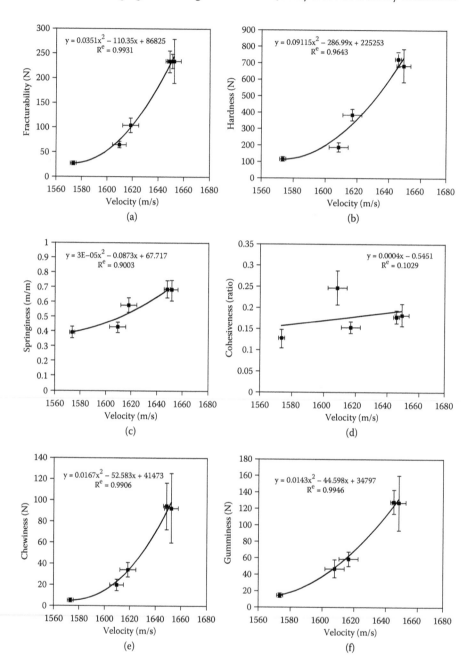

FIGURE 7.28 Relationship between ultrasonic velocity and texture parameters of cheeses.

Ultrasonic velocity is assumed to be related to the square root of the bulk modulus over the density of the solid material (Povey, 2001). Since the density differences of the cheese varieties are not significant, the ultrasound velocity is determined by the bulk modulus, which might be strongly interdependent on texture properties. Hence, the texture properties were quite well explained by a quadratic regression model.

The correlation between the ultrasonic attenuation coefficient and texture parameters showed relatively high variability. Even though the attenuation coefficient is assumed to be related to physical properties of a material, such as viscosity, density, and ultrasonic velocity, accurate measurement of the attenuation coefficient using a noncontact ultrasonic technique is not easy due to its sensitivity to uneven sample surface, tiny defects, porosities, and uneven component distribution in the sample materials.

Using a multivariate linear regression analysis, texture properties were estimated by ultrasound velocity (Figure 7.28). The regression model was developed with a coefficient of determination (R^2) of 0.9931 (SEE = 11.23N), 0.9634 (SEE = 75.05N), 0.9003 (SEE = 0.061m/m), 0.1029 (SEE = 0.049), 0.9906 (SEE = 5.686N), and 0.9946 (SEE = 5.28N) for fracturability, hardness, springiness, cohesiveness, chewiness, and gumminess, respectively. The results indicated that the texture properties of cheese were more significantly related to the ultrasound velocity than the attenuation coefficient.

7.5   PERSPECTIVES

Ultrasound technology was first developed about one hundred years ago and has become increasingly important for human life. It is widely applied to measurement of the internal quality of food and agricultural products, food process control, and medical diagnosis and treatment. Ultrasonic technology is a very effective means of evaluating the internal quality of food and agricultural products, including firmness and cavity. Recently, ultrasonic instrumentation and sensors have become smaller in accordance with advances in electronics and semiconductor technology. At present, micro-sized ultrasonic sensors such as capacitive micro-machined ultrasonic transducers (cMUTs) are being developed by applying microelectromechanical system (MEMS) technology (Kim et al., 2005). Hence, in the near future, new applications to the field of biosystems will appear based on developments in ultrasonic instrumentation and sensor technology.

REFERENCES

Bhardwaj, M. C. 2000. High transduction piezoelectric transducers and introduction of non-contact analysis. *e-Journal of Nondestructive Testing & Ultrasonics* (serial online) 1:1–21.

Castillo, M., P. Acevedo, and E. Moreno. 2003. KLM model for lossy piezoelectric transducers. *Ultrasonics* 41:671–79.

Cho, B., and J. M. K. Irudayaraj. 2003. A non-contact ultrasound approach for mechanical property determination of cheese. *Journal of Food Science* 68(7):2243–47.

Cho, B., J. M. K. Irudayaraj, and S. Omato. 2001. Acoustic sensor fusion approach for rapid measurement of modulus and hardness of cheddar cheese. *Applied Engineering in Agriculture* 17(6):827–32.

Fox, J. D., B. T. Khuri-Yakub, and G. S. Kino. 1983. High frequency wave measurements in air. *IEEE Ultrasonics Symposium* 1:581–92.

Haller, M. I., and B. T. Khuri-Yakub. 1992. 1–3 composites for ultrasonic air transducer. *IEEE Ultrasonics Symposium* 2:937–39.

Kervel, S. J. H., and J. M. Thijssen. 1983. A calculated scheme for the optimum design of ultrasonic transducers. *Ultrasonics* 21:134–40.

Kim, K. B., B. Ahn, H. W. Park, and Y. J. Kim. 2005. Fabrication and characterization of capacitive micromachined ultrasonic transducer. *Key Eng. Mater.* 297–300:2225–32.

Kim, K. B., M. S. Kim, H. M. Jung, S. Lee, J. G. Park, and G. S. Kim. 2006. Wavelet analysis of ultrasonic transmitted signal of apple during storage time. *Key Eng. Mater.* 321–23:1192–95.

Kim, K. B., M. S. Kim, S. Lee, and M. Y. Choi. 2004. Consideration of design parameters of ultrasonic transducer for fruit. *Review of Progress in Quantitative Nondestructive Evaluation* 24A:1047–53.

Kim, K. B., S. Lee, M. S. Kim, and B. K. Cho. 2009. Determination of apple firmness by nondestructive ultrasonic measurement. *Postharvest Biology and Technology 52*:44–48.

Kino, G. S. 1987. *Acoustic waves—Devices, imaging and analog signal processing*. Englewood Cliffs, NJ: Prentice-Hall.

Krimholtz, R., D. Leedom, and G. Matthaei. 1970. New equivalent circuits for elementary piezoelectric transducers. *Electronics Letter* 6:398–99.

Mizrach, A. 2000. Determination of avocado and mango fruit properties by ultrasonic technique. *Ultrasonics* 38:717–22.

Mizrach, A., U. Flitsanov, R. El-Batsri, and H. Degani. 1999. Determination of avocado maturity by ultrasonic attenuation measurements. *Scientia Horticulturae.* 80:173–80.

Mizrach, A., N. Galili, and G. Rosenhouse. 1989. Determination of fruit and vegetable properties by ultrasonic excitation. *Transactions of the ASAE* 32(6):2053–58.

Park, B., and A. D. Whittaker. 1990. *Determination of beef marbling score using ultrasound A-scan.* ASAE paper 90-6058. St. Joseph, MI: ASAE.

Povery, M. J. W., and J. Lamb. 2001. Elastic properties of food. In *Handbook of elastic properties of solids, liquids, and gases.* ed. M. Levy, H. Bass, R. Stern. Vol. 3. 129–145. San Diego, CA: Academic Press.

Sarker, N., and R. R Wolfe. 1983. Potential ultrasonic measurements in food quality evaluation. *Transactions of the ASAE* 26(2):624–29.

Shmulevich, I., N. Galili, and D. Rosenfield. 1996. Detection of fruit firmness by frequency analysis. *Transactions of the ASAE* 39(3):1047–55.

8 Hyperspectral and Multispectral Imaging Technique for Food Quality and Safety Inspection

Moon S. Kim, Kuanglin Chao, Diane E. Chan, Chun-Chieh Yang, Alan M. Lefcourt, and Stephen R. Delwiche

CONTENTS

Mention of a product or specific equipment does not constitute a guarantee or warranty by the U.S. Department of Agriculture and does not imply its approval to the exclusion of other products that may also be suitable.

8.1 INTRODUCTION

As consumer demands for high-quality and safe food products increase, agro-food industries are faced with overwhelming challenges to rapidly and objectively evaluate agricultural commodities to ensure that quality and safety needs are met (Kim et al., 2007). Especially, concerns about food safety are rising with every new outbreak of foodborne illness, both within the United States and worldwide (Armstrong et al., 1996; Mead et al., 1999). With the advancement and availability of optoelectronic imaging devices, applications of image-based machine vision techniques have become more prevalent as nondestructive inspection tools in the handling and processing of agricultural products (Schatzki et al., 1997; Wen and Tao, 2000; Chen and Tao, 2001; Aleixos et al., 2002; Shahin et al., 2002; Troop et al., 2005; Blasco et al., 2007). Conventional monochromatic or red, green, and blue (RGB)–based machine vision methods are typically used for rapid online sorting for basic physical quality attributes of agro-foods, such as size, shape, and color (Leemans and Destain, 1998; Miller and Drouillard, 2001). Chemical and biological properties of agro-food can often be assessed by spectroscopic methods. Imaging, compared to point source, spectroscopic techniques, allows for assessment of spatial attributes of agricultural commodities and is ideal for identification or detection of localized artifacts such as defects and contaminants (Kim et al., 2001).

Since the late 1990s, hyperspectral imaging, combining the advantages of spectroscopy and imaging, has found its way into agro-foods as a rapid, nondestructive means to assess safety and quality for human consumption (Kim et al., 2001; Mehl et al., 2004; Park et al., 2002; Lawrence et al., 2003; Lu, 2003; Gowen et al., 2007; Heitschmidt et al., 2007; Qin et al., 2008). Providing full spectrum data, often over one hundred spectral data points, for every pixel in the image of a food product enables spectral and spatial analysis for correlation to composition, contaminants, and physical attributes such as size and shape. However, high speeds and product volumes have presented significant challenges in implementing the use of hyperspectral imaging for rapid online food processing applications to improve real-time inspection across agro-food industries (Kim et al., 2001). Use of image data from a combination of only several key wavelengths (thus, multispectral imaging) and simpler processing methods has been particularly ideal for adaptation to online processing applications to address various safety and quality inspection needs (Kim et al., 2001, 2004; Mehl et al., 2002; Park et al., 2002; Chao et al., 2007). In this regard, hyperspectral imaging platforms have often served as a research tool. Numerous studies have reported the use of hyperspectral imagery data to determine a few spectral bands and algorithms (multispectral methods) for potential use in agro-food safety and quality inspection (Kim et al., 2002a, 2002b; Mehl et al., 2004; Liu et al., 2006; Chao et al., 2007; Cho et al., 2007; Lee et al., 2008; Qin et al., 2008; Won et al., 2009).

Fundamentally, there are two ways to acquire hyperspectral imaging data from an object: a band-sequential imaging method and a push-broom (line-scanning) imaging method. The band-sequential imaging method captures a full spatial scene at individual wavelengths to form a three-dimensional hyperspectral image data cube. The push-broom method captures a line of spatial information with full spectral data per spatial pixel, and the composites of a set of many spatial line scans form a hyperspectral image cube (Figure 8.1).

Over the past decade, researchers at the Agricultural Research Service (ARS), U.S. Department of Agriculture (USDA), have developed several versions of line-scan–based hyperspectral imaging systems capable of both visible to near-infrared reflectance and fluorescence methods. These line-scan hyperspectral imaging techniques have served dual purposes, as both basic laboratory-based research tools and online multispectral platforms to perform rapid inspection of foods. The line-scan speed (data acquisition) in conjunction with the speed of sample movement affects the spatial resolution of the target acquired by the line-scan hyperspectral imaging techniques. In the early 2000s, data acquisition and transfer rates were the limiting factors for online implementation of the technology, which at the time resulted in coarse spatial resolution of samples that were not adequate to address safety and quality inspection needs for agro-foods (Kim et al., 2001). Newer line-scan–based hyperspectral imaging approaches can now deliver high-speed online safety and quality inspection of food and agricultural products on high-throughput processing lines (Chao et al., 2007, 2010; Kim et al., 2007; Yang et al., 2008). This has become possible because of technology advancements as well as research and development for optimization of the system and methodologies. The improvement of low-light–sensitive, two-dimensional detectors (i.e., charged coupled device (CCD) and faster electronics for data transfer makes possible near-real-time processing with sufficient spatial resolution. In addition, in lieu of capturing the entire spectral range of hyperspectral image data, a few spectral bands (spectral regions of interests on the CCD) can be selectively captured and transferred to increase the line-scanning rate. The selections of desired multispectral bands are based on the software, and with this flexibility, a single platform can serve potentially many inspection needs. Even more recently, we have also developed additional methods to simultaneously capture multispectral line-scan images of fluorescence and reflectance, thus allowing

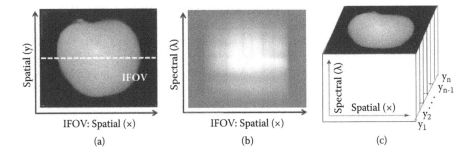

FIGURE 8.1 (a) Instantaneous field of view (IFOV) of a line scan. (b) IFOV dispersed onto the CCD as spectral and spatial directions. (c) Line-scan composition of hyperspectral image data cube.

simultaneous multiple inspection algorithms (multitasking) for different safety and quality problems (Kim et al., 2007, 2008).

Our research goals have been to develop image-based sensing methodologies and technologies to address food quality issues and safety concerns for food production and to aid in reducing food safety risks in food processing. Major research areas in recent years have included the development of portable low-cost imaging devices for *in situ* identification of contamination sites for use by food producer and processor operations, with the goal of commercial implementation for cleaning and sanitation inspection (Jun et al., 2010); automated online poultry inspection systems for detecting pathophysiological abnormalities in poultry, with the goal of commercial implementation as part of existing or new poultry processing systems (Chao et al., 2010); and image-based, rapid online techniques for simultaneous detection of fecal contamination and defects on fruits and vegetables (Kim et al., 2008).

In this chapter, details are provided on the most recent line-scan hyperspectral imaging platform that has served as a workhorse for our research endeavors, followed by case studies using the hyperspectral line-scan imaging technologies to determine several key spectral bands for detection of fecal contamination on apples; online poultry inspection; and multitasking apple inspection.

8.2 LINE-SCAN HYPERSPECTRAL IMAGING

We have developed a series of line-scan–based hyperspectral imaging systems (e.g., Kim et al., 2001; Vargas et al., 2005; Qin et al., 2008; Jun et al., 2009). Typically, the imaging platforms have been designed for acquisition of reflectance measurements in the range of approximately 400 to 1,000 nm, and for fluorescence measurements from approximately 420 to 700 nm (with UV-A excitation and, more recently, blue excitation). Another consideration was flexibility for the imaging of various sample sizes, in that the platform is able to accommodate small and large samples ranging from individual grains of wheat to whole chicken carcasses.

8.2.1 SENSING MODULE

Figure 8.2 shows an illustrative diagram of the transportable line-scan hyperspectral imaging system capable of both fluorescence and reflectance imaging. The sensing components include a low-light–sensitive electron-multiplying charge coupled device (EMCCD) camera (MegaLuca, Andor Technology, Inc., South Windsor, Connecticut). The EMCCD has $1,002 \times 1,004$ pixels and is thermoelectrically cooled to $-20°C$ via a Peltier device. The EMCCD is equipped with a 12.5 MHz pixel readout rate and 14-bit A to D digitizer. Image data are transferred to a PC via USB 2.0. Both vertical and horizontal pixels can be binned, and binning is achieved in the hardware prior to data transfer to PC.

An imaging spectrograph (400 to 1,000 nm, VNIR Hyperspec, Headwall Photonics, Inc., Fitchburg, Massachusetts) and a C-mount object lens (Schneider Optics, Van Nuys, California) are attached to the EMCCD. The instantaneous field

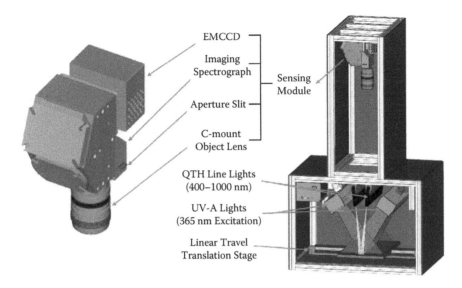

FIGURE 8.2 Schematics of the ARS laboratory-based line-scan hyperspectral imaging system and its critical sensing components.

of view (IFOV) is limited to a thin line by the spectrograph slit size: a range of aperture slits (10, 25, 60, and 100 μm width) are available and can be readily exchanged; typically, the 60 μm slit is used most often in our system. Note that the aperture slit size affects the IFOV, light throughput, and spectral resolution. Through the slit, light from the scanned line of the IFOV is dispersed by the holographic grating and projected onto the EMCCD. Therefore, for each line scan, a two-dimensional (spatial and spectral) image is created with the spatial dimension along the horizontal axis and the spectral dimension along the vertical axis of the EMCCD.

8.2.2 Illumination Sources

Two types of continuous wave (CW) light sources are used in the line-scan imaging system to provide illumination for fluorescence and reflectance imaging. For reflectance, the illumination source is a regulated DC 150-W quartz-tungsten halogen (QTH) lamp coupled with a pair of (bifurcated) fiber optic line lights (Fiber-Lite, Dolan-Jenner Industries, Inc., Lawrence, Massachusetts). These line lights are mounted to illuminate the line of the IFOV at backward and forward angles of approximately 5°. The imaging system and its components are enclosed in a light-tight casing to mitigate the effect of ambient light.

For fluorescence imaging, two high UV-A (320 to 400 nm, 365 nm peak) fluorescent lamp assemblies (Spectroline XX-15, Spectronics Corp., Westbury, New York) are used. Because the presence of the QTH fiber optic line lights within the confined space limited the options for placement of the UV fluorescent lamps, the UV lamps were mounted to illuminate the line of the IFOV at angles of approximately 35° from the vertical position, backward and forward. A long (>400 nm) wavelength pass filter

(Kodak Wratten Gelatin filter, #2A) was placed between the C-mount lens and the spectrograph to remove the second-order effects of the UV fluorescence excitation sources. Note that with UV-A excitation, a number of constituents in food products emit fluorescence in the visible range of the spectrum.

8.2.3   SOFTWARE INTERFACE

One of the key aspects of sensing system development is the versatility of interface software to control the imaging devices and acquire data. The interface software was developed using a software development kit provided by the camera manufacturer on a Microsoft (MS) Visual Basic (Version 6.0) platform in the MS Windows operating system. Because of the push-broom–based, line-scan imaging method coupled with the use of the dispersive spectrograph, the spectral information at each pixel location is presented in the vertical (traverse) direction of the EMCCD. With the aid of the software controls for the EMCCD, the system can be configured to acquire either hyperspectral images (with the entire span of the vertical spectral region) or multispectral images (with a few selected regions of interest in the vertical spectral direction using the random track acquisition mode). Vertical binning affects the spectral interval and, more importantly, the spectral resolution of the images. In general, reducing the data volume per line-scan image results in a higher number of lines scanned per second. However, the vertical pixel shift rate is also one of the limiting factors, along with the EMCCD exposure time, for the number of lines scanned per second.

8.2.4   SPECTRAL CALIBRATION

Spectral calibration of the EMCCD spectrograph was performed using a Hg-Ne spectral calibration lamp (Oriel Instruments, Stratford, Connecticut). To illuminate the entire line of the IFOV, the Hg-Ne pencil light was placed approximately 25 cm above (and at a 5° forward angle) a 30×30 cm polytetrafluoroethylene (Spectralon™) 99% diffuse white reflectance reference panel (SRT-99-120, Labsphere, North Sutton, New Hampshire) positioned horizontally level at the sample presentation distance. Note that Ne is the starter gas for the spectral Hg-Ne lamp, and immediately upon supplying the power to the lamp the dominant Ne lines were obtained, followed by the Hg lines.

Figure 8.3a shows a line-scan image of the Hg emission lines acquired with no pixel binning ($1,004 \times 1,002$ pixels) and with a 60 μm slit. The adjacent graph in Figure 8.3b shows the prominent Ne and Hg emission lines. The two emission lines observed between the 600th and 800th Y-pixel locations show the second-order effect of the 404.3 and 435.8 nm Hg emission lines. An optimal alignment of the EMCCD spectrograph resulted in the same vertical pixel locations for the individual Hg emission lines regardless of the horizontal pixel location (as seen in Figure 8.3a). If the EMCCD and spectrograph were to be misaligned, skewed emission lines (off the horizontal) would be observed. With the known Hg and Ne emission line wavelengths as the dependent variables and the corresponding vertical pixel locations as

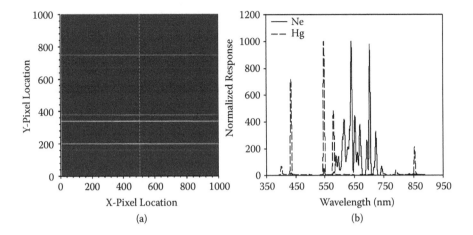

FIGURE 8.3   (a) A line-scan image of Hg lamp emission lines acquired with no pixel binning (1,004 × 1,002 pixels). The horizontal (X) axis represents the spatial dimension, and the vertical axis (Y) represents spectral dimension. (b) Hg and Ne emission spectra across the 501st horizontal pixel location. Note that the Ne emission line-scan image is not shown in (a).

the independent variables, a linear or polynomial regression fit is typically used to determine the wavelengths corresponding to the vertical pixel locations. The linear fit resulted in an r^2 value of 0.999, with the vertical pixels spaced by approximately 0.79 nm and the residual errors less than 1 nm. This observation suggested that a linear regression fit for wavelength calibration was sufficient.

The effects of spectral pixel binning on spectral resolution have been seldom discussed in the literature. Without any vertical spectral pixel binning, a hyperspectral image cube in the 400 to 1,000 nm range consists of over 800 channels; this may result in a redundancy of spectral information with the creation of unnecessarily large-size data files since biological materials do not exhibit such optical properties or fluorescence emissions as to warrant such spectrally large dimension data. Using the Hg emission line at 546.7 nm, we determined spectral resolution in full width at half maximum (FWHM) for no vertical pixel binning (1×), and then for 3×, 6×, and 10× vertical pixel binning: The spectral responses of individual binning selections were subjected to a cubic-spline interpolation to obtain spectral responses in a 0.1 nm increment; then the wavelength differences (FWHM) at 50% of the peak intensity levels prior to and after the peak wavelength were calculated. The FWHM were 4.4, 4.6, 6.0, and 9.4 nm for 1×, 3×, 6×, and 10× binning, respectively. Note that the data used to determine the spectral resolution were acquired with the use of a 60 μm spectrograph slit; the slit size in conjunction with the vertical binning affects spectral resolution.

We typically operate the imaging system in the laboratory with 6× vertical pixel binning, which results in spectral data in approximately 4.8 nm intervals with a spectral resolution of 6.0 nm. Thus, for fluorescence imaging, wavelengths span from approximately 420 to 700 nm across sixty wavebands. Hyperspectral reflectance images are captured from approximately 400 to 1,000 nm with 125 wavebands.

8.2.5 Spatial Resolution

During image acquisition, a sample is moved transversely through the illuminated IFOV by a programmable motorized translation stage (Velmex, Inc., Bloomfield, New York) that is controlled through the interface software. With each new line-scan acquisition, the translation stage is advanced by one incremental step. Depending on the nature of the samples being imaged, the step sizes may be as small as <0.1 mm for small target samples or when higher spatial resolution images are needed, or as large as several millimeters when fine spatial resolution is not necessary.

Myriad factors can affect the spatial resolution of the line-scan imaging system. Spatial resolution in the direction of the translation stage movement is dictated by the line-scan step increment. As shown in Figure 8.4, an ideal line-scan imaging would cover the entire sample surfaces. Undersampling by scanning with line-scan steps greater than the width of the IFOV will result in coarse resolution images and is likely to misrepresent samples with relatively small features. Spatial resolution in the transverse direction of the sample movement is a combined function of the detector pixel size/binning and the IFOV. Changing the distance between the object lens and the target will also affect the width and length of the IFOV.

After adjusting the length of the IFOV (object lens to the sample distance) to capture a full-scene length of approximately 80 mm, line-scan images of an Air Force standard target at 650 nm, shown in Figure 8.5, were acquired with an 0.08 mm step increment of the translation table to provide a near square-pixel size of 0.08×0.08 mm. The image on the right is an enlargement of the middle portion of the image of the standard target, showing that the individual dots, 0.25 mm in diameter and spaced 0.50 mm apart, can be clearly resolved. With these same spatial imaging parameters, hyperspectral images of two varieties of wheat kernels with crease side up and down were also acquired (650 nm reflectance image shown in Figure 8.6). In the sample image and the adjacent three-dimensional contour plots for the selected kernels, it is clear that various features (e.g., germ, crease, endosperm cheek) of individual wheat kernels can be observed.

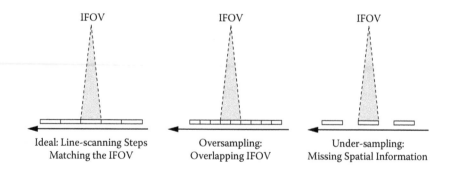

FIGURE 8.4   Three line-scanning scenarios that can affect the spatial resolution of line-scan image acquisition.

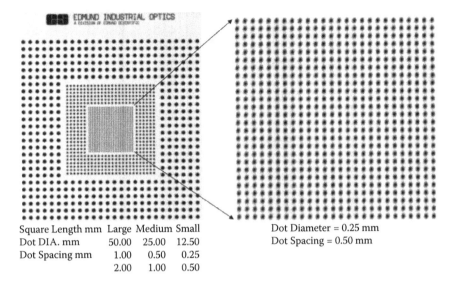

Square Length mm	Large	Medium	Small	Dot Diameter = 0.25 mm
Dot DIA. mm	50.00	25.00	12.50	Dot Spacing = 0.50 mm
Dot Spacing mm	1.00	0.50	0.25	
	2.00	1.00	0.50	

FIGURE 8.5 An Air Force standard target image at 650 nm. It was acquired with an 0.08 mm step increment of the translation table to provide a near square-pixel size of 0.08 × 0.08 mm. The image on the right is an enlargement of the middle portion of the standard target.

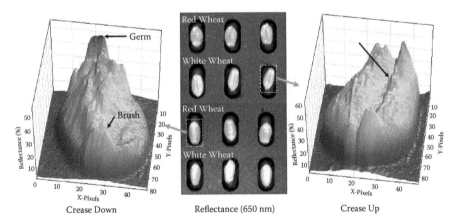

FIGURE 8.6 Reflectance image of red and white wheat kernels at 650 nm where hyperspectral image data were acquired with 0.08 × 0.08 mm pixel resolution. First and second rows show kernels with crease side up, and third and fourth rows are with crease side down. Adjacent three-dimensional plots illustrate reflectance intensity variations of various parts of wheat kernels.

8.2.6 DETECTION OF FECAL CONTAMINATION ON APPLES

Multivariate analysis of hyperspectral imagery, such as by principal component analysis (PCA), has been used as a systematic method for selecting key spectral bands to implement in rapid industrial multispectral imaging inspection applications (Kim et al., 2002a, 2002b; Mehl et al., 2002; Windham, 2003; Lee et al., 2008;

Qin et al., 2008). In addition, multispectral image fusion/ratio analysis methods can further enhance detection sensitivity (Kim et al., 2001, 2005). Below is a case study for the development of a method to detect fecal contamination on apples (Kim et al., 2004), for which a two-band fluorescence ratio was used to detect a range of dilutions of bovine feces on apples. The study was conducted using the laboratory-based hyperspectral imaging system with approximately 1 × 1 mm spatial resolution. The rationale for this work is that, in brief, contamination by animal fecal matter is recognized as a major culprit for occurrence of pathogenic *E. coli* O157:H7 in food products (Armstrong et al., 1996; Cody et al., 1999; Burnett et al., 2000).

Hyperspectral fluorescence images of apples artificially contaminated with spots of fecal matter were acquired, spanning the entire spectral region from 420 to 750 nm, and subjected to PCA. A total of ninety-six "tree run" (no wax or fungicide applied) Golden Delicious apples, exhibiting a range of the natural skin color variations typical of this cultivar and no visually identifiable physical defects, were randomly selected for this study. Fresh cow feces collected from USDA dairy farm facilities in Beltsville, Maryland, were mixed with distilled H_2O to prepare three dilutions: 1:2, 1:20, and 1:200 by weight. Using a pipette, three contamination spots—one 30 μl spot for each fecal dilution—were applied to one face of each apple, starting with the 1:2 dilution in one quadrant, followed by the 1:20 and 1:200 dilutions, respectively, on the adjacent quadrants, proceeding in the clockwise direction. Note that the 1:200 fecal dilution spots were not visible on the apple surfaces, and the 1:20 dilution resulted in relatively transparent contamination spots that were not easily discernable visually.

PCA was performed on the set of hyperspectral image data for all ninety-six apples. Subsequently, individual principal component (PC) score images were visually evaluated to determine those PC images that showed (1) all three feces contamination spots on individual apples, and (2) contrasting PC score values (gray levels) between the feces contamination spots and apple surfaces for potential discrimination of the two classes. Each PC image is a linear sum of images at individual wavelengths weighted by corresponding spectral weighing coefficients (eigenvectors). The wavelengths where weighing coefficients at local maximum and minimum indicate the dominant spectral regions. Additional transformations tested to discriminate fecal contamination over the range of normal apple surfaces were various two-band ratio images using the dominant wavelengths found by PCA. Individual ratio images were subjected to a nonparametric and unsupervised histogram threshold method by Otsu (1979) to test automated detection of fecal contamination on apples.

Figure 8.7a illustrates the first, second, fifth, and sixth PC score images obtained by PCA of the hyperspectral fluorescence images data for the fecal-contaminated Golden Delicious apples (only nine apples are shown, for illustrative purposes). The PC-1 image represents a composite of transformed information from all the spectral regions, which accounts for the largest variance. Subsequent PC images are ordered in terms of the sample variance size. Individual PC images exhibit unique features based on spectral variation of the image data. Prominent features observed in these images are the apple color variation and fecal contamination on apples. The PC-2, PC-5, and PC-6 images show all three (1:2, 1:20, and 1:200 dilution) fecal

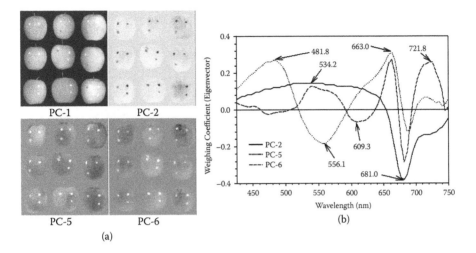

FIGURE 8.7 (a) First and the most significant principal component (score) images for classification of feces-contaminated spots. The results were obtained by subjecting the hyperspectral fluorescence images to 420 to 750 nm of feces-contaminated Golden Delicious apples. (b) Spectral weighing coefficients (eigenvectors) for the PC-2, PC-5, and PC-6 obtained from the PCA of hyperspectral fluorescence images of Golden Delicious apples treated with diluted cow feces.

contamination spots on the individual apples. In the PC-2 image, the fecal spots appear darker than the surrounding apple surfaces, while the opposite (lighter fecal spots) is observed in the PC-5 and PC-6 images. This observation suggests that the contrasts (in PC score values) between the fecal contamination spots and apple surfaces provide potential for classification or discrimination of the two classes, regardless of the concentrations of feces and the color variations that may occur across one apple or between different apples.

Spectral weighing coefficients (eigenvectors) for the PC-2, PC-5, and PC-6 images and the dominant wavelengths depicted as the local minimum and maximum are shown in Figure 8.7b. The wavelengths were identified to be 534 and 681 nm for PC-2; 481, 556, and 663 nm for PC-5; and 534, 609, 663, 681, and 722 nm for PC-6. The dominant wavelengths by the selected individual PCs shared a few common wavelengths, as well as slightly blue or red-shifted wavelengths from the naturally occurring emission peak wavelengths. Blue-green emissions from the samples with UV excitation are convoluted emissions from a mixture of many compounds where individual compounds exhibit broad emission characteristics with varying wavelength peak locations in nature. In general, the selected dominant wavelengths nearly coincided with the fluorescence emission maxima observed from the sample materials.

The PC images were constructed based on a linear combination (sum) of the original data weighted by corresponding weighing coefficients. The PC-2, PC-5, and PC-6 images can be approximated by linearly combining the original images at the selected dominant wavelengths (Kim et al., 2002a). For the PC-2 image, with the use of two dominant wavelength regions, a broad band in the blue-green and a band in

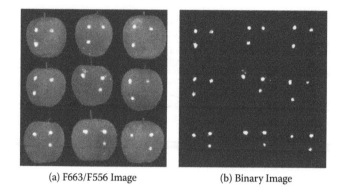

(a) F663/F556 Image   (b) Binary Image

FIGURE 8.8 (a) Fluorescence ratio image (F663/F556) of Golden Delicious apples contaminated with diluted cow feces. (b) Resultant binary image showing the detection of the feces contamination spots on apples obtained by applying automated histogram threshold values of 1.627 to the F663/F556 ratio image.

the red region centered at 681 nm, the darker feces contamination spots and lighter apple surfaces were the result of subtracting the red region from the blue-green region weighted by weighing coefficients. Another simple multispectral fusion method that may amplify the feature differences is the ratio of the two spectral band regions.

Two-band ratio images using all pair combinations of the PCA-selected wavelengths were created. Figure 8.8a illustrates the F663/F556 ratio image that yielded a minimal number of false positive pixels for feces contamination spots. To limit image analysis to only the apple surface areas, a mask image was created using a higher-intensity emission band for removing the image background areas surrounding the apples. Figure 8.8b shows the binary image resulting from the automated detection method that applied a threshold value of 1.627 by the Otsu method (Otsu, 1979) to the masked F663/F556 ratio image of the apples, clearly identifying the fecal contamination spots. Note that all the ratio images were subjected to the unbiased automated threshold method to determine the best ratio combination. In general, using a ratio of F663 to either of the selected blue-green or F609 band resulted in similar automated detection results with minimal false positives. All fecal spots on the apples were detected. On the center apple in Figure 8.8b, some false positive pixels near the 1:20 spot (upper right spot) were easily removed by using a spatial filter. The cluster of pixels near the 1:2 spot on the same apple (upper left) are not false positives, but resulted from smearing of the 1:2 fecal dilution that occurred when the spot was first applied. For detection of fecal contamination on apples, fluorescence methods have been found to be more sensitive in detecting diluted fecal matter or thin smear spots than reflectance methods (Kim et al., 2002b, 2007; Liu et al., 2006).

8.3   LINE-SCAN ONLINE POULTRY INSPECTION

Poultry broiler production has increased dramatically to meet rising consumer demand. U.S. poultry slaughter plants process over 8.8 billion broilers annually (USDA, 2009). Since 1957, USDA Food Safety and Inspection Service (FSIS) inspectors have conducted on-site organoleptic inspection of all chickens processed at U.S. poultry plants for indications of diseases or defects. To help maintain product safety and reduce the risk of food safety hazards for poultry, egg, and meat products, the USDA FSIS implemented the Hazard Analysis and Critical Control Point (HACCP) program in processing plants throughout the country (USDA, 1996). The HACCP-based Inspection Models Project (HIMP) has been tested in a small number of volunteer plants (USDA, 1997). HIMP performance standards require that chickens with infectious septicemia/toxemia conditions (septox) be removed from the processing lines. Commercial evisceration lines in the United States currently may operate at speeds up to 140 birds per minute (bpm); however, such high-speed processing lines require up to four inspection stations, each with an FSIS inspector to conduct bird-by-bird inspection at the individual limit of 35 bpm. These limiting factors on production throughput, combined with increasing chicken consumption and demand, place additional pressure on both chicken production and the food safety inspection system.

Since 1998, ARS researchers have been developing automated poultry inspection systems for high-speed online operations in the slaughterhouse plant environment (Chao et al., 2003, 2004, 2008, 2010; Yang et al., 2006, 2009). Visible/near-infrared (Vis/NIR) spectroscopy techniques were first developed that could effectively differentiate wholesome and unwholesome chicken carcasses and viscera (Chao et al., 2003). A preliminary laboratory imaging study by Yang et al. (2005) achieved classification accuracies of 95.7% for wholesome and 97.7% for unwholesome (systemically diseased) chicken carcasses using multispectral images at the 540, 610, and 700 nm wavebands. In this section, development of hyperspectral/multispectral line-scan imaging for high-speed online operation in a commercial processing environment in differentiating wholesome and unwholesome chickens is presented.

8.3.1 LINE-SCAN POULTRY INSPECTION SYSTEM

The fundamental design of the line-scan poultry inspection system, shown in Figure 8.9, was similar to that of the laboratory-based hyperspectral imaging system described earlier. The IFOV was illuminated by two pairs of high-power, broad-spectrum white light-emitting diode (LED) line lights (LL6212, Advanced Illumination, Inc., Rochester, Vermont). The effective spectral range for poultry inspection was in the visible region, and thus the white LED line lights were an adaptation specifically for the poultry inspection system. A thermoelectrically cooled EMCCD camera (PhotonMax 512b, Princeton Instruments, Roper Scientific, Inc., Trenton, New Jersey), containing a 512 × 512 pixel detector array cooled to an operating temperature of –70°C, was equipped with a 10 MHz 16-bit digitizer for high-speed low-light image acquisition.

As shown in Figure 8.9b, the distance from the camera lens to the shackle line was 914 mm. The angles of illumination before and after the IFOV were adjusted to maximize the reflectance intensity. Two line lights were mounted together on a single joint to the right of the camera, such that their angles of illumination were not independently adjustable. The other two line lights were similarly mounted to the left of the sensing module. The right and left pairs were separated by 115 mm, each pair

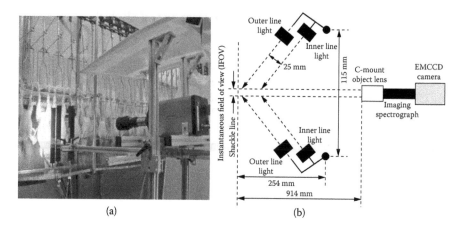

(a)   (b)

FIGURE 8.9 (a) A photograph of the hyperspectral/multispectral line-scan imaging system on a commercial chicken processing line. (b) Schematic of the poultry inspection system.

at a distance of 254 mm perpendicular to the shackle line (measured from shackle line to mounting joint). The inner line lights were positioned slightly behind the outer lights rather than with their forward faces flush, to allow for some forward movement of the target surface (up to 25 mm) so that the convex surface of a bird, even if pushed slightly more forward than normal, would still be adequately and uniformly illuminated. The outer lights were positioned a half head lower than the inner lights to avoid the interference that would have resulted if the adjacent LED lights had been mounted at equal height. The height of the IFOV for each scanned line was 178 mm. This linear distance was represented by 512 spatial points (0.35 mm per pixel).

This line-scan image size was reduced by binning the spectral dimension by four, reducing the total line-scan image size to 512 pixels × 128 spectral channels. However, because the white LEDs emitted light predominantly in the visible region, spectral channels outside the range of the LEDs—the first 19 and the last 54 spectral channels out of the 128 total channels—were discarded. Therefore, this poultry inspection imaging work was performed using only the remaining fifty-five spectral channels, for a line-scan image size of 512 × 55 pixels. Spectral calibration was performed using the method described earlier in the chapter—the fifty-five channels spanned from approximately 389 nm to 744 nm, with an average spectral bandwidth of 6.6 nm.

One of the significant advantages of the ARS line-scan imaging technique is the versatility to operate in either hyperspectral or multispectral mode, with the ability to select any of the available spectral channels through the camera control software. Initially, the system was operated in hyperspectral mode at the processing line, collecting hyperspectral images with fifty-five spectral channels (Figure 8.10) for later analysis to select wavebands and develop algorithms to differentiate wholesome and systemically diseased chickens. Implementation of the selected wavebands would thus only involve adjustments through software control settings. For online real-time inspection of poultry carcasses, only measurements for the spectral pixels

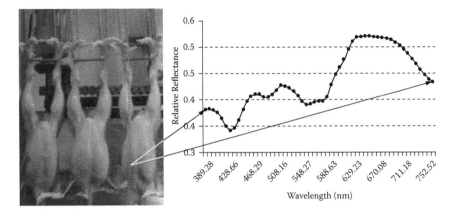

FIGURE 8.10 A photograph of poultry carcasses on a processing line. The plot shows a representative hyperspectral line-scan data of a pixel that was acquired by the full-spectrum (hyperspectral) acquisition mode of the system.

corresponding to the selected wavebands were digitized for the computer and used by the system in multispectral imaging mode. Both hyperspectral and multispectral line-scan images were acquired using a camera exposure time of 0.1 ms.

8.3.2 IN-PLANT ONLINE HYPERSPECTRAL IMAGING AND AUTOMATED MULTISPECTRAL INSPECTION

In-plant hyperspectral line-scan images of chickens were acquired on a 140 bpm commercial processing line. A total of 5,549 wholesome and 93 unwholesome chickens were imaged, their conditions identified by an FSIS veterinary medical officer who observed the birds as they approached the illuminated IFOV of the imaging system. Because a systemic unwholesome condition affects the entire body of a bird, it was not necessary to analyze the entire image of each chicken. The hyperspectral image data for the chicken carcasses were analyzed for region of interest (ROI) optimization and for selection of key wavelengths and development of classification algorithms for differentiating wholesome and unwholesome birds. Hyperspectral image analysis was performed using MATLAB software (MathWorks, Natick, Massachusetts).

Subsequently, the system was implemented to conduct automated multispectral online inspection for over 100,000 birds on a commercial processing line during two eight-hour shifts. To verify system performance, an FSIS veterinary medical officer identified wholesome and unwholesome conditions of birds immediately before they entered the IFOV of the imaging system, during several thirty- to forty-minute periods, for direct comparison with the classification results produced from the automated multispectral imaging inspection algorithms.

8.3.3 DEVELOPMENT OF POULTRY WHOLESOMENESS CLASSIFICATION METHODS

Image background removal was first performed by image masking using a 0.1 relative reflectance threshold value for the 620 nm waveband. For any pixel in a hyperspectral line scan, if its reflectance at 620 nm was below the threshold value, then the pixel was identified as background and its value at all wavebands was reassigned to zero. Within a bird image, the potential ROI area spanned from an upper border across the breast of the bird to a lower border at the lowest nonbackground spatial pixel in each line scan, or to the last (512th) spatial pixel if there were no background pixels present at the lower edge of the image.

For each potential ROI, the average relative reflectance spectrum was calculated across all ROI pixels for all wholesome chicken images, and the average relative reflectance spectrum was calculated across all ROI pixels for all unwholesome chicken images. The difference spectrum between the wholesome and unwholesome average spectra was calculated. This calculation was performed for all potential ROIs evaluated, which varied in size and were defined by the number of ROI pixels and their vertical coordinate locations within each line scan.

The optimized ROI was identified as being that which provided the greatest spectral difference between averaged wholesome pixels and averaged unwholesome pixels across all fifty-five wavebands. Using the optimized ROI, a key waveband was identified as being the waveband corresponding to the greatest spectral difference between averaged wholesome chicken pixels and averaged unwholesome chicken pixels, for differentiating wholesome and unwholesome chicken carcasses by relative reflectance intensity. In addition, wavebands at which local maxima and minima occurred were used to create two-waveband ratios for differentiating wholesome and unwholesome birds. The two-waveband ratio showing the greatest difference between average wholesome and average unwholesome chicken pixels was selected. Thus, multispectral online inspection used the selected key wavelength and the two-waveband ratio to differentiate between wholesome and unwholesome chicken carcasses operating in autonomous mode. For a more detailed description of methods for the determination of the ROI and multispectral fuzzy-logic–based detection, readers are referred to Yang et al. (2006) and Chao et al. (2010).

8.3.4   In-Plant Online Hyperspectral Imaging

A representative image of a chicken at 620 nm is shown in Figure 8.11a, with the starting point (SP) and ending point (EP). Within each line scan, possible ROI pixels begin at the SP-EP line and extend to the farthest nonbackground pixel below the SP-EP line, which in some cases may coincide with the lowermost pixel of the line-scan image. Parameters m and n indicate the location of the upper and lower ROI borders for ROIs under consideration in this study, as percentages of the pixel length between the SP-EP line and the farthest nonbackground pixel within each line-scan image. To optimize the ROI size and location, combinations of m and n, with values of m between 10 and 40% and values of n between 60 and 90%, were evaluated.

For each possible ROI, the average spectrum was calculated across all ROI pixels from the 5,549 wholesome chicken carcasses, and the average spectrum was calculated across all ROI pixels from the 93 unwholesome chicken carcasses. The difference between the average wholesome and average unwholesome value at each of the fifty-five bands was calculated (Figure 8.12). Because the 40 to 60% ROI (Figure 8.11b) showed the range with the greatest difference values between the

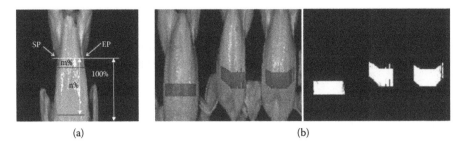

(a)   (b)

FIGURE 8.11 (a) Reference points defining the possible region of interest (ROI) areas for analysis: the starting point (SP), the ending point (EP). (b) Representative chicken image with the optimized ROIs and masked ROIs where intensity values in conjunction with fuzzy-logic algorithms were used to determine wholesomeness of the birds.

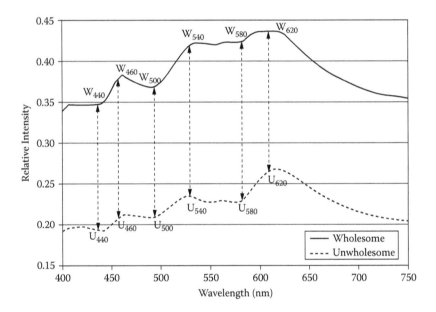

FIGURE 8.12 The averaged wholesome and unwholesome chicken spectra obtained during the hyperspectral analysis, highlighting possible key wavebands for two-waveband ratio differentiation of wholesome and unwholesome chickens.

average wholesome and average unwholesome spectra, this ROI was used in the waveband selection process. For this optimized ROI, consisting of the 40 to 60% region of each line-scan image, the 580 nm band showed the greatest difference between the average wholesome and average unwholesome spectra, and thus was selected as the key waveband to be used for intensity-based differentiation of wholesome and unwholesome chicken carcasses.

In addition, the spectra showed noticeable differences between the wholesome and unwholesome chickens in the three areas corresponding to 440 to 460 nm, 500 to 540 nm, and 580 to 620 nm peaks and valleys, and two-band ratios were investigated using these particular pairings (Figure 8.12). For these three two-waveband ratios (440 and 460 nm, 500 and 540 nm, and 580 and 620 nm), the average wholesome and average unwholesome ratio values were calculated. The differences in average ratio value between wholesome and unwholesome chickens were then calculated:

$$W_{440}/W_{460} - U_{440}/U_{460} = 0.003461$$

$$W_{500}/W_{540} - U_{500}/U_{540} = 0.038602$$

$$W_{580}/W_{620} - U_{580}/U_{620} = 0.115535$$

The last ratio, using the 580 and 620 nm wavebands, showed the greatest difference between the average wholesome and average unwholesome chicken spectra, and was thus selected for use in differentiation by the two-waveband ratio.

8.3.5   IN-PLANT ONLINE AUTOMATED MULTISPECTRAL INSPECTION

The optimized ROI and key wavebands (580 and 620 nm; Figure 8.13) determined from the hyperspectral data analysis were used for online multispectral inspection of over 100,000 chickens on a 140 bpm processing line during two eight-hour shifts at a commercial poultry plant. The inspection program specifically determined the 40 to 60% ROI (e.g., Figure 8.11b) for each line-scan image as it was acquired, which was clearly affected by the size and position of the bird, and thus could vary significantly

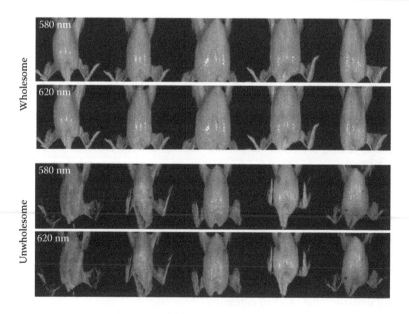

FIGURE 8.13   Representative multispectral images of wholesome and unwholesome chicken carcasses at two key spectral wavebands, acquired during in-plant testing of the ARS hyperspectral/multispectral poultry inspection system.

for different birds. For a bird whose body extended past the lower edge of the image, such as the first bird in Figure 8.11b, the total ROI was a rectangular area. In contrast, irregularly shaped ROIs resulted for birds positioned such that background pixels were present at the lower edge of the image.

The values for the 580 nm key waveband and 580/620 nm two-waveband ratio from the ROI data were used to build the fuzzy-logic membership functions for online multispectral classification. These functions were used to classify each ROI pixel within a line-scan image as either wholesome or unwholesome, by using each pixel's 580 nm intensity value and its ratio value, using 580 and 620 nm as inputs to obtain a decision output value D_0 between 0 and 1 for each criterion. The average of all the D_0 values calculated for all ROI pixels of a bird was used to determine a wholesome or unwholesome assignment by comparison with a threshold value of 0.6. Two D_0 values were calculated for each of these ROI pixels, one for the key waveband and one for the two-waveband ratio. Online multispectral inspection averaged the D_0 values for all ROI pixels for each bird, in order to classify the bird by comparison to the threshold value.

System verification was also performed for several thirty- to forty-minute periods within the inspection shifts, by an FSIS veterinary medical officer. This consisted of bird-by-bird observation of chicken carcasses on the processing line immediately before they entered the IFOV of the imaging system. The imaging system output was observed for comparison with the veterinary medical officer's identifications. Over four verification periods during inspection shift 1, the imaging system correctly identified 16,056 of 16,174 wholesome birds (99.27% correct) and 41 of 43 unwholesome birds (95.35% correct). Over six verification periods during inspection shift 2, the imaging system correctly identified 27,580 of 27,626 wholesome birds (99.83% correct) and 34 of 35 unwholesome birds (97.14% correct).

For multispectral inspection conducted on the 140 bpm processing line in this study, the imaging system acquired about thirty to forty line-scan images between the SP and EP for each chicken inspected. Previous testing of the imaging system (Chao et al., 2007) demonstrated similar performance in identifying wholesome and unwholesome birds on a 70 bpm processing line, with the multispectral classification based on about seventy to eighty line-scan images for each chicken. The imaging inspection system can be used as a tool to help poultry processors improve efficiency and lower the risks of cross-contamination on the processing line, by removing most unwholesome chicken carcasses earlier on the processing line, such that they are not presented for inspection and further unnecessary processing. The line-scan online inspection platform is also being adapted for online detection of fecal contamination on poultry carcasses using two other spectral bands at 520 and 560 nm with ratio algorithms (Park et al., 2007; Yoon et al., 2007).

8.4 LINE-SCAN MULTITASKING IMAGING FOR APPLE INSPECTION

For fruit and vegetable producers and processors, fecal contamination is only one detection problem among many safety and quality concerns affecting their products. Fruits with quality defects such as cuts, lesions, and rots are also safety concerns, since such defects can provide favorable ecological niches for bacterial growth (Mercier and Wilson, 1994; Burnett et al., 2000)—there have been numerous investigations to develop defect detection techniques (Leemans and Destain, 1998; Wen and Tao, 2000; Bennedsen and Peterson, 2004; Mehl et al., 2004; Kleynen et al., 2005; Throop et al., 2005; Lee et al., 2008). Rapid line-scan imaging is ideal for in-line processing inspection because a series of narrow spatial images can be captured as objects cross a linear field of view. Full-spectrum hyperspectral data can be selectively processed using multispectral algorithms for specific tasks in contaminant and defect detection and disease identification. However, to achieve such safety and quality assessments, both reflectance and fluorescence measurements at multiple wavelengths may be needed (Kim et al., 2007)—and to be of practical benefit, an effective product screening system must be able to address multiple inspection tasks. In this regard, a single imaging system needs to have the combination of flexibility in capturing fluorescence and reflectance, and selectivity and simultaneous acquisition capability for multispectral bands. The line-scan imaging technologies described in previous sections were further enhanced and specifically developed to simultaneously capture NIR reflectance and fluorescence, to address a combination of safety and quality inspection tasks (multitask inspection) for apples. In this section, an innovative approach to simultaneously capture multispectral fluorescence and reflectance from rapidly moving apples on a commercial apple-sorting machine, allowing multiple food processing and inspection tasks for safety and quality inspections, is presented.

8.4.1 ONLINE MULTITASKING LINE-SCAN IMAGING SYSTEM

Figure 8.14 illustrates critical components of the online multitasking line-scan imaging system. The system was designed and built to operate with a commercial

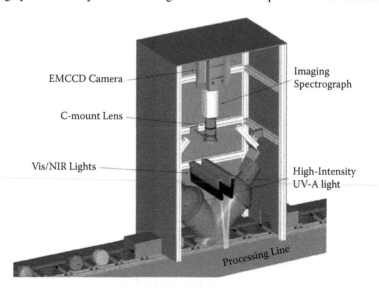

FIGURE 8.14 Schematics of the online multitasking line-scan imaging system on a commercial apple-sorting machine.

apple-sorting machine (FMC Corp., Philadelphia, Pennsylvania). The sorting machine loads apples onto a conveyor tray system to transport them for imaging. The speed of the conveyor is adjustable up to approximately five apples per second. Each rectangular tray, painted with nonfluorescent, flat-black paint, has a central circular depression to hold an apple in a steady position while in transit. For this system, an EMCCD imaging device (iXon, Andor Technology, Inc., South Windsor, Connecticut) with $1,004 \times 1,002$ pixels coupled with a 35 MHz (pixel readout rate), 14-bit digitizer PC board was used.

An imaging spectrograph (ImSpector V10E, Spectral Imaging Ltd., Oulu, Finland) and a C-mount lens (Rainbow CCTV S6X11, International Space Optics, S.A., Irvine, California) are attached to the EMCCD. The IFOV is limited to a thin line by the spectrograph aperture slit (50 µm). Through the slit, light from the scanned IFOV line is dispersed by a prism-grating prism device and projected onto the EMCCD. Therefore, for each line scan, a two-dimensional (spatial and spectral) image is created with the spatial dimension along the horizontal axis and the spectral dimension along the vertical axis of the EMCCD.

Two independent CW light sources are used for fluorescence and reflectance imaging. A micro-discharge high-intensity UV-A (320 to 400 nm) lamp with a diffuse filter (ML-3500, Spectronics Corp., Westbury, New York) is used for the fluorescence excitation source. For reflectance, the illumination is provided by a DC-regulated 50-W quartz halogen lamp coupled with a pair of bifurcated fiber optic line lights (Fiber-Lite, Dolan-Jenner Industries, Inc., Lawrence, Massachusetts). These line lights are mounted to illuminate the line of IFOV at 5° forward and backward angles. The imaging system and its components are enclosed in a light-tight casing to mitigate the effects of ambient light.

One of the technical challenges was to design a system that can simultaneously capture UV-A-induced fluorescence in the visible region and reflectance in the NIR region, at multiple wavelengths from a fast-moving object. A combined utilization of optical filters allowed the line-scan system to detect only fluorescence emissions in the visible and reflectance in the NIR portion of the spectrum. A NIR long-pass filter (>720 nm) was inserted in between the lamp aperture and the fiber optic bundle input to eliminate visible light illumination (thus allowing the sensors to detect fluorescence emissions from the samples), while allowing reflectance measurements only in the NIR region of the spectrum (720 to 1,000 nm). Due to the use of a grating spectrograph, a second-order effect of the fluorescence excitation was present in the longer-wavelength region starting at approximately 700 nm; to remove the second-order effects, a long-pass filter (>410 nm) was placed in front of the C-mount lens.

8.4.2 DETECTION OF FECAL CONTAMINATION AND DEFECTS

As illustrated in the earlier case study, a two-waveband fluorescence emission ratio is an effective method for detecting fecal contamination spots on apples. To further validate the spectral bands and methods for detection of fecal contamination and defects, hyperspectral images of samples at 10 nm spectral intervals from 400 to 700 nm for fluorescence and 450 to 1,000 nm for reflectance were acquired independently at a commercial processing speed. This online hyperspectral image study was conducted using another fast line-scan hyperspectral imaging system integrated with the above commercial apple-sorting machine operating at a processing line speed of over three apples per second (see Kim et al., 2007). As stated earlier, we have developed a series of line-scan–based hyperspectral imaging platforms with various lighting regimes, and readers should not confuse the main multitasking inspection system presented in this section with the system used for online hyperspectral imaging of apples.

The independent hyperspectral fluorescence study showed that a ratio of 660 nm to 530 nm (F660/F530) enhanced the response differences between the fecal spots and apples. Furthermore, the ratio method reduced the spatial heterogeneity of normal apple surfaces (Figure 8.15a). The second image in Figure 8.15a shows the binary image highlighting the regions of the fecal spots, obtained by subjecting the ratio image to a simple thresholding method using a global threshold value of 0.99. Based on the samples we investigated, a 100% detection rate (118 feces-treated apples) was achieved, with no false positives (0 out of 139 normal apples). Fluorescence ratio techniques have consistently demonstrated very high detection rates for animal fecal contamination on apples, even with slight changes in the center wavelengths of the bands used between the studies performed using a laboratory system and an online system.

For defect detection, a two-band NIR ratio using the 750 and 800 nm wavebands provided a significant difference in ratio values between normal apple surfaces and defect portions (Figure 8.15b). Note that because of the spectral resemblance of apple stem and calyx regions to defects such as rots, cuts, legions, and fungal growth, false positives are ever present. The presence of stem/calyx typically has been a problematic source for false positives in the detection of defects using machine vision techniques (Wen and Tao, 2000; Kim et al., 2002a; Mehl et al., 2004; Lefcourt et al., 2009). However, the two NIR ratio mean values for normal apples were significantly lower than for apples with defects, regardless of the presence of stem/calyx false positives. In addition to the presence of the stem/calyx, the presence of defects increased the mean ratio values and the spatial heterogeneity of the NIR ratio responses. Thus, using the means and coefficients of variation of the ratio values as simple classification model inputs, we achieved a defect detection rate of approximately 98%. For a detailed description of the above fluorescence and reflectance methods, see Kim et al. (2007).

8.4.3 MULTISPECTRAL BANDS FOR MULTITASK INSPECTION

Figure 8.15c shows key multispectral wavelengths needed for simultaneous detection of fecal contamination and defects on apples using fluorescence and reflectance, respectively. Relative intensity at individual wavelengths depends on the responsivity of the system and spectral responses of the samples. It is thus important to be able to adjust the system throughput to equilibrate the intensities at multispectral regions regardless of fluorescence and reflectance. With the aid of the software control for EMCCD, the line-scan imaging system can be configured to acquire hyperspectral (entire spectral dimension) images or several regions of interest in the vertical, spectral direction (multispectral/random track mode). The use of the system in the

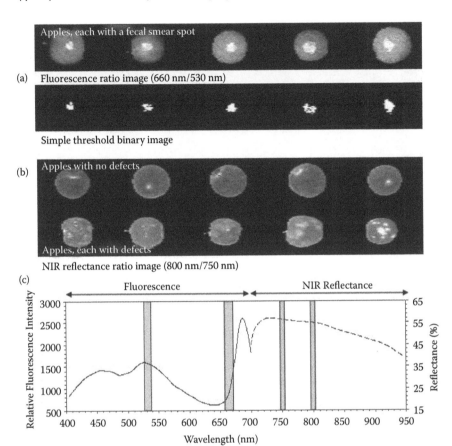

FIGURE 8.15 (a) Fluorescence ratio image at 660 nm over 530 nm for detection of fecal smear spots and simple threshold results. (b) NIR reflectance ratio image at 800 nm over 750 nm for detection of defects. With the presence of defects, the variations of the ratio values for the apple surfaces are greater than those for the apples with no defects. (c) Multispectral bands for simultaneous inspection for fecal contamination and defects superimposed on typical spectral responses, fluorescence in the visible and reflectance in the NIR, of a representative Golden Delicious apple.

multispectral mode reduces the data volume per line scan and can result in a higher number of lines scanned per second. This approach also allows selective changes in spectral bands (verticval pixel locations) and bandwidths (vertical pixel binning) without any hardware modifications in that each vertical ROI can be adjusted to match sample-dependent intensities regardless of fluorescence or reflectance. For instance, relative fluorescence intensity (fluorescence yield) shown in the fluorescence spectrum of a Golden Delicious apple at 660 nm (Figure 8.15c) is the lowest compared to the other key wavelength at 530 nm. A larger vertical ROI (pixel binning) centered at 660 nm increases the throughput, and thus allows matching of the relative intensity to that at other wavelengths. The relative intensity for fluorescence is generally lower than reflectance, although it also depends on the illumination

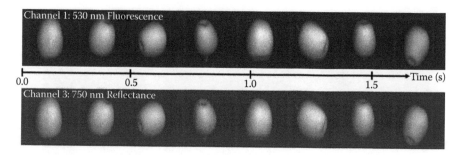

FIGURE 8.16   Fluorescence (530 nm) and reflectance (750 nm) images of Golden Delicious apples simultaneously acquired using the online multitasking imaging system. Timescale illustrates the processing line speed per second.

power. Relatively narrow spectral bands (smaller pixel binning) for reflectance compared to fluorescence can accommodate matching of the throughput from the two sensing methods. An additional benefit includes the better use of the A to D dynamic range for improvement of the signal-to-noise ratio of the data.

Figure 8.16 shows resultant multispectral fluorescence and reflectance images of normal Golden Delicious apples acquired using the online multitasking line-scan imaging system in multispectral mode. Although fluorescence at 530 and 660 nm and reflectance at 750 and 800 nm were simultaneously acquired, only the fluorescence image at 530 nm and reflectance image at 750 nm are shown for brevity. The line-scan images were acquired with 200 s exposure time at 50% electron multiplying gain set. The timescale in the figure indicates the sorting machine processing speed of approximately over four apples per second. There were a total of 333 lines per second and approximately 40 horizontal lines (pixels) per apple, which resulted in sufficient spatial resolution for image-based online inspection. The presented multi-task inspection approach in online applications may provide an economically viable means for a number of food processing industries to be able to adapt to operate and meet dynamic and specific inspection and sorting needs. Automated online inspection software is under development to test the real-time, multitask inspection of apples for decal contamination and defect detection. The most significant and important outcome is the technical demonstration of a single online inspection system that is capable of simultaneously acquiring multispectral images of both fluorescence and reflectance measurements at processing line speeds. Furthermore, the line-scan inspection system can potentially provide the capability for current sorting mechanisms, such as by size and color, in addition to sorting for fecal contamination and defects of food products.

8.5   SUMMARY

In this chapter, recently developed ARS line-scan hyperspectral-based sensing technologies to address agro-food safety concerns are presented, including two case studies using the laboratory-based hyperspectral imaging platforms. We envision that the line-scan spectral imaging technologies can deliver concurrent safety and quality inspection for a variety of agricultural products on high-throughput processing lines.

An online line-scan imaging system capable of both hyperspectral and multispectral reflectance imaging was developed to inspect freshly slaughtered chickens on a high-speed processing line for wholesomeness. During continuous in-plant operation in automated multispectral imaging inspection mode, the system inspected over 100,000 chickens on a 140 bpm processing line and accurately identified over 99% of wholesome chickens and over 96% of unwholesome chickens. These results demonstrated that with appropriate hyperspectral analysis methods and multispectral inspection algorithms for online operation, a line-scan spectral imaging system utilizing an EMCCD camera can perform food safety inspection tasks accurately and consistently while meeting the high-speed production requirements (e.g., at least 140 bpm) of commercial chicken processing. Efforts to commercialize the system are under way.

For fruit and vegetable producers and processors, fecal contamination detection is only one detection problem among many safety and quality concerns affecting their products. To be of practical benefit, an effective product screening system must be able to address multiple inspection tasks. A line-scan spectral imaging system was specifically developed to simultaneously capture NIR reflectance and fluorescence, and was applied to address a combination of safety and quality inspection tasks (multitasking) for apples on a commercial sorter operating at three to four apples per second. Reflectance images at 750 and 800 nm could identify defects such as cuts and rotted spots; fluorescence images at 530 nm and 660 nm detected fecal contamination and differentiated it from surface defects. Simultaneous fluorescence images and NIR reflectance images were enabled by using an NIR long-wavelength pass filter (750 nm) on the Vis/NIR lights to eliminate reflectance in the visible spectrum and a long-pass filter (450 nm) over the C-mount lens to remove second-order effects of the UV light. Our investigation demonstrated over 98% accuracy in online inspection of apples at speeds of three to four apples per second, for surface fecal contamination by fluorescence and for defect and disease detection by NIR reflectance.

Further research is being conducted to develop inspection systems suitable for commercial processing of other fresh produce, such as leafy greens. Effective detection of contamination by fecal matter, for example, is important due to its association with common bacterial causes of foodborne illness. Adaptable to a broad range of problems and commodities, the line-scan hyperspectral imaging platform will be critically useful for both research and commercial food safety and quality inspection applications.

REFERENCES

Aleixos, N., Blasco, J., Navarrón, F., and Moltó, E. 2002. Multispectral inspection of citrus in real-time using machine vision and digital signal processors. *Comput. Elec. Agric.* 33(2):121–37.

Armstrong, G. L., Hollingsworth, J., and Morris, J. G. 1996. Emerging foodborne pathogens: *Escherichia coli* O157:H7 as a model of entry of a new pathogen into the food supply of the developed world. *Epidemiol. Rev.* 18:29–51.

Bennedsen, B. S., and Peterson, D. L. 2004. Identification of apple stem and calyx using unsupervised feature extraction. *Trans. ASAE* 47(3):889–94.

Blasco, J., Aleixos, N., Gómez-Sanchis, J., and Moltó, E. 2007. Citrus sorting by identification of the most common defects using multispectral computer vision. *J. Food Eng.* 83(3):384–93.

Burnett, S. L., Chen, J., and Beuchat, L. R. 2000. Attachment of *Escherichia coli* O157:H7 to the surfaces and internal structures of apples as detected by confocal scanning laser microscopy. *Appl. Environ. Microbiol.* 66:4679–87.

Chao, K., Chen, Y. R., and Chan, D. E. 2003. Analysis of Vis/NIR spectral variations of wholesome, septicemia, and cadaver chicken samples. *Appl. Eng. Agric.* 19(4):453–58.

Chao, K., Chen, Y. R., and Chan, D. E. 2004. A spectroscopic system for high-speed inspection of poultry carcasses. *Appl. Eng. Agric.* 20(5):683–90.

Chao, K., Chen, Y. R., Early, H., and Park, B. 1999. Color image classification system for poultry viscera inspection. *Appl. Eng. Agric.* 15(4):363–69.

Chao, K., Yang, C. C., Chen, Y. R., Kim, M. S., and Chan, D. E. 2007. Hyperspectral/multispectral line-scan imaging system for automated poultry carcass inspection applications for food safety. *Poultry Sci.* 86:2450–60.

Chao, K., Yang, C. C., and Kim, M. S. 2010. Spectral line-scan imaging system for high-speed nondestructive wholesomeness inspection of broilers. *Trends Food Sci. Technol.* 21(3):129–37.

Chen, Z., and Tao, Y. 2001. Multi-resolution local multi-scale contrast enhancement of x-ray images for poultry meat inspection. *Appl. Optics* 40(8):1195–2000.

Cho, B., Chen, Y. R., and Kim, M.S. 2007. Multispectral detection of organic residues on poultry processing plant equipment based on hyperspectral reflectance imaging technique. *Comput. Elec. Agric.* 57(2):177–89.

Cody, S. H., Glynn, M. K., Farrar, J. A., Cairns, K. L., Griffin, P. M., Kobayashi, J., Fyfe, M., Hoffman, R., King, A. S., Lewis, J. H., Swaminathan, B., Bryant, R. G., and Vugia, D. J. 1999. An outbreak of *Escherichia coli* O157:H7 infection from unpasteurized commercial apple juice. *Ann. Intern. Med.* 130:202–9.

Gowen, A. A., O'Donnell, C. P., Cullen, P. J., Downey, G., and Frias, J. M. 2007. Hyperspectral imaging—An emerging process analytical tool for food quality and safety control. *Trends Food Sci. Technol.* 18(12):590–98.

Heitschmidt, G. W., Park, B., Lawrence, K. C., Windham, W. R., and Smith, D. P. 2007. Improved hyperspectral imaging system for fecal detection on poultry carcasses. *Trans. ASABE* 50(4):1427–32.

Jun, W., Kim, M. S., Cho, B., Millner, P., Chao, K., and Chan, D. 2010. Microbial biofilm detection on food contact surfaces by macro-scale fluorescence imaging. *J. Food Eng.* 99:314–22.

Jun, W., Kim, M. S., Lee, K., Millner, P., and Chao, K. 2009. Assessment of bacterial biofilm on stainless steel by hyperspectral fluorescence imaging. *Sensing Instrum. Food Quality Safety* 3:41–48.

Kim, M. S., Chen, Y. R., Cho, B., Lefcourt, A. M., Chao, K., and Yang, C. C. 2007. Hyperspectral reflectance and fluorescence line-scan imaging for online quality and safety inspection of apples. *Sensing Instrum. Food Quality Safety* 1(3):151–59.

Kim, M. S., Chen, Y. R., and Mehl, P. M. 2001. Hyperspectral reflectance and fluorescence imaging system for food quality and safety. *Trans. ASAE.* 44(3):721–29.

Kim, M. S., Lee, K., Chao, K., Lefcourt, A. M., Jun, W., and Chan, D. E. 2008. Multispectral line-scan imaging system for simultaneous fluorescence and reflectance measurements of apples: Multitask apple inspection system. *Sensing Instrum. Food Quality Safety* 2:123–29.

Kim, M. S., Lefcourt, A. M., Chen, Y. R., and Kang, S. 2004. Uses of hyperspectral and multispectral laser induced fluorescence imaging techniques for food safety inspection. *Key Eng. Mater.* 273:1055–63.

Kim, M. S., Lefcourt, A. M., Chen, Y. R., Kim, I., Chan, D. E., and Chao, K. 2002a. Multispectral detection of fecal contamination on apples based on hyperspectral imagery. Part I. Applications of visible and near-infrared reflectance imaging. *Trans. ASAE* 45(6):2027–37.

Kim, M. S., Lefcourt, A. M., Chen, Y. R., Kim, I., Chao, K., and Chan, D. 2002b. Multispectral detection of fecal contamination on apples based on hyperspectral imagery. Part II. Application of fluorescence imaging. *Trans. ASAE.* 45(6):2039–47.

Kim, M. S., Lefcourt, A. M., Chen, Y. R., and Yang, T. 2005. Automated detection of fecal contamination of apples based on multispectral fluorescence image fusion. *J. Food Eng.* 71(1):85–91.

Kleynen, O., Leemans, V., and Destain, M.-F. 2005. Development of a multi-spectral vision system for the detection of defects on apples. *J. Food Eng.* 69(10):41–49.

Lawrence, K. C., Park, B., Windham, W. R., and Mao, C. 2003b. Calibration of a pushbroom hyperspectral imaging system for agricultural inspection. *Trans. ASAE* 46:513–21.

Lawrence, K. C., Windham, W. R., Park, B., and Buhr, R. J. 2003a. A hyperspectral imaging system for identification of fecal and ingesta contamination on poultry carcasses. *J. Near-Infrared Spectrosc.* 11(4):269–81.

Lee, K., Kang, K., Delwiche, S., Kim, M. S., and Noh, S. 2008. Correlation analysis of hyperspectral imagery for multispectral wavelength selection for detection of defects on apples. *Sensing Instrum. Food Quality Safety* 2(2):90–96.

Leemans, V., and Destain, M.-F. 1998. Defect segmentation in 'golden delicious' apples by using colour machine vision. *Comput. Elec. Agric.* 20(2):117–30.

Lefcourt, A. M., Narayanan, P., Tasch, U., Kim, M. S., Reese, D., Rostamian, R., and Lo, M. Y. 2009. Orienting apples for imaging using their inertial properties and random apple loading. *Biosyst. Eng.* I04:64–71.

Liu, Y., Chen, Y. R., Kim, M. S., Chan, D. E., and Lefcourt, A. M. 2006. Development of simple algorithms for the detection of fecal contaminants on apples from visible/near infrared hyperspectral reflectance imaging. *J. Food Eng.* 81:412–18.

Lu, R. 2003. Detection of bruises on apples using near-infrared hyperspectral imaging. *Trans. ASAE* 46(2):523–30.

Mead, P. S., Slutsker, L., Dietz, V., McCaig, L. F., Bresee, J. S., Shapiro, C., Griffin, P. M., and Tauxe, R. V. 1999. Food-related illness and death in the United States. *Emerg. Infect. Dis.* 5:607–25.

Mehl, P. M., Chao, K., Kim, M. S., and Chen, Y. R. 2002. Detection of contamination on selected apple cultivars using reflectance hyperspectral and multispectral analysis. *Appl. Eng. Agric.* 18:219–26.

Mehl, P. M., Chen, Y. R., Kim, M. S., and Chan, D. E. 2004. Development of hyperspectral imaging technique for the detection of apple surface defects and contaminations. *J. Food Eng.* 61(1):67–81.

Mercier, J., and Wilson, C. L. 1994. Colonization of apple wounds by naturally occurring microflora and introduced *Candida olephila* and their effect on infection by *Botrytis cinerea* during storage. *Biol. Control* 4:138–44.

Miller, W. M., and Drouillard, G. P. 2001. Multiple feature analysis for machine vision grading of Florida citrus. *Appl. Eng. Agric.* 17(5):627–33.

Otsu, N. 1979. A threshold selection method from gray-level histogram. *IEEE Trans. Syst. Man Cybernet.* 9:62–66.

Park, B., Lawrence, K. C., Windham, W. R., and Buhr, R. J. 2002. Hyperspectral imaging for detecting fecal and ingesta contamination on poultry carcasses. *Trans. ASAE* 45:2017–26.

Park, B., Windham, W. R., Lawrence, K. C., and Smith, D. P. 2007. Contaminant classification of poultry hyperspectral imagery using a spectral angle mapper algorithm. *Biosyst. Eng.* 96(3):323–33.

Qin, J., Burks, T., Kim, M. S., Chao, K., and Ritenour, M. A. 2008. Citrus canker detection using hyperspectral reflectance imaging and PCA-based image classification method. *Sensing Instrum. Food Quality Safety* 2(3):168–77.

Schatzki, T. F., Haff, R. P., Young, R., Can, I., Le, L.-C., and Toyofuku, N. 1997. Defect detection in apples by means of x-ray imaging. *Trans. ASAE* 40(5):1407–15.

Shahin, M. A., Tollner, E. W., McClendon, R. W., and Arabnia, H. R. 2002. Apple classification based on surface bruises using image processing and neural networks. *Trans. ASAE* 45(5):1619–27.

Throop, J. A., Aneshansley, D. J., Anger, W. C., and Peterson, D. L. 2005. Quality evaluation of apples based on surface defects: Development of an automated inspection system. *Postharvest Biol. Technol.* 36(3):281–90.

USDA. 1996. Pathogen reduction: Hazard Analysis and Critical Control Point (HACCP) systems: Final rule. *Fed. Reg.* 61:38805–989.

USDA. 1997. HACCP-based Inspection Models Project (HIMP): Proposed rule. *Fed. Reg.* 62:31553–62.

USDA. 2009. *Poultry—Production and value.* 2008 summary. Washington, DC: USDA.

Vargas, A. M., Kim, M. S., Tao, Y., Lefcourt, A. M., Chen, Y. R., Luo, Y., Song, Y., and Buchanan, Y. 2005. Detection of fecal contamination on cantaloupes using hyperspectral fluorescence imagery. *J. Food Science.* 70(8):E471–76.

Wen, Z., and Tao, Y. 2000. Dual-camera NIR/MIR imaging for stem-end/calyx identification in apple defect sorting. *Trans. ASAE* 43(2):446–52.

Windham, W. R., Smith, D. P., Park, B., Lawrence, K. C., and Feldner, P. W. 2003. Algorithm development with visible/near-infrared spectra for detection of poultry feces and ingesta. *Trans. ASAE* 46:1733–38.

Yang, C. C., Chao, K., Chen, Y. R., and Early, H. L. 2005. Systemically diseased chicken identification using multispectral images and region of interest analysis. *Comput. Elec. Agric.* 49:255–71.

Yang, C. C., Chao, K., Chen, Y. R., Kim, M. S., and Chan, D. E. 2006. Development of fuzzy logic-based differentiation algorithm and fast line-scan imaging system for chicken inspection. *Biosyst. Eng.* 95:483–96.

Yang, C. C., Chao, K., and Kim, M. S. 2009. Machine vision system for online inspection of freshly slaughtered chickens. *Sensing Instrum. Food Quality Safety* 3(1):70–80.

Yoon, S. C., Lawrence, K. C., Park, B., and Windham, W. R. 2007. Optimization of fecal detection using hyperspectral imaging and kernel density estimation. *Trans. ASABE* 50(3):1063–71.

9 Electronic Nose for Detection of Food Flavor and Volatile Components

Bongsoo Noh

CONTENTS

9.1 INTRODUCTION

Detecting food flavor and volatile components is important for quality control of food. Quality control of product fragrance and flavor is based on the comparison of instrumental analysis and sensory evaluation. Until now, gas chromatography (GC), mass spectroscopy (MS), and sensory evaluation have been used for analysis

235

of flavor. The combination of gas chromatography and GC-MS is the most popular technique for the identification of volatile compounds in foods because the separation achieved by the GC system can be complemented by MS with a high sensitivity. The ability to identify the molecules on the basis of their fragmentation patterns at one part per billion sensitivity can often be determined by GC-MS.

The electronic nose is an instrument that analyzes gaseous mixtures discriminating different organoleptic properties of different foodstuffs. Those properties include qualities, origins, defects, and concentrations of undesirable components. Analytical methods allow chemical patterns to be discriminated and individual components to be identified and quantified. Pretreatments for isolation of specific constituents are not necessary. The human nose is still the most efficient for sensory evaluation; however, an organizing team of trained judges as a panel involves considerable expense. The electronic nose is a comparatively simpler, cheaper device and is faster to apply.

The use of the electronic nose has been proposed in recent years in the food industry as an analytical tool. The electronic nose can be used to detect various kinds of food, for example, meat, vegetables, fish, rice grains, coffee beans, wine, beer, or other beverages, to evaluate their freshness, measure the shelf life, and check chemical and microbial contaminations of foods. Manufacturers adopt this technology for product development and quality and safety controls.

In this chapter the electronic nose system and its use for quality control of food will be described. A more detailed description will be given regarding several sensor types, with their pattern recognition or fingerprints. And finally, specific application for detection of food flavor and volatile components will be explored.

9.2   SYSTEM OF THE ELECTRONIC NOSE

The electronic nose has been designed to mimic the human olfactory system. There are a number of olfactory sensors, such as a phospholipids layer in the human nose. Human olfactory sensors are not specific, but are sensitive to odor compounds. The smelled aroma of foodstuff is sent to the brain to analyze the information transferred by the entire olfactory sensor in the human nose. The stages of the recognition process in the electronic nose system are similar to those of human olfaction. These are performed for identification, comparison, quantification, and other applications. While the electronic nose provides objective opinions, a hedonic scale evaluation in sensory analysis is related to subjective opinions because of a specificity of the human nose. The electronic nose consists of a sampling system (odor handling and delivery system), sensor array, electronic circuitry, and pattern-recognizing software. Sensors can be divided into three categories according to the type of sensitive material used: inorganic crystalline materials (e.g., semiconductors and metal oxides), organic materials, and polymers. This part of the chapter describes the sensor for the electronic nose and data analysis by a pattern recognition that has been popularly utilized.

9.2.1   Gas Sensors

Gas sensors can be divided into five categories: metal oxide sensor, conducting polymer sensor, quartz crystal microbalance sensor, surface acoustic wave sensor, and fingerprint mass spectrometry. The main features of these sensors are described below.

9.2.1.1   Metal Oxide Sensors (MOSs)

MOSs are inorganic and operate at high temperatures, around 400°C. Oxygen in the air reacts with lattice vacancies and removes electrons from a conducting band. Odor molecules react irreversibly with the oxygen species, liberating electrons and lowering the measured resistance of the sensor. MOSs can detect aldehydes, alcohols, hydrogen sulfide, ammonia, hydrocarbon, and air contaminants. In general, this type of sensor is to be replaced every ten months or one and a half years due to its destabilization.

9.2.1.2   Conducting Polymer (CP) Sensors

These are made of pyrrole, indole, or derivatives of similar materials, electrochemically deposited on a silicon substrate. The sensor tends to swell in the presence of odors and thus change resistance. CP sensors are very responsive to polar molecules, thus making them complementary to the function of metal oxides. CP sensors operate at room temperature and display rapidly reversible reactions leading to rapid discrimination. A less desirable feature of CP is its interference with water and humidity. In addition, it is about three orders of magnitude less sensitive than MOSs. Its tendency to drift and destabilize is a function of its organic nature.

9.2.1.3   Quartz Crystal Microbalance (QCM) Sensor

The sensor element is a quartz resonator coated with an organic material similar to the stationary phase of a GC column. The resonant frequency of the sensor changes as aromas adsorb and desorb from the coating, changing the mass of the resonator, and hence its frequency. This sensor depends on an organic interface. Like all sensors, the material from which it is made will determine its lifetime and drift characteristics.

9.2.1.4   Surface Acoustic Wave (SAW) Sensor

The SAW resonator frequency is measured using a 20 ms gated reciprocal counter that produces five hundred readings in 50 s. This translates into an array of a five-hundred-sensor reading with each chemical sensor assigned to a unique retention time and frequency reading for a chemical. The sensor responses are nearly orthogonal with minimum overlap. A virtual sensor array with any number of chemical sensors needed for any odor, fragrance, or smell can be created and saved for late retrieval. Chemical sensor space can be defined mathematically by assigning unique retention time slots to each sensor in the electronic nose system.

9.2.1.5   Fingerprint Mass Spectrometry

The quadrupole instruments for electronic olfaction, sometimes called fingerprint mass spectrometry, are similar in design to the technology used for GC-MS. The only difference comes from the absence of the GC module used to isolate the volatile molecules before their detection by the mass spectrometer. In fingerprint mass spectrometry, the entire aroma enters the quadrupole module without isolation of specific molecules. The resulting fingerprint is from an aroma ensemble, just as it is presented to humans. A mass spectrometer consists of an ion source, a mass-selective analyzer, and an ion detector. To detect molecules, the mass spectrometer uses the difference in mass-to-charge ratios (m/z) of ionized atoms or molecules to separate them from each other. Mass spectrometry is useful for quantifying atoms or molecules and also for determining chemical and structural information about molecules. Molecules have distinctive fragmentation patterns that provide structural information to identify structural components. The mass spectrum obtained from fingerprint mass spectrometry can be linked with few molecules identified with the GC-MS technique. The electronic nose based on GC with the SAW sensor or mass spectrometry quantifying constituents is much advanced from the first generation of electronic nose that was used only for qualitative analyses.

9.2.2 SENSOR ARRAYS

Each sensor of the sensor array in an electronic nose is sensitive to a volatile component of foodstuffs. Sensor array formats depend upon different volatile molecules of a food product and provide an electronic signal that can be utilized effectively as a fingerprint of the volatile molecules associated to the product (Gardner and Bartlett, 1999). The sensor array is exposed to a volatile odor through suitable odor handling and a delivery system that ensures a constant exposure rate of each sensor. The response signals of a sensor array are conditioned and processed through a suitable circuitry and fed to an intelligent pattern recognition engine for classification, analysis, and declaration. The most complicated parts of an electronic olfaction process are odor capture and associated sensor technology. Any sensors that respond reversibly to chemicals in the gas or vapor phase have the potential to participate in a sensor array of an electronic nose. The array of sensors consists of several nonspecific sensors. The sensors are sensitive to biological or chemical materials, such as flavors or odor. It is not enough to recognize and find out quality exactly with simple information through nonspecific sensors. Nonetheless, the difference between a bad sample and a good one can be discriminated by pattern recognition and a statistical analysis. A stimulus of volatile compound generates a characteristic fingerprint from a sensor array, which provides information about pattern recognition. A fingerprint, which is global information, of the samples to be classified is sometimes used to construct a database and train a pattern so that unknown odors can subsequently be classified or identified. An artificial neural network is utilized for training known data.

9.2.3 DATA ANALYSIS AND PATTERN RECOGNITION

9.2.3.1 Pattern Recognition

In order to interpret the volatile pattern rapidly, the data collected must be combined with a chemometric analysis based on multivariate analysis such as principal component analysis (PCA) or discriminant function analysis (DFA). A specific interesting application comes from researches conducted in order to evaluate the capability of an electronic nose in the characterization of volatile components in food. Interpretation of data with multiple variables such as several sensors or mass fragments for multiple samples requires the use of statistical interpretation methods. Chemometric techniques provide a way of presenting the data in an understandable format. With suitable training sets designed for specific needs, it is possible to quantify, discriminate, and identify an unknown sample. Data acquisitions of sample are stored in data files and the files are used to build a library necessary to train the electronic nose. A library is a matrix whose rows include the samples, and the columns contain the sensors or fingerprint mass spectrometry ion intensities. Multivariate statistics are used in the analytical field as a statistic treatment method. The statistical method used is based on PCA, which is a means of reducing the degree of complexities of spectral data into less complex data and just the most significant components. It provides a view into a two-dimensional map that can group sets of data into population groups. Product identification using a discriminated analysis involves the prior training of the system with known samples to build an identification model based on DFA. This training leads to prediction and classification of unknowns. As an alternative method to recognize a pattern, a unique approach by a GC-SAW sensor based on the electronic nose is two-dimensional olfactory images, called VaporPrint™, without statistical analysis. It displays the forest of trees in its entirety. This approach provides a high-resolution image that changes an olfactory response to a visual response. It is produced by a polar plot chromatogram, using a retention time as the angular variable and the SAW detector response as a radial variable. A gallery of VaporPrint images for the heated coconut oil with different heating times at 160°C is shown in Figure 9.1 (Han et al., 2006).

9.2.3.2   Determination of Concentration Levels Using Linear Regression

A modeling method that simultaneously uses all information contained in a predictor matrix is used to predict a specific property, including sample categories. This model is based on a partial linear regression to build a calibration graph using samples with known concentrations.

Accepted-nonaccepted samples are a supervised method allowing the identification of samples based on the simple independent modeling class analogy (SIMCA) model. Each group is represented on a plan that allows the projection of unknown samples recognized as belonging to a territory into the same representative group (accepted or nonaccepted).

9.2.3.3   Artificial Neural Networks (ANNs)

In order to analyze complex data and recognize patterns, an ANN could be applied to a set of inputs and produce the desired set of outputs after training by a set of input and output vectors to adjust the weights. The ANN is trained for chemical vapor recognition. Operation consists of propagating the sensor data through the network. Since this is simply a series of vector-matrix multiplications, unknown chemicals can be rapidly identified in the field.

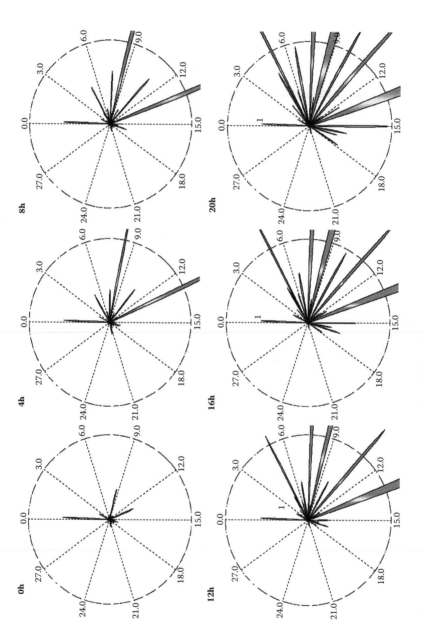

FIGURE 9.1 Derivative patterns of the heated coconut oil at 160°C using the electronic nose based on GC with a SAW sensor and VaporPrint image program. (Data from Han, K. Y., Oh, S. Y., Kim, J. H., Youn, A. R., and Noh, B. S., *Kor. J. Food Sci. Technol.*, 38: 16–21, 2006.)

9.3 APPLICATION OF ELECTRONIC NOSE

The electronic nose has been applied to various fields in the food industry. Several applications are being discussed in this chapter. Other review papers will be helpful for more information (Peris and Escuder-Gilabert, 2009).

9.3.1 RANCIDITY OF OILS

Oxidation of lipids is a common and frequently undesirable chemical change that can impact flavor, volatile components, and nutritional quality of foods, and can compromise the odor. The response of an electronic nose can be utilized effectively as a fingerprint for the off-flavor volatile molecules associated to rancidity. Major volatile components of soybean oil, including hexanal, were measured by an electronic nose. As shown in Table 9.1, several sensors typically reacted well on hexanal,

TABLE 9.1

Changes of Sensitivity by the Portable Electronic Nose for Major Volatile Components in Soybean Oil

| Component | Amount (µL) | Sensitivity[a] of Sensor | | | | | |
		TGS825	TGS824	TGS880	TGS822	TGS800	TGS813
Hexanal	0.1	1.0050	0.8334	0.4342	0.4383	0.9105	0.9279
	1	1.0132	0.8792	0.2894	0.3087	0.8698	0.8991
	10	0.9607	0.3640	0.0612	0.0781	0.3839	0.4512
Pentanal	0.1	0.9983	0.8064	0.3411	0.3432	0.8339	0.8753
	0.5	0.9977	0.7814	0.2111	0.2164	0.8299	0.7842
	1	0.9563	0.7232	0.1279	0.1331	0.7876	0.6821
1-Pentanol	0.1	1.0065	0.8233	0.6177	0.6149	0.9492	0.9390
	1	0.9954	0.8370	0.3553	0.3544	0.8983	0.9127
	10	0.9912	0.8295	0.1060	0.1041	0.8277	0.8858
m-Xylene	0.1	0.9881	0.8210	0.9431	0.9093	0.8988	0.9191
	1	0.9994	0.8238	0.8854	0.8094	0.9009	0.9019
	10	0.9864	0.8222	0.8773	0.7549	0.8912	0.9291
n-Decane	1	0.9948	0.9862	0.9599	0.9518	0.9316	0.9799
	10	0.9892	0.9648	0.8312	0.8325	0.9443	0.9815
	100	0.9999	0.9208	0.7337	0.7575	0.9233	0.9653
Benzene	0.1	1.0148	0.8381	0.7485	0.6266	0.8701	0.9225
	1	0.9973	0.8343	0.7041	0.4967	0.8866	0.9327
	10	0.8537	0.8310	0.5358	0.2787	0.8462	0.9296
n-Octane	1	0.9983	0.8260	0.9099	0.8986	0.8898	0.9415
	10	1.0055	0.8154	0.8398	0.7245	0.8735	0.9298
	100	1.0082	0.8269	0.6369	0.5036	0.8730	0.9171

Source: Data from Yang, Y. M., Han, K. Y., and Noh, B. S., *Food Sci. Biotechnol.*, 9: 146–50, 2000.

[a] Sensitivity is expressed by R_{gas}/R_{air} in 325 ml bottle.

pentanal, and 1-pentanol, among several volatile components (Yang et al., 2000). The responses of the sensors were correlated with the concentration of volatile components, implying that the electronic nose could distinguish rancid oil from less oxidized oils. Lipid oxidation of soybean oil was intentionally accelerated by adding a copper ion as a prooxidant and raising the heating temperature. This metal ion caused lipid oxidation to accelerate. The higher the concentration of copper sulfate added to soybean oil during thermal treatment, the faster the lipid oxidation happened. When copper sulfate was not added to soybean oil, the change of sensitivity was small. After twenty-five hours of heating, the sensitivity (R_{gas}/R_{air}) of sensors decreased with the addition of 0.5% copper sulfate. These results exhibited correlation among acid value, peroxide value, and iodine value of the heated oil (Figure 9.2). Thus, the degree of lipid oxidation could be explained by the sensitivity of the metal oxide sensor of the electronic nose. The data at various heating times were analyzed by the PCA. The PCA was operated by synthesizing the data by removing the redundant information. As could be inferred, the sensor array showed good performance in the

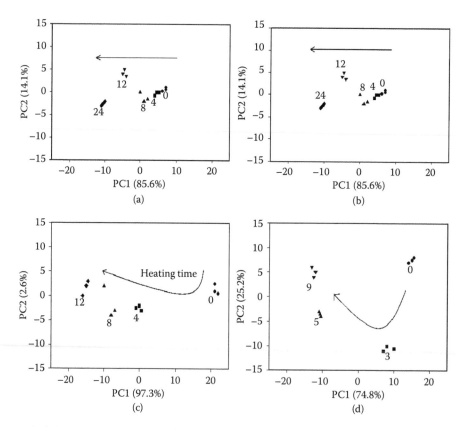

FIGURE 9.2 Principal component analysis of sensitivity obtained by the portable electronic nose, acid value, iodine value, and peroxide value of the heated soybean oil at 200°C with various copper sulfate concentrations. (Data from Yang, Y. M., Han, K. Y., and Noh, B. S., *Food Sci. Biotechnol.*, 9: 146–50, 2000.)

recognition of the heated oil with 0.25 and 0.5% copper sulfate, while a little lipid oxidation occurred to the heated oil without copper sulfate. As the concentration of copper sulfate increased, more lipid was oxidized. The samples were well separated by PCA. A PCA of the lipid oxidation data matrix showed a characteristic development of rancid flavor dependent on heating time. As heating time increased, the PCA plot extended from the right side (positive value of PC1) through the middle to the left side (negative value of that) in Figure 9.2. This means that the first principal component score (PC1) correlated with the heating time or degree of lipid oxidation. The PCA technique and analysis of the electronic nose was very useful for evaluating the quality of the oil. Lee et al. (1998) showed the characterization of fatty acids composition in vegetable oil by gas chromatography and chemometrics. They also found that PCA or DFA was very useful for many aspects of the vegetable oil industry, including pattern recognition or primary evaluation of category similarity, detection of adulterants, and quality control. Lipid autoxidation of rapeseed oil was also subjected to an accelerated storage test for 12 days at 60°C, and to an ambient temperature storage test in which it was stored in retail plastic bottles for up to six months (Mildner-Szkudlarz et al., 2008). Sensory evaluations, Solid Phase Micro Extraction-GC-MS analyses, and chemical analyses were conducted. The PCA of electronic nose data stored at an elevated temperature was related to the PCA of sensory evaluation. Prediction models based on partial least squares of electronic nose data could predict the sensory quality changes during storage at an elevated temperature. Peroxide and anisidine values were well predicted on the basis of the electronic nose and sensory evaluation. Therefore, a rancidity analysis for routine quality control of oil could be replaced by the application of an electronic nose.

9.3.2 Freshness of Milk and Soymilk

As a good predictor of shelf life, the determination of volatiles in milk using a dynamic headspace capillary GC, followed by multivariate interpretation of gas chromatographic peak area, was reported (Vallejo-Cordoba et al., 1995). It has also been successfully applied for the classification of abused milk samples to determine the cause of off-flavors (Marsili and Miller, 1998). However, significant problems exist, which include tedious procedures and potential errors associated with tracking a large number of chromatographic peaks generated by volatiles in the milk, even when automatic peak recognition software is applied. Furthermore, the operation of GC instrumentation, peak tracking, identification, and area quantization require a skilled person. It is also not easy to use a dynamic headspace capillary GC for quality control (Marsili, 2000). Therefore, a more rapid method that would provide a direct measurement of flavor quality, resulting in reliable estimates of the milk freshness, and could be handled with ease by untrained quality control personnel is sought. To determine the shelf life of milk, an electronic nose based on MOSs was applied. Milk was stored at room temperature and at a constant temperature of 5°C. Microorganisms produced undesirable flavors with increasing time of storage. They were monitored by the electronic nose with a high sensitivity. There was a good correlation between sensitivity by the electronic nose and acidity or pH during storage (Yang et al., 1999). The repeated monitoring provides good training for learning by computer. The data at different times were trained by the ANN, and the storage time of unknown milk was predicted. Not only milk but also dairy products can be applied for determining shelf life (Ko et al., 2000). Soymilk was also stored at different temperatures for forty-five days. The quality of the soymilk was measured by the MOS-based electronic nose. The obtained data by the electronic nose were analyzed by PCA. Significant differences in volatile profiles were found during storage (Figure 9.3). PC1 is correlated with the freshness of soymilk. As storage time increased, the PCA plot extended from the positive value of PC1 through the middle to the negative value (Park et al., 2002). It showed a good correlation between PC1 and storage time in the bar graph (Figure 9.4). The analysis of soymilk quality was recognized by the ANN with 90.0 and 97.5% probability levels at 20 and 35°C (Table 9.2).

9.3.3 Evaluation of Adulteration

Adulteration of food products, involving replacement of expensive ingredients with cheaper substitutes, could be potentially very lucrative for a vendor or raw material supplier. Sesame oils have a mild odor and a pleasant taste. They don't require winterization. They have been popular as a cooking oil in many Asian and African countries, and more expensive than other edible vegetable oils. Thus, adulteration with lower-quality or cheaper oils may generate considerable profits from the economic point of view. There is still a problem to sell adulterated sesame oil on the market. Adulterations are easily carried out with corn germ oil, soybean oil, sunflower oil, maize oil, and perilla oil (Mildner-Szkudlarz and Jeleń, 2008; Son et al., 2009). In order to evaluate an adulterated oil, the electronic nose can be used because of fast analysis and effective techniques for quality control of oils. An electronic nose

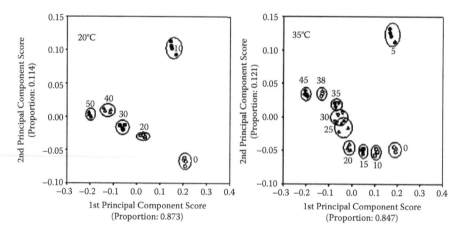

FIGURE 9.3 Principal component analysis of sensitivity by the electronic nose for soymilk at various temperatures. The storage time (day) of the sample was indicated with a number. (Data from Park, E. Y., Kim, J. H., and Noh, B. S., *Food Sci. Biotechnol.*, 11(3): 320–23, 2002.)

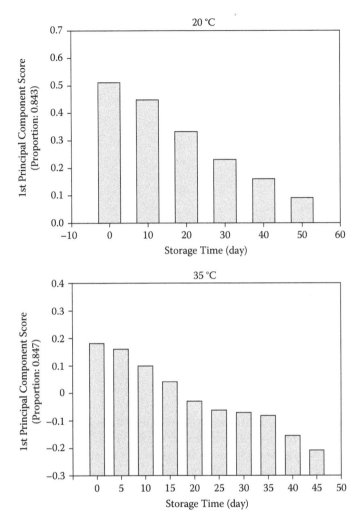

FIGURE 9.4   The relationship between storage time and the first principal component score during storage of soymilk at 20 and 35°C.

based on MOSs was used to classify sesame oil with different adulteration levels and predict the percentage of adulteration. More than one hundred samples were treated, and the obtained signals were analyzed by several feature extraction methods and pattern recognition techniques. A high correlation ($r = 0.99805$) between predicted concentrations and measured values was found (Zheng and Wang, 2006). An electronic nose based on a mass spectrometer system (SmartNose200) was also used to investigate authentic sesame oil (Figure 9.5). The obtained intensities of each fragment in the adulterated sesame oil by the electronic nose were analyzed by discriminant function analysis. A DFA plot showed clear separations of a pure sesame oil and other oils. The added concentration of perilla oil to sesame oil was

TABLE 9.2

Classification of Storage Day for Soymilk Stored at Different Temperatures Using the Electronic Nose with Artificial Neural Network System

Actual Time (day)	Predicted Time (day)				Probability	Average
	10	**20**	**30**	**40**		
(20°C) 10	10/10				100%	
20		7/10	3/10		70%	
30			9/10	1/10	90%	
40				10/10	100%	90%
	5	**15**	**30**	**40**		
(35°C) 5	10/10				100%	
15		10/10			100%	
30			10/10		100%	
40			1/10	9/10	90%	97.5%

Source:　Data from Park, E. Y., Kim, J. H., and Noh, B. S., *Food Sci. Biotechnol.*, 11(3): 320–23, 2002.

correlated with the result of DFA (Figure 9.6). The electronic nose based on the MS system could be used as an efficient method for the discrimination of the purity of oil and quality control of food products.

9.3.4　Irradiation of Foods

When foodstuffs are irradiated, many volatile compounds are formed, which may be useful as markers for irradiation treatment (Raffi, 1998). The percentage of radiolytic products from lipids, such as volatile hydrocarbons, aldehydes, or butanones, is directly linked to the chemical composition of the lipids. When fatty acids are exposed to a high-energy radiation, they undergo preferential cleavages, which, unlike other treatments, occur in the ester carbonyl region. Two types of hydrocarbons are predominantly produced through the irradiation of fatty acids: one that has one carbon fewer than the parent fatty acid (C_{n-1}), and another that has two carbons fewer and an additional double bond at position 1 (C_{n-2},1-ene) (Schreiber et al., 1994). Meat generally contains palmitic, stearic, oleic, and linoleic acids. Pentadecane ($C_{15:0}$) and 1-tetradecene ($C_{14:1}$) from palmitate, heptadecane ($C_{17:0}$) and 1-hexadecene ($C_{16:1}$) from stearate, 8-heptadecene ($C_{17:1}$) and 1,7-hexadecadiene ($C_{16:2}$) from oleate, and 6,9-heptadecadiene ($C_{17:2}$) and 1,7,10-hexadecatriene from linoleate are expected to be detected in irradiated meat. Therefore, an electronic nose can be applied for the detection of these types of hydrocarbons. As the concentration of these hydrocarbons increased, the ratio of resistances in the electronic nose decreased (Table 9.3). Several sensors generally reacted well with these hydrocarbons. The responses of the sensor were correlated with the concentration of hydrocarbons, an indication that the electronic nose could detect irradiated meat. Irradiated meat at 1 ~ 10 kGy levels was classified by PCA (Figure 9.7). The

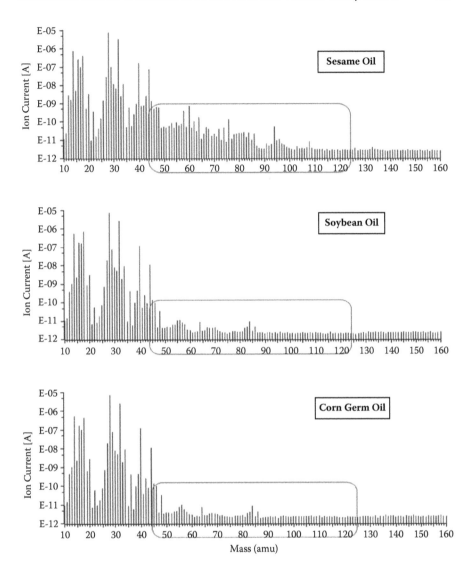

FIGURE 9.5   Mass spectrum of sesame oil, soybean oil, corn germ oil, and mixed oils (85% sesame oil + 15% other vegetable oil) by MS-based electronic nose. (Data from Son, H. J., Hong, E. J., Ko, S. H., Choi, J. Y., and Noh, B. S., *Food Eng. Progress*, 13: 275–81, 2009.)

PCA of irradiated meat showed characteristic development of off-flavor at different levels of irradiation doses (Han et al., 2001). As the irradiation dose increased, the PCA plot extended from the right side (positive value of the PC1) through the middle to the left side (negative value of the PC1), an indication that the PC1 was correlated with the irradiation dose or degree of off-flavor. An ANN analysis for identifying the irradiation of meat was applied to unknown samples (Table 9.4).

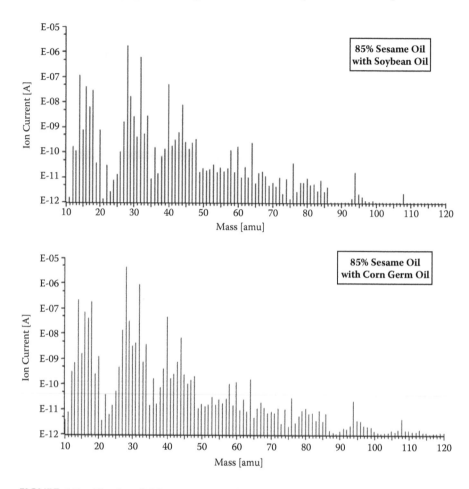

FIGURE 9.5 (Continued) Mass spectrum of sesame oil, soybean oil, corn germ oil, and mixed oils (85% sesame oil + 15% other vegetable oil) by MS-based electronic nose. (Data from Son, H. J., Hong, E. J., Ko, S. H., Choi, J. Y., and Noh, B. S., *Food Eng. Progress*, 13: 275–81, 2009.)

The known data of irradiated meat were used as a database to learn the neural network. On the basis of the database, input patterns and target patterns were built for a back-propagation learning algorithm. After supervised learning, unknown sample data were input into the learned neural network system and analyzed to detect irradiation doses. The output patterns analyzed by the neural network analysis system are used to determine irradiation doses of unknown samples. The correct probabilities, which predicted the irradiation dose of unknown samples in the electronic nose, are as shown in Table 9.4. Unknown irradiated samples were detected well. The electronic nose could detect off-flavor from irradiated and rancid samples. Therefore, the storage temperature would be important in the detection of irradiated meat.

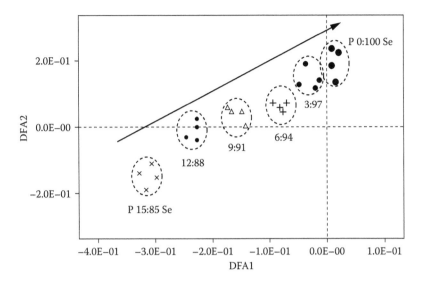

FIGURE 9.6 Discriminant function analysis of the electronic nose data for mixed oils at different ratios of concentration. Mixed oils with perilla oil (P) and different commercial sesame oil (Se). The arrow indicates increasing concentration of sesame oil in mixed oil. (Data from Son, H. J., Kang, J. H., Hong, E. J., Lim, C. R., Choi, J. Y., and Noh, B. S., *Kor. J. Food Sci. Technol.*, 41: 609–14, 2009.)

TABLE 9.3
Changes in the Ratio of Resistance of Sensor in the Electronic Nose to Hydrocarbons

Component	Amount (µl)	Ratio of Resistance of Sensor[a]					
		TGS825	**TGS824**	**TGS880**	**TGS822**	**TGS800**	**TGS813**
n-Pentadecane	1	0.73014	0.78999	0.71862	0.76959	0.76872	0.71735
	10	0.67854	0.70319	0.64164	0.68766	0.69998	0.66122
	100	0.67063	0.70518	0.66649	0.68481	0.70799	0.66687
Tetradecene	1	0.72376	0.74242	0.68968	0.74617	0.74432	0.68336
	10	0.57627	0.57740	0.61100	0.55542	0.59784	0.62012
	100	0.34362	0.35199	0.48192	0.29067	0.38081	0.51668
n-Heptadecane	1	0.73344	0.78778	0.78152	0.73658	0.73903	0.74520
	10	0.62898	0.64745	0.62712	0.61279	0.63210	0.64337
	100	0.43741	0.45261	0.59964	0.39772	0.38114	0.61624
1-Hexadecene	1	0.67606	0.68707	0.62406	0.66193	0.67860	0.62131
	10	0.58482	0.56003	0.53997	0.54287	0.57983	0.56243
	100	0.56303	0.54260	0.62817	0.52084	0.58310	0.64708

Source: Data from Han, K. Y., Kim, J. H., and Noh, B. S., *Food Sci. Biotechnol.*, 10(6): 668–72, 2001.
[a] Ratio of resistance is expressed by R_{gas}/R_{air}.

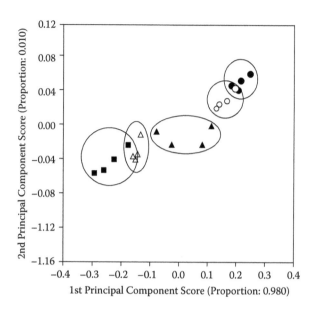

FIGURE 9.7 Principal component analysis of ratio of resistance by the electronic nose to volatile compounds of meat irradiated with different doses. ●, 0 kGy; ○, 1 kGy; ▲, 3 kGy; △, 5 kGy; ■, 10 kGy. (Data from Han, K. Y., Kim, J. H., and Noh, B. S., *Food Sci. Biotechnol.*, 10(6): 668–72, 2001.)

TABLE 9.4

Classification of Meat Irradiated with Different Irradiation Dose by Artificial Neural Network Program with Ratio of Resistance in the Electronic Nose

Irradiation	Predicted Dose (kGy)					Correct
Actual Dose (kGy)	0	1	3	5	10	Probability (%)
0	8/8					100
1		8/8				100
3		1/8	6/8	1/8		75
5				7/8	1/8	87.5
10				2/8	6/8	75

Source: Data from Han, K. Y., Kim, J. H., and Noh, B. S., *Food Sci. Biotechnol.*, 10(6): 668–72, 2001.

9.3.5 RIPENESS OF FRUIT AND VEGETABLES

Monitoring and controlling ripeness is important for management of fruit and vegetable harvesting as well as consumer preference. Many methods to monitor fruit ripeness cannot be applied to the packaging line and require destruction of the sample for analysis. Judging or prediction of ripeness of fruits and vegetables is mainly carried out based on practical experience. Sometimes poor-quality fruits and vegetables are provided to consumers when large quantities of them are inspected and a critical decision is needed in a short time. The electronic nose has been proved to be a fast tool for fruit ripeness assessments. An electronic nose, consisting of twenty-one MOS arrays, assesses the ripeness state of pink lady apples through their shelf life. To evaluate the electronic nose performance, fruit quality indicators such as firmness, starch index, and acidity were also obtained to compare results from both techniques. Apples were harvested at their optimal dates so that the electronic nose and fruit quality measurements could be performed on the fruit samples during their ripening process (Saevels et al., 2004). Fuzzy networks and ANNs were used for a good classification for optimal time of quality. Good correlations between electronic nose signals and classical parameters of fruit quality (firmness and acidity) were found in partial least squares modeling. During postharvest ripening volatile organic compounds (VOCs) such as aldehydes and esters formed, and the relative ratio of these compounds changed. Especially trans-2-hexenal can serve as an indicator compound because its concentration increases significantly during the ripening. An electronic nose based on QCM sensor arrays has been used to discriminate VOCs, which discriminate types of apples as a qualitative identification. Since the correlation between vapor concentration and the response of the QCM sensor is linear, even in a limited concentration range, quantitative identifications are possible. Applications of the electronic nose have been devoted to monitor the ripeness and shelf life of fruits and vegetables (Table 9.5). An MS-based electronic nose to monitor changes in apple fruit volatility during shelf life has been studied. The volatile profile changes were evaluated during storage. A specific electronic nose with ten different MOSs was able to detect clearer differences in volatile profiles, and showed well the discrimination of tomato with different ripeness states: unripe, half ripe, fully ripe, and overripe (Gómez et al., 2006). Each group was clearly distinguishable from other groups by linear discriminant analysis (LDA). These data were also analyzed by PCA. When the electronic nose was performed with LDA, better classification rates were observed than with performance by PCA. It is possible to classify the different tomato maturity states by LDA or PCA.

TABLE 9.5

Application of Electronic Nose to Fruit and Vegetable Processing

Sample	Detection System	Purpose	Reference
Apple	MOS-based e-nose	Ripeness states	Brezmes et al. (2001)
	MS-based e-nose	Shelf life	Saevels et al. (2004)
Tomato	QCM	Quality	Berna et al. (2004)
Peach	MOS-based e-nose	Ripeness	Benedetti et al. (2008)
Mandarin	MOS-based e-nose	Shelf life	Gómez et al. (2007)

9.3.6 Freshness of Fish Products

A number of different VOCs are generated during storage of foods. The freshness of foodstuffs will be predicted when the electronic nose detects off-flavor or off-odor from degradation by bacterial processes or chemical reactions. A significant release of volatile compounds in fish (Natale et al., 2001), oysters (Tokusoglu and Balaban, 2004), and shrimp (Luzuriaga et al., 2007) occurs at market distribution or during storage. Investigation by an electronic nose mainly deals with freshness of fish and fish products. An electronic nose for fish products is composed of sensors each oriented toward a certain gas (NH_3, H_2S, NO, CO, SO_2). Electronic nose data were analyzed by the means of PCA, PLS, DFA, etc. Salmon fillets were investigated for quality assessment under various storage conditions. The fillets were stored at –20, 4, and 10°C for fourteen days. Changes in bacteria and histamine over time were analyzed by an electronic nose. Comparing sensory properties with electronic nose data might be valuable in evaluating fish quality. Salmon was smoked under four different smokehouses in Europe and stored in different packaging (vacuum and modified atmosphere) for up to four weeks at 5 and 10°C. The quality changes of them were monitored by fish nose (MOS sensor array). The response of gas sensors correlated well with the results from the sensory evaluation of spoilage odor and microbial counts. It demonstrated that the electronic nose could detect volatile microbiologically produced compounds causing spoilage odor in cold-smoked salmon during storage as a fast method for quality control. This might be easily extended to other types of fish or fish products.

9.3.7 Detection of Microorganisms

The electronic nose can be applied to a technology for the early detection of undesirable off-odors and microbial contaminants in dairy and bakery products. The EU-funded project (electronic nose) found that the new technology can be used to quickly detect bacteria, yeasts, filamentous fungi, and off-odors. A Cranfield University team used different types of electronic nose systems, based on CP sensor arrays or MOS arrays, for the rapid and early cost-effective detection of undesirable, harmful contaminants, toxins, and taints in the dairy and bakery product industries. Magan and his team (Magan and Evans, 2001) are now in the final stages of completing electronic nose trials in the food industry together with a cost-benefit analysis in areas of the dairy and bakery industries. It is possible to differentiate and detect fungal contaminants in different cheese matrices. Fungal volatile production profiles can be used to discriminate mycotoxigenic strains from nonmycotoxigenic strains of some spoilage molds. In addition, MOS arrays were effective in detecting bacterial and mold contaminations twenty-four to forty-eight hours prior to visible spoilage in bread. Abramson et al. (2005) monitored ergosterol content, odor volatiles, microfloral infection, and level of major mycotoxins using an electronic nose. The MOS array was used to monitor odor volatile evolution. Several of the twelve sensors were able to track odor volatile changes. Signals from these chemosensors showed a good correlation with ochratoxin A formation (r = 0.84 ~ 0.87), a citrinin formation (r = 0.83), and a ergosterol production (r = 0.98). The data indicated that an effective chemosensor for odor volatiles presaging the appearance of mycotoxins in moist durum wheat needs to contain only several types of chemosensors.

9.3.8   QUALITY OF CONTROL FOR OTHER PRODUCTS

Electronic noses, launched commercially in 1995, are computerized tabletop units with sensors that detect odor molecules. The uptake of the electronic nose technology by the food industry has been slow due to its capital cost, but in recent years researchers in both the EU and United States have revealed that the noses can be useful tools to save costs through their ability to detect the quality of products. Commercial coffees are blends, which contain coffees of various origins for economic reasons. It is of great importance for the producers the availability of analysis and control techniques. There exists a rich literature on the characterization of coffee reporting results from the analysis of a fraction of its chemical profiles, such as the volatile fraction of green or roasted beans or phenolic compounds. Up to seven hundred diverse molecules have been identified in the headspace. Their relative abundance depends on the type, provenance, and manufacturing of the coffee. It is to be noticed that none of these molecules can alone be identified as a marker. On the contrary, one has to consider the whole spectrum, as, for instance, the gas chromatographic profile. The electronic nose is normally used to discriminate different classes of similar odor-emitting products. In particular, the electronic nose already serves to distinguish different coffee blends and coffee roasting levels. The numerous varieties of coffee beans contain a wide range of qualities. While the varieties of green coffee beans can generally be distinguished by their appearance, the visual criterion is impossible after the roasting process. An electronic nose based on GC with a SAW sensor can be used for evaluation of the classification of origin and blended commercial brands in roasted coffee beans. The origins of four similarly roasted ground coffee beans and eight blended commercial brands were rapidly classified by the VaporPrint image program. Those would be easily understandable over the aroma image pattern by VaporPrint. Analysis of GC by an expert might be replaced by this olfactory image. The reactions and chemical transformations during a fermentation process to an optimum limit are vital for producing superior quality tea. It is critical that the tea leaf be allowed to ferment only up to the desired limit so that the complex series of chemical changes within the leaf are accomplished optimally. Conventionally, length of fermentation is subjectively estimated by the human senses of smell and vision. Experts can sense conversion of a grassy smell to a floral smell of leaves after fermentation. Evaluation of good quality is very subjective, depending upon a skilled person. When humans are tired or sick, evaluation of tea quality can be delayed. A colorimetric approach is also used at times, where fermentation completion time is determined based on color. A specially designed electronic nose has been successfully used to monitor a volatile emission pattern in a fermentation process over the passage of time. Through prolonged experimentation with various fermentation processes and climatic variations, smell changes during the process may be reliably detected repeatedly by the electronic nose.

Application of the electronic nose has been expanded dramatically for a variety of areas in the food industry. The electronic nose can also be used to predict sensorial descriptors of food quality as determined by a panel. In some instances, it can be used to augment or replace panels of human experts. In other cases, electronic noses can be used to reduce the amount of analytical chemistry. There are a few disadvantages associated with data processing. An example would be the need of a lot of data for ANN analysis. Regardless of these problems, the electronic nose is promising, as it can fulfill niche analyses and be expanded to many fields of food industries.

REFERENCES

Abramson, D., Hulasare, R., York, R. K., White, N. D. G., and Jayas, D. S. 2005. Mycotoxins, ergosterol, and odor volatiles in durum wheat during granary at 16% and 20% moisture content. *J. Stored Products Res.* 41:67–76.

Benedetti, S., Buratti, S., and Spinardi, S. 2008. Electronic nose as a nondestructive tool to charaterise peach cultivars and to monitor their ripening stage during shelf-life. *Postharvest Biol. Technol.* 47:181–88.

Berna, A. Z., Lammertyn, J., Saevels, S., Natale, C. D., and Nicolaï, B. M. 2004. Electronic nose systems to study shelf life and cultivar effect on tomato aroma profile. *Sensors Actuators B* 97:324–33.

Brezmes, J., Llobet, E., Vilanova, X., Orts, J., Saiz, G., and Correig, X. 2001. Correlation between electronic nose signals and fruit quality indicators on shelf-life measurements with pink lady apples. *Sensors Actuators B* 80:41–50.

Gómez, A. H., Hu, G., Wang, J., and Pereira, A. G. 2006. Evaluation of tomato maturity by electronic nose. *Comput. Electron. Agric.* 54:44–52.

Gómez, A. H., Wang, J., Hu, G., and Pereira, A. G. 2007. Discrimination of storage shelf-life for mandarin by electronic nose technique. *LWT* 40:681–89.

Han, K. Y., Kim, J. H., and Noh, B. S. 2001. Identification of the volatile compounds of irradiated meat by using electronic nose. *Food Sci. Biotechnol.* 10(6):668–72.

Han, K. Y., Oh, S. Y., Kim, J. H., Youn, A. R., and Noh, B. S. 2006. Discrimination of the heated coconut oil using the electronic nose. *Kor. J. Food Sci. Technol.* 38:16–21.

Gardner, J. W., and Bartlett, P. N., 1999. *Electronic noses: Principles and applications.* Oxford: Oxford University Press.

Ko, S. H., Park, E. Y., Han, K. Y., Noh, B. S., and Kim, S. S. 2000. Development of neural network analysis program to predict shelf-life of soymilk by using electronic nose. *Food Eng. Progress* 4:193–98.

Lee, D. S., Noh, B. S., Bae, S. Y., and Kim, K. 1998. Characterization of fatty acids composition in vegetable oils by gas chromatography and chemometrics. *Anal. Chim. Acta* 358:163–75.

Luzuriaga, D. A., Korel, F., and Balaban, M. Ö. 2007. Odor evaluation of shrimp treated with different chemicals using an electronic nose and a sensory panel. *Aquat. J. Food Product Technol.* 16:57–75.

Magan, N., and Evans, P. 2001. Volatile as an indicator of fungal activity and differentiation between species and the potential use of electronic nose technology for early detection of grain spoilage. *J. Stored Products Res.* 36:319–40.

Marsili, R. T. 2000. Shelf-life prediction of processed milk by solid phase microextraction, mass spectrometry, and multivariate analysis. *J. Agric. Food Chem.* 48:3470–75.

Marsili, R. T., and Miller, N. 1998. Determination of the cause of off-flavors in milk by dynamic headspace GC/MS and multivariate data analysis. In *Food flavor formation, analysis and packing influences*, ed. C. Mussian, E. Contis, C. T. Ho, T. Parliament, A. Spanier, and F. Shaidi, 159–71. Amsterdam: Elesvier Science Publishers.

Mildner-Szkudlarz, S., Henryk, H., and Zawirska-Wojtasiak, J. R. 2008. The use of electronic and human nose for monitoring rapeseed oil autoxidation. *Eur. J. Technol.* 110:61–69.

Mildner-Szkudlarz, S., and Jeleń, H. 2008. The potential of different techniques for volatile compounds analysis coupled with PCA for the detection of the adulteration of olive oil with hazelnut oil. *Food Chem.* 110:751–61.

Natale, C. D., Olafsdottir, G., Einarsson, S., Martinelli, E., Paolesse, R., and D'Amico, A. 2001. Comparison and integration of different electronic noses for freshness evaluation of cod-fish fillets. *Sensor Actuators B Chem.* 77:572–78.

Park, E. Y., Kim, J. H., and Noh, B. S. 2002. Application of the electronic nose and artificial neural network system to quality of the stored soymilk. *Food Sci. Biotechnol.* 11(3):320–23.

Peris, M., and Escuder-Gilabert, L. 2009. A 21st century technique for food control: Electronic noses. *Anal. Chim. Acta* 638:1–15.

Raffi, J. 1998. Identifying irradiated foods. *Trends Anal. Chem.* 17:226–33.

Saevels, S., Lammertyn, J., Berna, A. Z., Veraverbeke, E. A., Natale, C. D., and Nicolaï, B. M. 2004. An electronic nose and a mass spectrometry-based electronic nose for assessing apple quality during shelf life. *Postharvest Biol. Technol.* 31:9–19.

Schreiber, G. A., Schulzki, G., Spiegelberg, A., Helle, N., and Bögl, K. W. 1994. Evaluation of a gas chromatographic method to identify irradiated chicken, pork, and beef by detection of volatile hydrocarbons. *J. Am. Oil Chem. Soc. Int.* 77:1202–17.

Son, H. J., Hong, E. J., Ko, S. H., Choi, J. Y., and Noh, B. S. 2009. Identification of vegetable oil-added sesame oil by a mass spectrometer-based electronic nose. *Food Eng. Progress* 13:275–81.

Son, H. J., Kang, J. H., Hong, E. J., Lim, C. R., Choi, J. Y., and Noh, B. S. 2009. Authentication of sesame oil with addition of perilla oil using electronic nose based on mass spectrometry. *Kor. J. Food Sci. Technol.* 41:609–14.

Tokusoglu, O., and Balaban, M. 2004. Correlation of odor and color profiles of oysters with electronic nose and color machine vision. *J. Shellfish Res.* 23:143–48.

Vallejo-Cordoba, B., Arteaga, G. E., and Nakai, S. 1995. Predicting milk shelf-life based on artificial neural networks and headspace gas chromatographic data. *J. Food Sci.* 60:885–88.

Yang, Y. M., Han, K. Y., and Noh, B. S. 2000. Analysis of lipid oxidation of soybean oil using the portable electronic nose. *Food Sci. Biotechnol.* 9:146–150.

Yang, Y. M., Noh, B. S., and Hong, H. K. 1999. Prediction of shelf life for milk using portable electronic nose. *Food Eng. Progress* 3(1):45–50.

Zheng, H., and Wang, J. 2006. Electronic nose and data analysis for detection of maize oil adulteration in sesame oil. *Sensors Actuators B* 119:449–55.

10 Biosensors for Evaluating Food Quality and Safety

Namsoo Kim and Yong-Jin Cho

CONTENTS

10.1 INTRODUCTION

Biosensors are normally composed of transducers and biological components (Figure 10.1), which are converted into physicochemical parameters such as ions, electrons, heat, mass, and light through specific reactions between biological components attached to the transducer surface and analytes that produce electric signals (Sethi, 1994). Biosensors are classified according to the transduction mechanisms and biological components used. The four main categories based on the transduction mechanisms include piezoelectric, optoelectronic and electrochemical biosensors, and biothermisters. In addition, the biological components used in biosensors include enzymes, microbial, immuno-, and nucleic acids, biochips (protein, DNA, and cell chips), and biomimetic sensors, which utilize artificial biorecognition elements (electronic nose and electronic tongue). Due to the advancements in biomicroelectromechanical system technology, transducers used for biosensors have gradually been miniaturized, which has resulted in the development of microfluidics technology. Moreover, in the case of some biosensors, the biological components are not immobilized onto the transducer surface, but rather are present in the vicinity of the transducer surface. Therefore, it is worth mentioning one definition of biosensors that was introduced by Cranfield Biotechnology Center (UK): "A biosensor is a compact analytical device incorporating a biological or biologically derived sensing element either integrated within or intimately associated with a physicochemical transducer."

The four major applications of biosensors include clinical diagnostics, industrial biotechnology, environment/wastewater, and food and beverage. The global

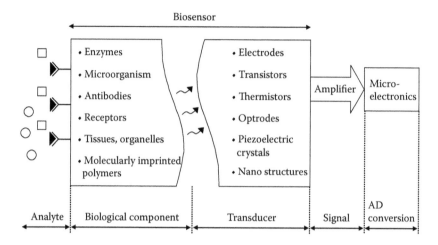

FIGURE 10.1 Schematic representation of biosensor signal transduction.

biosensor market increased from $0.7 billion in 1997 to $8.9 billion in 2005. It is expected that this increasing trend will continue due to the increasing demand in the bionanotechnology sector. Currently, commercialized biosensors, including glucose biosensors, have been marketed by many companies, such as Roche Diagnostics, Lifescan, Bayer, and Abbot. Although most biosensors have been developed and marketed for medical purposes, the biosensor market in the food industry has also increased by fifteen-fold during the same period described above, amounting to $150 million. The need for food biosensors to meet the increasing demand for new methods to evaluate food safety and functionality is expected to continually increase in the future. Some companies and institutions that are developing biosensors for food analysis include ATI Orion (United States), Cranfield Institute of Technology (UK), Biometry (Germany), Fuji Electric Co. (Japan), Molecular Devices Co. (United States), Oriental Electric Co. (Japan), Solea-Tacussel (France), Technicon, Inc. (United States), Toyo Jozo (Japan), and Universal Sensors (United States). Until now, the majority of marketed food biosensors have been targeted for the measurement of glucose, L-lactate, galactose, and alcohol (Warsinke, 1997).

The use of biosensors in the food industry may include nutrient analysis, detection of natural toxins and antinutrients, food process monitoring by measuring enzyme activity and microbial contamination, and rapid detection of genetically modified organisms (GMOs). Proteins, vitamin B complex, essential amino and fatty acids, and hazardous residual materials comprising pesticides and antibiotics can be measured using biosensors through enzymatic and immune reaction. Fungal toxins like aflatoxin; secondary metabolites including flavonoid compounds, which show functional activities; antinutrients such as trypsin inhibitor and lectin related with cytoplasmic membrane reactions are also measurable by complex formation between antigens and antibodies over biosensor transducers. The degree of microbial contamination is an important index of food wholesomeness. Until now, microbial examination has conventionally been done using culturing methods, which normally require 24 to 72 h. These methods, however, are complex and time consuming, and are strongly dependent on well-furnished laboratory facilities and expert personnel. The development of biosensors that use DNA and antibodies as sensing probes has made it possible to shorten the time needed to measure and detect microorganisms of hygienic importance, such as *Escherichia coli* and *Salmonella typhimurium*. The application of these biosensors to detect food poisoning bacteria and index microorganisms will be increasing in the near future. The inactivation of index enzymes such as polyphenol oxidase (PPO), lipoxygenase, and alkaline phosphatase (ALP) occurs by various heat treatment procedures during food processing. Based on this fact, biosensors that measure the activities of these enzymes can be devised to optimize on-site heat treatment procedures in the food industry. In particular, ALP and PPO activity are good indices for evaluating the pasteurization of milk and blanching of vegetables. As shown in the case of soybean and soybean products, consumer interest in GMOs has been increasing. Therefore, the necessity for rapid and high-sensitivity detection for the presence of GMOs in agricultural products and processed foods seems to be very high. Currently, biosensors and biochips targeted for GMO detection are under development in the levels of marker gene and protein, together with techniques for the miniaturized immobilization of biological components and nano-delivery of reaction mixtures. These efforts will lead to high-sensitivity GMO detection (Minunni et al., 2001).

During the last decade, our group has studied various types of food biosensors for the evaluation of fish and meat freshness, discriminant analysis of taste, rapid detection of microorganisms for food sanitation, and determination of hazardous residual substances. In the main text of this chapter, we will describe this work in more detail.

10.2   BIOSENSORS TO MEASURE BIOCHEMICAL FRESHNESS INDICES

Freshness of fish and meat is conventionally determined by visual observation, evaluation of physicochemical and sensory properties, and measurement of freshness indices. Of these factors, measurement of the freshness indices is the most important for evaluating the overall freshness of fish and meat. The generally recognized freshness indices include volatile basic nitrogen (VBN), trimethylamine (TMA), biogenic amines including histamine and cadaverine, and biochemical freshness indices such as K_i and H, which are expressed as the relative ratios of ATP-degradative compounds such as hypoxanthine (H_x), inosine ($H_x R$), and inosine 5'-monophosphate (IMP) (Ng et al., 1983).

The postmortem changes are as follows: adenosine 5'-triphosphate (ATP), adenosine 5'-diphosphate (ADP), and adenosine 5'-monophosphate (AMP) are rapidly degraded after death and are converted to IMP, $H_x R$, and H_x. In contrast, IMP starts to increase 5 to 12 h after death and decreases thereafter. $H_x R$ and H_x increase when IMP starts to decrease (Kim et al., 1988). Since consumers normally buy fish and meat around 24 h after death, the concentrations of ATP, ADP, and AMP are negligible. Therefore, K_i and H, which are defined below, only evaluate the relative concentrations of IMP, $H_x R$, and H_x in postmortem fish and meat.

$$K_i \text{ value} = \frac{H_x + H_xR}{H_x + H_xR + IMP} \times 100\% \tag{10.1}$$

$$H \text{ value} = \frac{H_x}{H_x + H_xR + IMP} \times 100\% \tag{10.2}$$

In the case of fish, the H value is a more appropriate biochemical freshness index of H_xR-forming species such as cod, mackerel, snapper, and skipjack, which accumulate H_xR during ATP degradation (Luong et al., 1992).

During the endogenous biochemical events from IMP to H_xR, H_x, and uric acid, three enzymes, 5'-nucleotidase (5'-NT), nucleoside phosphorylase (NP), and xanthine oxidase (XOD), are involved, together with the liberation of hydrogen peroxide (H_2O_2). Therefore, biosensor systems making use of oxygen (O_2) and H_2O_2 electrodes can be constructed to determine these biochemical freshness indices. Our group has developed a biosensor system that measures K_i and H by introducing a serially connected three-enzyme reactor that differs in column length combination (Park and Kim, 1999).

Figure 10.2 shows a schematic representation of the biosensor system containing three-enzyme reactors and three oxygen electrodes for the separate determination of H_x, H_xR, and IMP. One-enzyme (XOD), two-enzyme (NP and XOD), and three-enzyme (5'-NT, NP, and XOD)-immobilized Chitopearls (chitosan porous

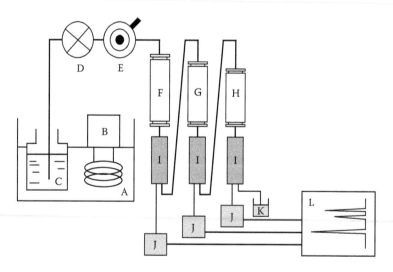

FIGURE 10.2 Schematic diagram of the biosensor system using three-enzyme reactors for the detection of H_x, H_xR, and IMP: (A) water bath, (B) circulator, (C) buffer solution, (D) peristaltic pump, (E) injector, (F) H_x column, (G) H_xR column, (H) IMP column, (I) oxygen electrode, (J) oxygen electrode adaptor, (K) waste solution, (L) multirecorder. (Reprinted from I.-S. Park and N. Kim, *Analytica Chimica Acta*, 394:201–10, 1999, with permission from Elsevier.)

beads) were packed into columns that had an internal diameter of 3 mm, and the column lengths were varied to degrade H_x, H_xR, and IMP at the corresponding F, G, and H columns. Here, columns F, G, and H were designated as the H_x, H_xR, and IMP columns. The maximum sensor signals for the H_x, H_xR, and IMP columns were obtained at pH values of 7.5, 8.0, and 9.5, respectively, and alkaline shifts in the pH optimum due to immobilization were observed for the H_xR and IMP columns. The sensor signals for the H_x and H_xR columns were maximal at 35°C, and the sensor signals for the IMP column were maximal at 30°C (Park et al., 2000a).

By using 0.05 M phosphate buffer (pH 7.5) as the eluent, the flow type biosensor system for freshness measurements was operated at a flow rate of 0.3 ml/min. First, various column length combinations of the H_x, H_xR, and IMP column were compared to determine the optimum column lengths of the three-enzyme reactors (Table 10.1). From these experiments the optimal length combination of the serially connected three enzyme reactors was determined to be 15-12-7 cm. Under these conditions, when only H_x was loaded into the system, it was completely degraded at the H_x column. Likewise, H_xR was only decomposed at the H_xR column. This result clearly showed that a separate analysis of the major components related to the biochemical freshness indices is possible with three-enzyme reactors having the length combination of 15-12-7 cm.

TABLE 10.1

Comparison of the Accuracy of the Three-Enzyme Reactors at Different Length Combinations

Column length (cm)		H_x (3.73 mM) Major peak Cm (mM)	H_x (3.73 mM) Minor peak	H_xR (1.86 mM) Major peak	H_xR (1.86 mM) Minor peak
Reactor F	10	8.70 (3.15a)			
G	10		3.30 (0.58a)	3.52 (1.42b)	
H	7				1.00 (0.17b)
Reactor F	12	9.90 (3.35a)			
G	10		0.60 (0.37a)	4.00 (1.62b)	
H	7				0.70 (0.05b)
Reactor F	12	10.20 (3.60a)			
G	12		0.40 (0.10a)	5.20 (1.79b)	
H	7				0.30 (0.02b)
Reactor F	15	11.40 (3.73a)			
G	12		0.20 (0.00a)	5.05 (1.83b)	
H	7				0.05 (0.00b)

[a] Calculated from an H_x calibration curve.
[b] Calculated from an H_xR calibration curve.

The application of the biosensor system with three-enzyme reactors having the length combination of 15-12-7 cm on the freshness evaluation of fish fillet was performed together with a comparative high-performance liquid chromatography (HPLC) analysis using a reversed-phase μBondapak C_{18} column. In this comparative analysis, a fish extract was prepared by the homogenization of a 5 g sample with 25 ml of 0.05 M phosphate buffer (pH 7.5) for 5 min, followed by the successive filtration with Whatman no. 1 filter paper and a 0.45 μm polyvinylidene fluoride (PVDF) syringe filter. After ATP-degradative compounds from various fish extracts were measured by the two methods, the analytical results and biochemical freshness indices deduced from them were compared and correlated (Figure 10.3). A good correlation with a correlation coefficient (r) of 0.9713 was obtained between the biosensor method and HPLC in regards to the contents of H_x, H_xR, and IMP. The two biochemical freshness indices, K_i and H, were also similar between the two methods.

The extracts of beef loins and chickens, which were bought on the day of sample treatment and preserved for 7 days at 4°C after purchase, were prepared according to the procedure described above. The extracts were then analyzed by the three-enzyme reactors having the length combination of 15-12-7 cm. Compared to the H_x contents in the control samples, those in the beef loins and chickens undergoing cold storage were conspicuously high, whereas the IMP contents in the beef loins and chickens decreased abruptly, resulting in a conspicuous increase in K_i and H in both meats (Table 10.2). It was also presumed that chickens are more susceptible to deterioration during cold storage, and K_i and H are good indices of meat deterioration during cold storage.

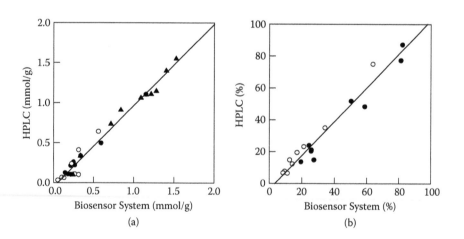

FIGURE 10.3 Correlations between the biosensor system and HPLC with respect to ATP-degradative compounds (a) and fish freshness indices (b). Symbols: (a) (•) H_x, (o) H_xR, (▲) IMP; (b) (•) K_i value; (o) H value. (Reprinted from I.-S. Park and N. Kim, *Analytica Chimica Acta*, 394:201–10, 1999, with permission from Elsevier.)

TABLE 10.2

Determination of H_x, H_xR, and IMP in the Beef Loins and Chickens, and the Resulting K_i and H Values

Sample[a]	Control[b]					After 7 days of storage at 4°C				
	H_x (μmol/g)	H_xR (μmol/g)	IMP (μmol/g)	K_i value (%)	H value (%)	H_x (μmol/g)	H_xR (μmol/g)	IMP (μmol/g)	K_i value (%)	H value (%)
Loin	1.149	0.198	0.994	57.5	49.1	2.156	0.222	1.248	65.6	59.5
Loin	1.258	0.194	1.158	55.6	48.2	2.033	0.210	1.050	68.1	61.7
Loin	1.536	0.290	1.306	58.3	49.0	1.203	0.134	0.382	77.8	70.0
Loin	1.265	0.182	0.573	71.6	62.6	2.005	0.225	0.328	87.2	78.4
Chicken	1.139	0.473	1.884	46.1	32.6	1.308	0.741	1.185	63.4	40.4
Chicken	0.718	0.335	1.655	38.9	26.5	1.938	0.332	0.570	79.9	68.2
Chicken	0.637	0.349	1.882	34.4	22.2	3.164	0.480	0.806	81.9	71.1
Chicken	0.890	0.368	1.632	43.5	30.8	1.980	0.259	0.206	91.6	81.0

Source: Reprinted from I.-S. Park, Y.-J. Cho, and N. Kim, *Analytica Chimica Acta*, 404:75–81, 2000, with permission from Elsevier.

[a] Treated from different samples.

[b] Delivered from a market and treated immediately.

10.3 BIOMIMETIC SENSOR FOR DISCRIMINANT TASTE EVALUATION AND FERMENTATION MONITORING

It is expected that instrumental methods for taste evaluation could offer an efficient method for taste quantification and grouping compared to conventional sensory evaluation. Thus, various types of taste sensors have been developed (Birch and Ogunmoyela, 1980; Ikezaki et al., 1991; Akiyama et al., 1996; Oohira and Toko, 1996; Toko and Fukusaka, 1997). In this context, the mechanism of taste sensation by the human tongue is a useful reference in the preparation of a taste sensor. Taste sensation starts with the perception by the biological membrane of gustatory cells in the taste buds on the tongue, and a taste substance perceived by the membrane brings about an electric signal transmitted along the nerve fiber to the brain, which results in taste sensation (Toko, 1998). Therefore, a significant amount of research has been devoted to manufacturing artificial taste sensors that mimic the transduction mechanism of the human gustatory system, and various polymer membrane-based taste sensors exploiting membrane components like phospholipids have been developed.

In addition to the taste sensors described above, another approach uses ionophores and ion exchangers as the electroactive materials for the preparation of polymer membranes, which are required to handle food samples mixed with various components like ions and salts. These biosensors normally determine membrane potential, and the optimal membrane potential has been reported to be 59.1 mV per concentration decade in a semilogarithmic plot at 25°C, irrespective of the electroactive materials used (Atkins, 1996). Our group developed an eight-channel taste evaluation system that employed cation- and anion-selective ionophores, and ion exchangers as electroactive materials, and applied it to the discriminant analysis of various food products and to monitor *Kimchi* fermentation.

The polymer membranes used for the eight-channel sensor array contained the electroactive material, Bis(2-ethylhexyl)sebacate as a plasticizer to enhance the electrochemical reaction during the perception of taste substances and to provide membrane flexibility, and polyvinyl chloride as the polymer matrix in a ratio of 1:66:33 with 1 ml of tetrahydrofuran. The polymer membrane was prepared by casting the above mixture into a glass cylinder holder that had an internal diameter of 2.5 cm. After solvent evaporation, the polymer membrane with a uniform surface and an approximate thickness of 300 μm was obtained. When required, the polymer membrane was taken from a storage bottle, punched in 5 mm diameter, and installed into a sensitive area of a Phillips electrode body. Out of the eight electroactive materials used, 4-tert-butylcalix[4]arene tetraacetic acid tetraethyl ester (calix[4]arene), monensin decyl ester (MDE), valinomycin, nonactin, and tridodecylamine (TDDA) were cation selective. Tridodecylmethylammonium chloride (TDMA), tri-*n*-octylmethylammonium chloride (TOMA), and *meso*-tetraphenylporphyrin manganese (III) chloride (Mn-porphyrin) were anion selective (Kim et al., 2005). Figure 10.4 shows a schematic representation of the eight-channel taste evaluation system. It consisted of an eight-channel sensor array, a double-junction Ag/AgCl reference electrode, a sixteen-channel high-input impedance amplifier, and a multichannel A/D converter. The sensor array and reference electrode were fixed into a round acryl frame, and the resulting frame was inserted into the sample during measurement. The A/D converter

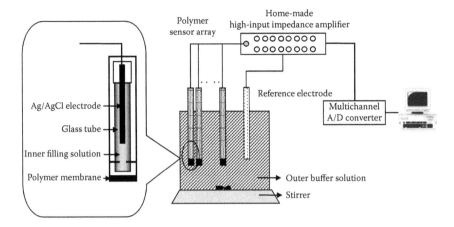

FIGURE 10.4 Schematic diagram of the taste evaluation system. (Reprinted from N. Kim, K.-R. Park, I.-S. Park, Y.-J. Cho, and Y. M. Bae, *Biosensors and Bioelectronics*, 20:2283–91, 2005, with permission from Elsevier.)

control software was developed with LabView, and white noise was removed by an averaging method for digital filtering to improve the signal-to-noise ratio.

Each cation- and anion-selective polymer membrane in the sensor array was determined for concentration dependency against an artificial taste substance (Figure 10.5). To do this, solutions of NaCl, KCl, NaBr, and lactic acid were made at concentrations of 10^{-6}, 10^{-5}, 10^{-4}, 10^{-3}, 0.01, and 0.1 M, and were used as the analytes to measure membrane potential. Out of the cation-selective polymer membranes, the calix[4]arene membrane showed an increased membrane potential just after the electrode was placed into different NaCl solutions, and the steady-state responses were obtained within 50 s. The lower detection limit for Na$^+$ ions was 10^{-6} M, and the membrane potentials at increasing Na$^+$ ion concentrations from 10^{-5} to 0.1 M were in the range of –219.0 to 60.6 mV, with a slope of 76.7 mV/decade. In the case of the MDE membrane, the concentration dependency at varying Na$^+$ ion concentrations was less pronounced than that of the calix[4]arene membrane. The valinomycin membrane electrode showed excellent linearity with a slope of 59.4 mV/decade at varying K$^+$ ion concentrations. The nonactin membrane, however, showed a less conspicuous concentration dependency, and thus responded to K$^+$ ions at a concentration of 10^{-4} M with a slope of 61.9 mV/decade. As the electroactive material responding to H$^+$ ions, TDDA is reportedly known to be involved in sour taste sensation (Schulthess et al., 1981). In our study with the TDDA membrane electrode, a concentration-dependent response to lactic acid was observed with a nearly Nernstian slope of 60.7 mV/decade. From the anion-selective polymer membranes used, the response of the TDMA membrane, which had quaternary amine functionality, was measured at different NaBr concentrations. In this case, the membrane potentials decreased according to the increase in analyte concentration, which indicated an anion-selective response of the TDMA membrane. The TOMA membrane electrode, which also contained quaternary amines, showed a linear response to bromide ions at a concentration of 10^{-5} M, with a slope of 51.9 mV/decade. This

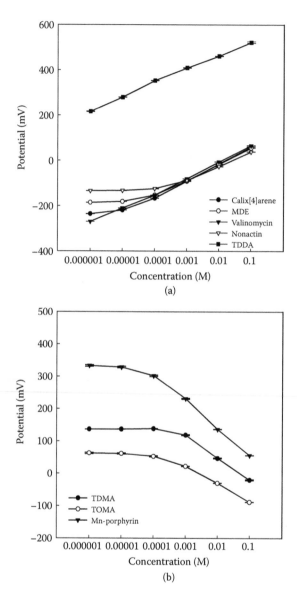

FIGURE 10.5 Changes in membrane potential through polymer membranes at various concentrations of artificial taste substances: (a) cation-selective polymer membranes and (b) anion-selective polymer membranes. The artificial taste substances for calix[4]arene, MDE, valinomycin, nonactin, and TDDA membrane were NaCl, NaCl, KCl, KCl, and lactic acid, respectively. That for TDMA, TOMA, and Mn-porphyrin membrane was NaBr. Measurements were conducted in duplicate and the error bars were inserted. (Reprinted from N. Kim, K.-R. Park, I.-S. Park, Y.-J. Cho, and Y. M. Bae, *Biosensors and Bioelectronics*, 20, 2283–91, 2005, with permission from Elsevier.)

concentration-dependent response was significantly better than that of the TDMA membrane. At a NaBr concentration of 10^{-5} M, the Mn-porphyrin membrane electrode showed responses to NaBr, and the total change in membrane potential and the slope in the linear region were 272.7 mV and 71.0 mV/decade, respectively. Overall, the concentration-dependent slopes in the linear region varied from 51.9 to 76.7 mV/decade for the eight polymer membrane electrodes tested, which corresponded to 87.8 to 129.8% of the theoretical value of 59.1 mV/decade (Pungor, 1998). This deviation, which was observed for all membranes except the valinomycin and TDDA membrane, might be due to the analyte properties such as lipophilicity, structural properties of electroactive materials, a possible interferant effect, and the change in buffering capacity of the basal solution (Craggs et al., 1974; Pungor, 1998; Oh et al., 1999; Bae, 2000).

A meaningful application of the taste sensor might be discriminant taste analysis (Imamura et al., 1996; Toko, 1998). Therefore, a discriminant analysis was performed using the eight-channel taste sensor array for six groups of beverages marketed in Korea: *Sikhye*, *Sujunggwa*, tangerine juice, *Ume* juice, ionic drink, and green tea (Park et al., 2004a). The responses of the individual polymer membrane electrodes constituting the sensor array against twenty-four samples belonging to six beverage groups were determined (Figure 10.6). The significance of the taste sensor responses was verified using a Statistical Analysis System (SAS) in the 95% confidence interval (Table 10.3). All membrane electrodes in the sensor array showed positive averaged potentials against tangerine juice, *Ume* juice, and ionic drink. However, *Sikhye* showed a negative averaged potential against the calix[4]arene membrane electrode; *Sujunggwa* against TDMA, TOMA, and calix[4]arene membrane electrodes; and green tea against TDMA, TOMA, calix[4]arene, MDE, and TDDA membrane electrodes. In contrast, the difference in averaged sensor responses for each beverage group increased in the following order: ionic drink, green tea, tangerine juice, *Sujunggwa*, *Ume* juice, and *Sikhye*. As shown in Table 10.3, all polymer membranes in the sensor array showed significant differences between tangerine juice and green tea, *Ume* juice and green tea, and ionic drink and green tea. In the case of *Ume* juice and the ionic drink, however, only the Mn-porphyrin membrane showed significant differences. For *Sikhye* and *Sujunggwa*, the Mn-porphyrin and valinomycin membrane showed significant differences. By using the membrane potentials obtained by the sensor array, principal component analysis (PCA) was carried out in three-dimensional planes (Figure 10.7). Using this approach, the six beverage groups were successfully discriminated.

The taste evaluation system described above was also applied to the discriminant analysis of three groups of alcoholic liquor marketed in Korea: *Maesilju*, *Soju*, and beer. In this analysis, the first, second, and third principal components were responsible for most of the total data variance, and the analyzed liquor samples were well classified into two-dimensional planes by PCA composed of the first-second and first-third principal components from the potential responses of the sensor array (Kim, 2005). In the future, more beverages and liquor, including soybean milk and *Macgulri*, could be discriminated by statistical analysis or an artificial neural network from the potential responses of the taste sensor array after further development of novel electroactive materials that act on ion groups or specific ions.

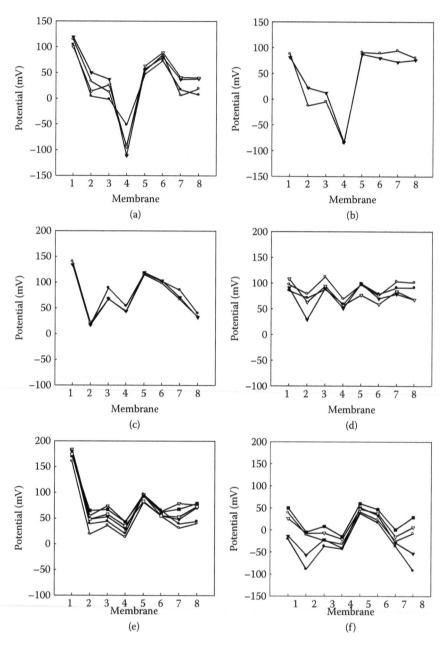

FIGURE 10.6 Responses of the individual taste sensors in the sensor array to various beverage groups: *Sikhye* (a), *Sujunggwa* (b), tangerine juice (c), ionic drink (d), *Ume* juice (e), and green tea (f). Electroactive materials: 1, Mn-porphyrin; 2, TDMA; 3, TOMA; 4, calix[4] arene; 5, valinomycin; 6, nonactin; 7, MDE; 8, TDDA. All symbols in each beverage group represent different commercial products. (Reproduced from K.-R. Park, Y.-M. Bae, I.-S. Park, Y.-J. Cho, and N. Kim, *J. Kor. Soc. Appl. Biol. Chem.*, 47:300–6, 2004a, with permission from the Korean Society for Applied Biological Chemistry.)

TABLE 10.3

Response of Each Taste Sensor in the Sensor Array to Various Marketed Beverages

Electroactive material			Response (mV)			
	Sikhye	Sujunggwa	Tangerine juice	Ume juice	Ionic drink	Green tea
Mn-porphyrin	109.25[c]	73.67[d]	138.00[b]	175.00[a]	97.80[c]	16.00[e]
TDMA	25.25[ab]	−18.67[ab]	18.33[ab]	45.67[a]	41.00[a]	−34.00[b]
TOMA	18.25[b]	−16.67[b]	75.00[a]	55.33[a]	86.00[a]	−16.40[b]
Calix[4]arene	−87.25[c]	−65.67[c]	47.00[a]	30.50[a]	48.20[a]	−30.40[b]
Valinomycin	55.25[c]	77.00[b]	117.00[a]	90.50[b]	92.80[b]	47.60[c]
Nonactin	81.75[b]	73.00[b]	101.00[a]	59.33[b]	67.00[b]	31.80[c]
MDE	25.25[a]	45.67[a]	74.33[a]	53.00[a]	76.80[a]	−21.00[b]
TDDA	26.00[ab]	39.00[ab]	35.00[a]	63.83[a]	75.60[a]	−23.60[b]

Source: Reproduced from K.-R. Park, Y.-M. Bae, I.-S. Park, Y.-J. Cho, and N. Kim, *J. Kor. Soc. Appl. Biol. Chem.*, 47:300–6, 2004a, with permission from the Korean Society for Applied Biological Chemistry.

[a–c] Means within the same row with different superscripts are significantly different at $p < 0.05$ by ANOVA.

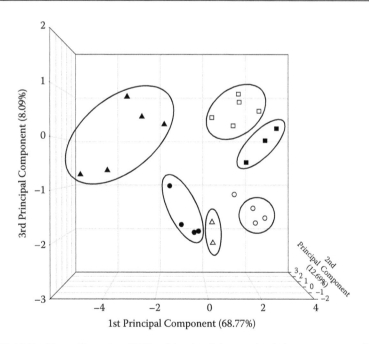

FIGURE 10.7 Three-dimensional PCA of the signal data on the six beverage groups. Symbols: •, *Sikhye*; △, *Sujunggwa*; ○, tangerine juice; ■, ionic drink; □, *Ume* juice; ▲, green tea. (Reproduced from K.-R. Park, Y.-M. Bae, I.-S. Park, Y.-J. Cho, and N. Kim, *J. Kor. Soc. Appl. Biol. Chem.*, 47:300–6, 2004a, with permission from the Korean Society for Applied Biological Chemistry.)

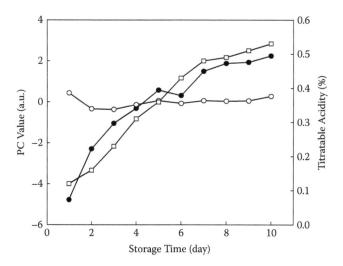

FIGURE 10.8   Changes in the values of PCs and titratable acidity of the analytical samples prepared from *Backkimchi* during storage for ten days at 4°C in the case of the cation-selective polymer membranes. Symbols: (•) PC1, (o) PC2, (ɔ) titratable acidity. (Reprinted from N. Kim, K.-R. Park, I.-S. Park, Y.-J. Cho, and Y. M. Bae, *Biosensors and Bioelectronics*, 20:2283–91, 2005, with permission from Elsevier.)

Kimchi, which is a traditional Korean pickle fermented by lactic acid bacteria, is now consumed worldwide because it contains many functional properties, including anticarcinogenic activity (Choi et al., 1997). Therefore, this taste evaluation system was used to monitor *Kimchi* fermentation because it normally contains a high content of organic acids and salts. In *Kimchi* fermentation, titratable acidity is conventionally used as the ripening index and is adjusted to around 0.4% before low-temperature marketing at 0 to 4°C. During *Kimchi* fermentation at 4°C, the time-dependent changes in the first and second principal components were tracked together with the titratable acidity using cation-selective polymer membrane electrodes. As shown in Figure 10.8, a better correlation between the first principal component and titratable acidity was found than that obtained with the all-polymer membrane electrodes in the sensor array. This might be attributable to the fact that H+ ions were dissociated from the organic acids, including lactic acid, during fermentation. In the case of anion-selective polymer membranes, however, the first principal component was less correlated with titratable acidity during fermentation. Based on this result, the biomimetic sensor system for taste evaluation seems to be capable of monitoring pickle fermentation and holds promise for use by the food industry to measure the fermentation process in real time on a computer.

10.4   BIOSENSORS FOR FOOD COMPONENT ANALYSIS

Biosensors are useful for on-site analysis and monitoring of food components, which are important indices of quality during fermentation, storage, and manufacturing. These components might include sugars like glucose and lactose, organic acids such as L-lactic acid and L-ascorbic acid (vitamin C), and chelating agents like ethylene-diaminetetraacetic acid (EDTA) (Pfeiffer et al., 1990; Park et al., 1995; Volotovsky and Kim, 1998a). Our group has developed electrochemical biosensors for food component analysis based on a needle type three-layer membrane electrode and ion-sensitive field effect transistor (ISFET).

Lactic acid is produced during pickle fermentation by lactic acid bacteria such as *Leuconostoc mesenteroides* and *Lactobacillus plantarum*, which are also responsible for its characteristic taste and flavor. Acid production, routinely measured by titratable acidity and pH, has been used as an important index of the ripening of pickles, including *Kimchi*. However, there is accumulating evidence that the lactic acid content is more correlated with the sensory properties during pickle ripening than titratable acidity and pH (Kim et al., 1986; Park et al., 1993). The needle type enzyme membrane H_2O_2 electrode was prepared by merging a working platinum electrode and Ag/AgCl reference electrode into insulating tubing and heat-shrink-able vinyl tubing after individual preparation with a platinum and silver wire having a diameter of 0.5 mm and heat-shrinkable fluorinated ethylene propylene copolymer (Kim, 1997). The enzyme membrane was prepared using three layers: an inner layer for interference elimination including Nafion or cellulose actetate, a lactate oxidase (LOD)–immobilized middle layer, and a outer protective layer containing ethyl cellulose, cellulose acetate, or cellulose triacetate. LOD catalyzes lactic acid oxidation over the electrode surface as follows:

$$\text{Lactic acid} + \text{LOD (FAD}^+) \rightarrow \text{pyruvic acid} + \text{LOD (FADH}_2) \qquad (10.3)$$

$$\text{LOD (FADH}_2) + O_2 \rightarrow \text{LOD (FAD}^+) + H_2O_2 \qquad (10.4)$$

$$H_2O_2 \rightarrow 2H^+ + O_2 + 2e^- \qquad (10.5)$$

When the electrochemical properties of the lactate sensor containing the middle enzyme layer were evaluated (Figure 10.9), a current response increasing sigmoi-dally against 4.99 mg/dl of L-lactate was observed at increasing potential. However, a significant inhibition current due to ascorbic acid oxidation was also found at a similar potential range (Scholze et al., 1991). When the cyclic voltammogram up to 700 mV against 8.32 mg/dl of lactic acid was compared to that without the addition of analyte, a clear distinction was observed. A current-time recording and the corresponding calibration curve indicated a linear relationship between analyte concentration and current response up to a lactic acid concentration of 4.99 mg/dl. Since only 1.8 to 3.0 min was needed to reach a steady-state current response after analyte addition, this lactate sensor was demonstrated to be a rapid analytical tool to measure lactic acid in food samples.

As shown in Figure 10.9, a H_2O_2-based electrode is susceptible to interference from reducing compounds such as vitamin C, cysteine, and tocopherol. In the needle type lactate sensor, interference was eliminated through the inner layer by exploiting the bulkiness and electrical properties of the polymer. As shown in Table 10.4, ascorbic acid interference was almost completely eliminated by using an inner layer coating of 2% acetonic or 5% ethanolic polymer solution. Ascorbic acid interference was

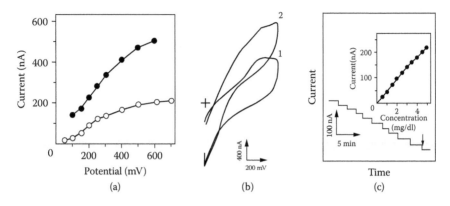

(a)   (b)   (c)

FIGURE 10.9 Electrochemical properties of a needle type lactate sensor. (a) Effects of the measuring potential on the current response against 4.99 mg/dl of lactic acid (o) and the inhibition current against 4.99 mg/dl of ascorbic acid (•). (b) Cyclic voltammogram of the biosensor at lactic acid concentrations of 0 (1) and 4.99 (2) mg/dl, respectively. (c) Current-time recording and the corresponding calibration curve. Arrow indicates point of analyte addition. (From N. Kim, *Foods Biotechnol.*, 6:113–17, 1997, with permission from the Korean Society of Food science and Technology.)

TABLE 10.4

Removal of L-Ascorbic Acid Interference in Phosphate Buffer (pH 7.0) by Inner Layer Coating with a Polymer Solution

	Removal of L-ascorbic acid interference (%)[a]				
Dip coating (times)	Nafion[b]	Cellulose acetate[c]	Ethyl cellulose[c]	Palmitic acid[d]	Lauric acid[d]
1	60.6	47.0	0	-[e]	-
2	95.7	57.8	-	52.4	0
3	97.0	73.0	-	-	-
4	97.9	90.7	-	-	-
5	-	96.0	67.6	64.3	14.0
8	-	-	83.2	-	34.8
10	-	-	78.4	-	-
15	-	-	91.9	-	-

Source: N. Kim, *Foods Biotechnol.*, 6:113–17, 1997, with permission from the Korean Society of food science and Technology.

[a] Determined by comparing the current response of the electrode with or without inner layer coating.

[b] One percent solution dissolved in the mixture of low chain aliphatic alcohol.

[c] Two percent acetonic solution.

[d] Five percent ethanolic solution.

[e] Not determined.

removed by 97 to 98% after Nafion coating three or four times. Cellulose acetate was also as effective as Nafion. In the case of ethyl cellulose, the coating times needed to be increased to obtain a similar degree of interference elimination. On the other hand, fatty acids including lauric and palmitic acid were shown to be less effective in eliminating the interfering current caused by reducing compounds. By casting an outer layer over the enzyme-containing middle layer, the dynamic range of the L-lactate sensor was significantly increased from 15–25 to 150–250 mg/dl of lactic acid (Kim, 1997).

The needle type lactate sensor based on H_2O_2 electrode showed a pH optimum around 9 to 10 and temperature optimum of 45°C. Current responses were stable over forty days and the response to lactic acid was quite specific. When applied to lactic acid measurements in *Kimchi* and yogurt, the measured lactic acid contents were accurate, 187.4 ± 4.1 and 734.1 ± 34.5 mg/dl, and similar to the values measured using a spectrophotometric lactate kit (Kim et al., 1996).

Ascorbic acid is an important quality index of fruit beverages, including fruit juice, because of its strong antioxidative activity. Therefore, it is frequently selected as a functional material for use in creating functional foods with antioxidative activity. During storage processes including low-temperature preservation, the oxidation of ascorbic acid to dehydroascorbic acid might occur, which greatly deteriorates juice quality. Many studies have been conducted to develop methods of detecting ascorbic acid in food matrices. Some of the approaches developed depend on an efficient separation of ascorbic acid from all other substances, after which the analyte is measured using a suitable detector. On the other hand, enzyme-based biosensor analysis makes use of the specificity of enzyme reactions to measure ascorbic acid. For example, peroxidase reduces H_2O_2 to water in the presence of ascorbic acid together with the liberation of dehydroascorbic acid as follows (Volotovsky and Kim, 1998b):

$$\text{Ascorbic acid} + H_2O_2 \xrightarrow{\text{Peroxidase}} \text{Dehydroascorbic acid} + 2H_2O. \quad (10.6)$$

Although phenolic compounds, diamines, and leuco dyes are also possible hydrogen donors, vitamin C is likely to be the only peroxidase substrate in fruit beverages present at relatively high concentrations up to several millimoles/liter (Miller and Rice-Evans, 1997; Uchiyama et al., 1997). Another reason for using peroxidase in a biosensor system is that it causes a local pH change in the biomembrane immobilized over a pH-sensitive potentiometric transducer, ISFET.

Figure 10.10 shows the design of the ISFET biosensor used for vitamin C determination. The sensor chip (3 × 10 mm) contains two identical ISFETs covered with or without the biomembrane, which was formed by depositing a solution containing 5% peroxidase, 5% bovine serum albumin (BSA), and 10% glycerol, followed by cross-linking with saturated glutaraldehyde vapor for 30 min at room temperature and drying in air for 15 min after rinsing. Consumption of ascorbic acid during enzyme-catalyzed H_2O_2 reduction caused a local pH increase in the biomembrane of ISFET 1, and the differential signal between the working and reference ISFETs was amplified and recorded. Measurements were conducted at room temperature in a glass cell having the nominal capacity of 1.5 ml filled with a solution containing phosphate or citrate buffer and 100 mM NaCl.

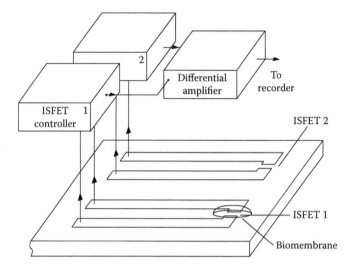

FIGURE 10.10 Setup of the ISFET-based biosensor. (Reprinted from V. Volotovsky, Y. J. Nam, and N. Kim, *Sensors and Actuators B: Chemical*, 42:233–37, 1997, with permission from Elsevier.)

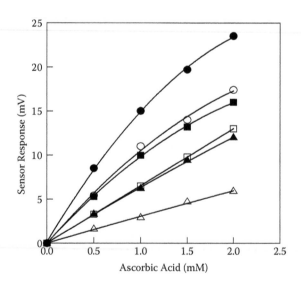

FIGURE 10.11 Sensor performance in phosphate (filled symbols) and citrate (open symbols) buffers (pH 6.0). Buffer concentrations were 5 (•,○), 10 (■,□), and 20 mM (▲,△), respectively. (Reprinted from V. Volotovsky and N. Kim, *Analytica Chimica Acta*, 359:143–48, 1998, with permission from Elsevier.)

TABLE 10.5

Comparison of Analytical Methods

	Vimatin C Concentration (mM)	
Beverage	HPLC	Biosensor
Del Monte 100% orange juice	2.57 ± 0.08	2.72 ± 0.10
Lotte Samkang 100% orange juice	2.42 ± 0.05	2.34 ± 0.07
Haejo Miin carbonated drink	1.91 ± 0.04	1.60 ± 0.06
Gerber mixed juice (apple, cherry)	1.40 ± 0.01	1.68 ± 0.07
Capri Sonne "Safari" drink	0.89 ± 0.02	1.12 ± 0.02

Source: Reprinted from V. Volotovsky and N. Kim, *Analytica Chimica Acta*, 359, 143–48, 1998, with permission from Elsevier.

Vitamin C was determined using the ISFET biosensor (Figure 10.10) at a preoptimized H_2O_2 concentration of 1 mM dissolved in the two reaction buffers described above. The degree of response of a potentiometric biosensor has been shown to depend on the buffering capacity of the bulk solution. Thus, as expected, the buffer concentration determined the magnitude of the response and linear dynamic range (Figure 10.11).

Some fruit beverages from a local market were diluted with 10 mM phosphate buffer containing 100 mM NaCl by fourfold so that the ascorbic acid content was lower than 1 mM (Cooke and Moxon, 1982). The contents of vitamin C in the diluted samples were then determined using the ISFET biosensor after pH adjustment to 6.0, which is the optimal pH for peroxidase activity. The results of this analysis were compared to the analytical results by HPLC. As shown from Table 10.5, the biosensor response was quite reproducible and a relatively good correlation was obtained between the two methods. When ascorbic acid measurements in the two buffer systems were compared, a good correlation was also obtained. However, in the citrate buffer system, the pH adjustment to 6.0 was not required because of the relatively high buffering capacity of the citrate buffer. Therefore, the citrate buffer system was recommended in the case of fruit beverages that have a high vitamin C content, such as orange juice. In contrast, the phosphate buffer system was more approximate for analysis of beverages with a low ascorbic acid concentration since the sensor response is higher than in the citrate buffer system under these conditions.

Chelating agents (sequestrants) maintain the integrity of many food products by removing trace metal catalysts in fats and oils, and preventing deteriorative changes leading to discoloration, off-flavor, and unacceptable odors in shellfish (Volotovsky and Kim, 1998a). EDTA, which has a good chelating property at relatively low concentrations, is the most prominent sequestrant used in food. As the use of EDTA and other sequestrants is strictly regulated by the Food and Drug Administration or related foreign agencies, it is important to develop biosensors to measure these substances.

By exploiting the EDTA-induced restoration of urease activity inhibited by copper ions (II), we prepared a potentiometric ISFET biosensor that could measure EDTA concentration. In this device, urease was immobilized onto the film of a positively charged polymer, poly(4-vinylpyridine-*co*-styrene) (PVPy), on the sensitive area of

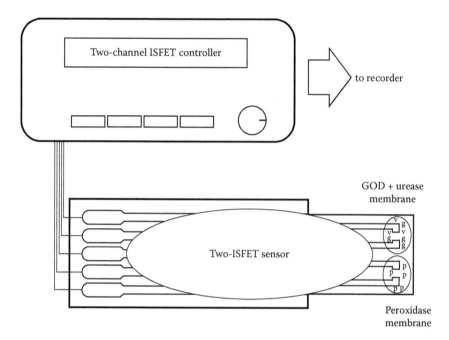

FIGURE 10.12 Experimental setup of a two-ISFET sensor for multicomponent analysis. (Reprinted from V. Volotovsky and N. Kim, *Sensors and Actuators B: Chemical*, 49:253–57, 1998, with permission from Elsevier.)

an ISFET, and this restored the catalytic activity of urease, which is inhibited by Cu^{2+} ions in the presence of EDTA. This biosensor was able to measure EDTA concentration in the range of 5 to 500 ppm (Volotovsky and Kim, 1998a).

By using a two-ISFET multienzyme biosensor as depicted in Figure 10.12, it was possible to measure glucose, vitamin C, and citric acid. To prepare the peroxidase biomembrane, a solution containing 5% peroxidase, 5% BSA, and 10% glycerol in 10 mM phosphate buffer (pH 6.0) was deposited onto the sensitive area of one ISFET, followed by cross-linking with saturated glutaraldehyde vapor for 30 min. A solution including 10% glucose oxidase (GOD) and 5% urease in the same buffer solution was separately deposited onto the other ISFET, using the same procedures described above. The resulting biomembranes were covered with a 0.5% ethanolic solution of PVPy and dried in air for 5 min. The proposed biosensor was able to measure glucose, vitamin C, and citric acid in fruit juices, and the dynamic ranges for glucose, ascorbic acid, and citric acid were 1–10, 0.25–2, and 5–100 mM, respectively (Volotovsky and Kim, 1998c).

10.5 LABEL-FREE BIOSENSORS FOR MICROBIAL DETECTION

The methods commonly used for microbial detection with biosensors include nucleic acid hybridization, complex formation between antigens and antibodies, and binding of ligand and receptor. Out of these approaches, complex formation between antigens and antibodies has been the most widely used method for the development of immunosensors, which allows for high-sensitivity detection of microbes of hygienic importance, such as *Salmonella typhimurium* and *Escherichia coli*. Conventionally, biosensors using this detection method adopt experimental formats similar to those used by immunoassays, including enzyme-linked immunosorbent assay (ELISA). A labeled immunosensor exploits the competitive binding between analytes and analytes labeled with an enzyme or chromogen (tracer) to antibodies immobilized on the transducer surface (direct competitive), or the competitive binding between analytes and labeled antibody to antigens immobilized on the transducer surface (indirect competitive). On the other hand, a label-free immunosensor measures the changes in physical parameters, such as frequency response and refractive index after immune binding, without the need for a tracer or labeled antibody for signal transduction. The former can be divided into fiber optic, particle-based, and membrane-based immunosensors, and has the merits of intrinsic high sensitivity due to the use of labels like quantum dot and fluorescent silica nanoparticle, diversity in system setup, and expansibility to other domains like cell imaging. Since the latter does not use labels for signal transduction, it normally requires simple protocols and is easy to operate. Simultaneously, it can be highly sensitive, depending on selection of the appropriate analysis formats. Representative label-free immunosensors include quartz crystal microbalance (QCM), surface plasmon resonance (SPR), optical grating coupler (OGC), and scanning angle reflectometry and ellipsometry (Huetz et al., 1995; Elwing, 1998; Homola et al., 1999; Benesch et al., 2000; Vörös et al., 2002). Our group has developed direct-binding QCM and OGC immunosensors for *S. typhimurium* and *E. coli* by immobilizing thiolated antibodies to gold electrodes on a 9 MHz AT-cut QCM through self-assembled monolayers (SAMs), or by the immobilization of microbial antibodies over the activated quartz grating on the waveguide layer of an OGC via amino-, epoxide- and thiol-terminal silane surface.

To measure *S. typhimurium*, an anti-*Salmonella* antibody raised against the common structural antigen of the *Salmonella* spp. was thiolated with four thiolation cross-linkers comprising sulfosuccinimidyl 6-[3-(2-pyridyldithio)propionamido]hexanoate (sulfo-LC-SPDP), N-succinimidyl 3-(2-pyridyldithio)propionate (SPDP), 3,3'-dithiobis(sulfosuccinimidylpropionate) (DTSSP), and dimethyl-3,3'-dithiobispropionimidate·2HCl (DTBP) (Park and Kim, 1998). The procedure used to immobilize the antibodies to the QCM surface is depicted in Figure 10.13. Sulfo-LC-SPDP and SPDP react with primary amines and sulfhydryls in the antibody molecules, which introduces 2-pyridyl disulfide groups on the antibody molecules. The disulfide bonds are cleavable with a reducing agent like dithiothreitol. The thiolated antibody–gold complex SAM was spontaneously formed by chemisorption upon incubation with the thiolated antibody solution.

The QCM sensors immobilized with various thiolated antibodies were mounted inside a dip holder in a batch type reaction cell having a capacity around 30 ml, and the resonant frequencies after injecting 7.0×10^7 CFU/ml of *S. typhimurium* suspension were measured (Figure 10.14). The frequency shift (ΔF), which was defined as F_1 (the steady-state baseline resonant frequency before the addition of microbial suspension) minus F_2 (the steady-state baseline resonant frequency after the addition of microbial suspension), caused by complex formation between *Salmonella* and the immobilized antibodies was at the highest for sulfo-LC-SPDP thiolation, slightly exceeding 1,400 Hz. Therefore, thiolation through sulfo-LC-SPDP was determined

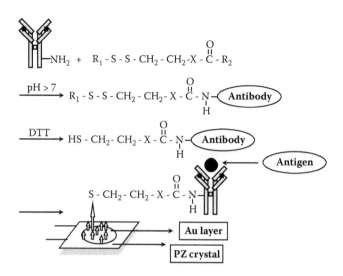

FIGURE 10.13 Mechanism of antibody adsorption with the thiolation cross-linkers onto the gold electrode of the quartz crystal. R_1 and R_2 are leaving groups. X is the residue remaining after antibody immobilization. DTBP, none; DTSSP, none; SPDP, none; sulfo-LC-SPDP, $-CO-NH-(CH_2)_5-$. (Reprinted from I.-S. Park and N. Kim, *Biosensors and Bioelectronics*, 13:1091–97, 1998, with permission from Elsevier.)

to be the best immobilization method to detect *S. typhimurium*. It was also presumed that sulfo-LC-SPDP might facilitate the binding of the microorganism to the immobilized antibody by reducing steric hindrance through its extended spacer arm (Park and Kim, 1998). When the frequency shifts from *Salmonella* adsorption during the course of antibody immobilization via sulfo-LC-SPDP were compared, a meaningful sensor response only occurred when the sensor chip was prepared using the procedure outlined in Figure 10.13. The frequency shift obtained with the bare chip and that obtained without cross-linker treatment were 30 and 50 Hz, respectively. These values, however, were negligible compared to the frequency shift when the chip was prepared using the thiolation cross-linker and antibody.

When the time-dependent resonant frequencies of the QCM sensor were measured at varying *S. typhimurium* concentrations in the range of 5.6×10^6 to 1.8×10^8 CFU/ml (Figure 10.15), the degree of the frequency shift from complex formation between the antigen and antibody, and the elapsed time were strongly dependent on the microbial density used. At cell densities higher than 1.4×10^7 CFU/ml, the resonant frequency began to decrease after 5 min and the steady-state frequency responses were acquired after 30 to 60 min. The *Salmonella* concentrations and frequency shifts obtained showed a linear relationship in a double-logarithmic scale in the cell concentration range from 2.0×10^6 to 1.8×10^8 CFU/ml (Figure 10.16). When a market milk spiked with 6.6×10^6 and 1.6×10^7 CFU/ml of *S. typhimurium* suspension was individually added to the above batch type reaction cell, frequency shifts far exceeding those due to the milk matrix were obtained. This result clearly indicated that the antibody QCM sensor can be applied to *Salmonella* detection in liquid food samples, including milk (Park et al., 2000b).

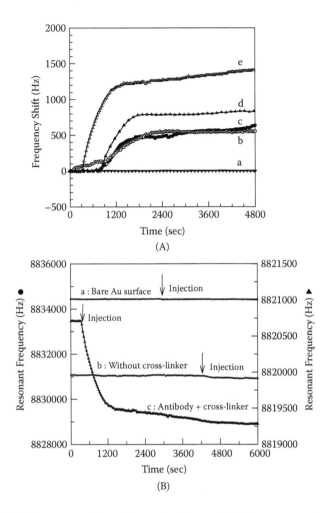

FIGURE 10.14 Frequency shifts of the antibody-immobilized quartz crystals according to the thiolation methods (A) and typical response profiles of the antibody-coated crystals during *Salmonella* adsorption (B). Symbols: (A) (a) without cross-linker, (b) DTSSP, (c) DTBP, (d) SPDP, (e) sulfo-LC-SPDP; (B) (a) the crystal with bare gold surface, (b) the crystal coated without sulfo-LC-SPDP, (c) the crystal immobilized with the cross-linker. The *Salmonella* suspension containing 7.0×10^7 CFU/ml cells was injected into the reaction cell and the frequency shift was measured. (Reprinted from I.-S. Park and N. Kim, *Biosensors and Bioelectronics*, 13: 1091–97, 1998, with permission from Elsevier.)

A flow type QCM immunosensor system for the detection of *E. coli*, which is an important index microorganism in the food industry, was also devised (Kim and Park, 2003). Figure 10.17 shows the arrangement of the sensor system. It was constructed with a micro-dispensing peristaltic pump, an injector, a flow cell installed with the sensor chip, an oscillator circuit, a quartz crystal analyzer, a potentiostat, and an IBM-compatible PC. The system was connected through capillary tubing and controlled by operating software. The reaction buffer of the system was 0.1 M

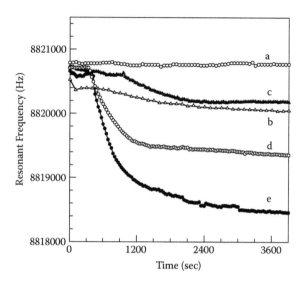

FIGURE 10.15   Typical responses of the antibody-immobilized sensor prepared by sulfo-LC-SPDP thiolation. The concentrations of the *Salmonella* suspension added were 0 (a), 5.6 × 10⁶ (b), 1.4 × 10⁷ (c), 7.0 × 10⁷ (d), and 1.8 × 10⁸ (e) CFU/ml. (Reprinted from I.-S. Park and N. Kim, *Biosensors and Bioelectronics*, 13:1091–97, 1998, with permission from Elsevier.)

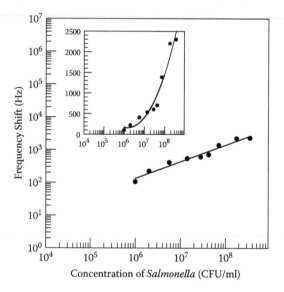

FIGURE 10.16   Relationship between *S. typhimurium* concentration and the frequency shift of the antibody-immobilized sensor. (Reprinted from I.-S. Park and N. Kim, *Biosensors and Bioelectronics*, 13:1091–97, 1998, with permission from Elsevier.)

sodium phosphate buffer (pH 7.2), as used by the batch type sensor system. The system operation was performed in stop-flow mode. That is, the reaction buffer was initially supplied to the system through the micro-dispensing pump. When the baseline frequency was stabilized, the sample was added through a 500 µl injection loop until the sample reached the flow-through cell. At this moment, the pump was stopped until a steady-state frequency shift was attained. The pump was then turned on and 1.2 M NaOH was supplied through the injection loop to remove bound bacterial cells, followed by flushing with the reaction buffer.

The specificity of the QCM sensor was evaluated in terms of detecting *E. coli*, which is related to food hygiene. To do this, cultured suspensions of *Enterobacter aerogenes* (one major coliform bacteria) and *S. typhimurium* and *Listeria monocytogenes* (food poisoning bacteria) were added at a cell density of 1.5×10^7, 1.9×10^7, and 3.5×10^7 CFU/ml, respectively, to the flow line of the sensor system at the optimized flow rate of 0.188 ml/min. The resulting sensor responses were compared to that obtained by the addition of 5.5×10^7 CFU/ml of *E. coli*. As shown in Table 10.6,

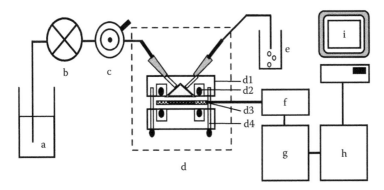

FIGURE 10.17 Schematic diagram of the apparatus used for the flow type antibody sensor based on QCM analysis. Symbols: (a) buffer, (b) peristaltic pump, (c) injector, (d) flow-through cell, (d_1) acryl holder, (d_2) O-ring, (d_3) quartz crystal, (d_4) joint, (e) waste, (f) oscillator circuit, (g) quartz crystal analyzer, (h) potentiostat, (i) PC. (Reprinted from N. Kim and I.-S. Park, *Biosensors and Bioelectronics*, 18:1101–7, 2003, with permission from Elsevier.)

TABLE 10.6
Specificity of the Flow Type Antibody Sensor

Microorganism	Frequency shift (Hz)	Cell concentration (CFU/ml)
Escherichia coli	281.1 ± 61.2	5.5×10^7
Enterobacter aerogenes	54.8 ± 15.6	1.5×10^7
Salmonella typhimurium	24.5 ± 22.3	1.9×10^7
Listeria monocytogenes	32.1 ± 14.7	3.5×10^7

Source: Reprinted from N. Kim and I.-S. Park, *Biosensors and Bioelectronics*, 18:1101–7, 2003, with permission from Elsevier.

the frequency shifts obtained by the addition of interfering bacteria were in the range of 24.5 ~ 54.8 Hz. These values were conspicuously lower than that attained by the addition of *E. coli*, which clearly indicated a good selectivity of the QCM sensor.

The flow type antibody sensor was applied to *E. coli* detection in various food samples, including drinking water, milk, dumpling, and beef. To prepare analytical samples, 20 g of solid sample was added to a Stomacher bag, followed by the addition of 200 ml of sterile distilled water. The bag was shaken vigorously for 10 min in the Stomacher. After setting still, the presence of coliform bacteria and *E. coli* in the supernatant was assessed by viable cell count based on lauryl tryptose agar and *E. coli* Petrifilm™ identification. Liquid samples were used as they were or after dilution with sterile distilled water, followed by viable cell count and *E. coli* Petrifilm identification. However, since all samples were *E. coli* negative, samples were spiked with cultured *E. coli* suspensions and enriched by inoculating 10^2 CFU/ml of cultured *E. coli* before microbial detection with the QCM sensor (Kim and Park, 2003). As shown in Table 10.7, the frequency shifts obtained with nonspiked samples ranged from 6.0 to 50.0 Hz, which might indicate a background effect, whereas spiking significantly increased the sensor response to values of 48.0 to 487.0 Hz, depending on sample type and spiked *E. coli* concentration. Similar results were found in the enriched samples. Sensor responses amounting to 52.0 to 487.0 Hz were attained according to the sample types used. The conspicuous differences in frequency shift between the spiked or enriched samples and negative control clearly indicated that the QCM sensor was capable of *E. coli* detection in

---

TABLE 10.7

Detection of the Spiked and Enriched *E. coli* in Various Food Samples with the Flow Type Antibody Sensor

Sample	Spiked *E. coli* (CFU/ml)	Frequency shift (Hz) Not spiked	Frequency shift (Hz) Spiked	Enriched *E. coli* (CFU/ml)	Frequency shift (Hz) Not enriched	Frequency shift (Hz) Enriched
Drinking water	1.3×10^7	27.9 ± 8.8^a	156.0 ± 22.2	1.0×10^6	27.9 ± 8.8	68.0 ± 32.2
Milk	1.0×10^7	21.1 ± 13.5	88.0 ± 23.6	1.0×10^6	21.1 ± 13.5	52.0 ± 23.1
Laver-rolled cooked rice	6.0×10^5	6.0 ± 4.9	48.0 ± 14.3	1.3×10^8	6.0 ± 4.9	476.0 ± 24.8
Dumpling	2.0×10^6	24.0 ± 14.3	72.0 ± 21.1	6.3×10^7	25.0 ± 14.3	291.0 ± 34.4
Boiled fish paste	1.0×10^8	26.0 ± 18.1	487.0 ± 32.1	1.6×10^8	26.0 ± 18.1	487.0 ± 36.6
Beef	4.0×10^7	33.7 ± 7.6	292.0 ± 17.3	1.1×10^7	51.0 ± 17.4	114.0 ± 12.5
Shaved ice	3.2×10^7	50.0 ± 9.2	266.0 ± 23.1	_^{b}	-	-
Pork	-	-	-	4.4×10^7	45.0 ± 11.2	356.0 ± 21.7
Pacific cod	-	-	-	2.2×10^8	26.0 ± 9.8	412.0 ± 25.4

Source: Reprinted from N. Kim and I.-S. Park, *Biosensors and Bioelectronics*, 18:1101–7, 2003, with permission from Elsevier.

^{a} Mean ± S.D. (n=5).

^{b} Not determined.

the spiked or enriched food samples. Considering a total analysis time shorter than 24 h was required for sample preparation, spiking or enrichment, and measurement, compared to the analysis time of around 3 to 4 days required for conventional culturing methods, the sensor seemed to provide a rapid and powerful screening tool for *E. coli* detection (Kim and Park, 2003).

OGC biosensors based on optical waveguide lightmode spectroscopy (OWLS) have been recently developed in the sector of integrated optics, exploiting the science of light guided in structures smaller than the wavelength of light (Vörös et al., 2002). This method has been reported not only to have superior intrinsic sensitivity, but also to be convenient and versatile, especially for biological applications comprising protein-DNA interactions, receptor-ligand interactions in biomembranes, biomaterial-surface-induced blood coagulation and thrombosis, interactions with cells such as cell adhesion and spreading, cellular response to toxic compounds, and monitoring of environmental pollution. As a way to improve the sensitivity for microbial detection, a label-free OGC immunosensor for *S. typhimurium* was developed (Kim et al., 2007a).

As a transducer, OGC is composed of a glass substrate and a waveguide film over which the diffraction grating has the surface relief depth of ~20 nm, the grating periodicity of 2,400 lines/mm, and the grating area dimensions of ~2 mm in length and 16 mm in width. The OGC immunosensor used for the detection of *S. typhimurium* is depicted in Figure 10.18. The system was operated in the flow mode at a flow rate of 194 μl/min except during antibody immobilization. This system consisted of a microdispensing peristaltic pump, an injector, and the main unit was composed of a sensor

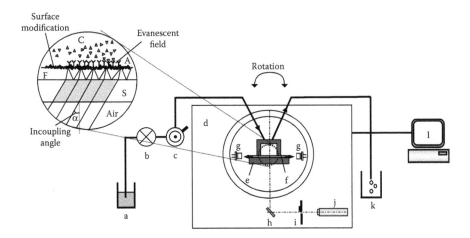

FIGURE 10.18 Schematic diagram of the OGC immunosensor system. The instrument was equipped with a flow-through cell to introduce the liquid sample to the grating part of the sensor surface in a reproducible manner. Symbols: (a) buffer solution, (b) peristaltic pump, (c) injector, (d) main unit, (e) sensor holder, (f) OGC sensor chip, (g) photodiode, (h) beam mirror, (i) shutter, (j) laser source, (k) waste, (l) PC, (A) adlayer, (C) bulk solution, (F) waveguide film, (S) glass substrate. (Reprinted from N. Kim, I.-S. Park, and W.-Y. Kim, *Sensors and Actuators B: Chemical*, 121:606–15, 2007, with permission from Elsevier.)

holder attached to a flow-through cuvette over the sensor chip, two photodiodes, a beam mirror, a shutter, a He-Ne laser emitting a monochrome (λ = 632.8 nm), and a PC. The operating principles of this system are as follows. The monochromatic light diffracted by the optical grating is propagated via total internal reflection inside the waveguide film (F). At a well-defined angle α, the phase shift during one internal reflection equals zero and a guided mode is excited, which generates an evanescent field penetrating into the bulk solution (C). The time-dependent change in refractive index at the surface caused by the formation of an adlayer (A) owing to bound microbial cells can be monitored in real time by the precise measurement of incoupling angles in the transverse electric and magnetic mode (Kim et al., 2007a).

The responses of the OGC sensor chips, which were prepared by the immobilization of an anti-*Salmonella* antibody using different immobilization protocols, were compared after the addition of *S. typhimurium* at concentrations ranging from 1.3×10^1 to 1.3×10^8 CFU/ml. The antibody immobilizations to amino-, epoxide-, and thiol-terminal silane functionalized waveguide surfaces were conducted using 3-aminopropyltriethoxysilane (APTS), 3-glycidoxypropyltrimethoxysilane (GOPS), 3-mercaptopropyltrimethoxysilane (MTS)-N-γ-maleimidobutyryloxysuccinimide ester (GMBS), and MTS-GMBS-streptavidin-biotinylated antibody protocols, respectively (Bhatia et al., 1989; Polzius et al., 1996; Kim et al., 2007a). For maximum data fitting, the relationships between the microbial density and change in surface coverage were plotted in a double-logarithmic scale (Figure 10.19). Linear

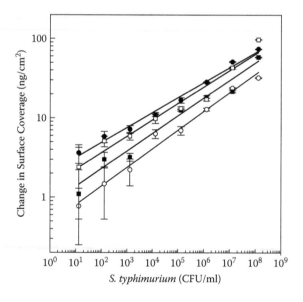

FIGURE 10.19 Relationship between *Salmonella* concentration and the change in surface coverage according to the immobilization methods. Symbols: (•) APTS protocol, (o) GOPS protocol, (■) MTS-GMBS protocol, (□) MTS-GMBS-streptavidin-biotinylated antibody protocol. Changes in surface coverage at an adjacent measuring time were measured and the error bars were inserted. (Reprinted from N. Kim, I.-S. Park, and W.-Y. Kim, *Sensors and Actuators B: Chemical*, 121:606–15, 2007, with permission from Elsevier.)

equations were acquired for the tested concentrations of *S. typhimurium* with the *r* values over 0.9744, and the best data convergence was observed when the APTS protocol was used. The LOD was determined to be 1.3×10^3 CFU/ml, irrespective of the antibody immobilization methods, and the sensitivity obtained was comparable to those obtained from SPR and piezoelectric immunosensors against *S. paratyphi* and *S. typhimurium* (Fung and Wong, 2001; Kim et al., 2003; Oh et al., 2004). On the other hand, the dynamic detection range varied from 1.3×10^3 to 1.3×10^8 CFU/ml.

When a repetitive measurement using an OGC sensor chip prepared by the APTS protocol was conducted at a *S. typhimurium* concentration of 6.3×10^8 CFU/ml after a regeneration procedure using 10 mM HCl, the sensor responses were quite reproducible during ten cycles of measurement, with a mean surface coverage change of 113.70 ng/cm^2 and a coefficient of variability (CV) of 2.87% (Figure 10.20). After ten repetitive cycles, the sensor response was reliable enough for further measurement.

The specificity of the direct-binding OGC immunosensor prepared by the APTS protocol was evaluated according to immobilization steps and immobilized proteins (Figure 10.21). In the former case, the sensor responses during each step of antibody immobilization were determined by injecting a *S. typhimurium* suspension at a cell density of 6.3×10^6 CFU/ml into the flow-through cell over the sensor chip immobilized with the anti-*Salmonella* antibody. The responses of the bare chip and sensor chip treated only with glutaraldehyde were 3.31 and 3.49 ng/cm^2, respectively. In contrast, the sensor chip treated with glutaraldehyde and immobilized with the antibody had a response of 36.05 ng/cm^2. The addition of the reaction buffer, 4 mM Tris-HCl buffer (pH 7.2), as a negative control, caused a surface coverage change of 2.86 ng/cm^2. Based on these results, it was inferred that the sensor response only occurred in the case of the antibody-immobilized sensor chip, which indicated that

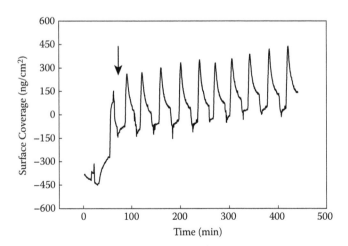

FIGURE 10.20 Sensor responses during ten repetitive measurements after regeneration with 10 mM HCl. The arrow indicates the start of the first measurement. (Reprinted from N. Kim, I.-S. Park, and W.-Y. Kim, *Sensors and Actuators B: Chemical*, 121:606–15, 2007, with permission from Elsevier.)

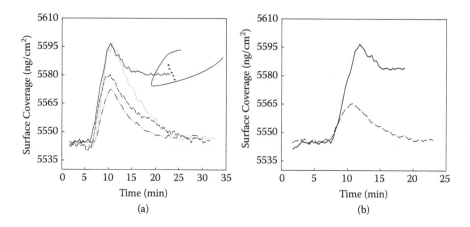

FIGURE 10.21 Comparison of the sensor responses according to the immobilization steps (a) and immobilized proteins (b). (a) (_) sensor surface immobilized with the anti-*Salmonella* antibody, (…) sensor surface treated with glutaraldehyde, (– –) bare sensor surface, (– ·· –) sensor surface immobilized with the antibody and injected with the reaction buffer. (b) (_) sensor surface immobilized with the anti-*Salmonella* antibody, (– –) sensor surface immobilized with BSA. (Reprinted from N. Kim, I.-S. Park, and W.-Y. Kim, *Sensors and Actuators B: Chemical*, 121:606–15, 2007, with permission from Elsevier.)

nonspecific adsorption did not take place (Park and Kim, 1998). In the latter case, the responses of the OGC sensor chips coated with the antibody and BSA, which was evidently not related with the immune response, were compared. As shown in Figure 10.21, no sensor response was observed in the case of BSA immobilization, which also indicated the specificity of the *Salmonella* biosensor (Zourob et al., 2003). The responses of the OGC immunosensor against similar concentrations of *L. monocytogenes*, *E. coli*, and *Pseudomonas aeruginosa* were 0.25 to 2.32% of that obtained with 6.3×10^8 CFU/ml of *S. typhimurium*. These results demonstrated that the OGC immunosensor holds promise for use as a rapid screening tool for hazardous food poisoning bacteria and index microorganisms related with food hygiene.

10.6   FOOD BIOSENSORS MEASURING HAZARDOUS RESIDUAL MATERIALS

The hazardous residual materials that are occasionally present in food matrices include antibiotics, pesticides, heavy metals, and cyanide and its glycosides. These materials can be measured for the purpose of quantification or screening (Water and Haagsma, 1991; Kijk, 1994; Norma et al., 2002). For quantification, instrumental methods such as HPLC, gas chromatography (GC), GC–mass spectrometry (MS), and LC-MS have been routinely used (Li et al., 2002). Recently, the importance of rapid screening methods for hazardous residual materials possibly present in food has been increasing in the sector of food safety management, and rapid screening methods such as thin-layer chromatography (TLC), *Bacillus megaterium* disc assay, ELISA, and fluorescence immunoassay have been developed to monitor possible residual materials (Park et al., 2004b). These methods, however, are complicated and laboratory intensive and are not easily applicable to on-site inspection. As a new screening method, biosensing of the residual materials described above has been developed. As mentioned before (Mellgren et al., 1996), biosensors are suitable for high-sensitivity detection of residual materials in real time, together with convenience and rapidity, which make them appropriate for on-site analysis. On the other hand, the presence of hazardous residual materials in raw food sources like fish can be also assessed by detecting specific biomarkers, including vitellogenin and zona radiata protein for endocrine disruptors, metallothionein for heavy metals, and cytochrome $P_{450}1A$ for pesticides (Arukwe et al., 2001; Lacorn et al., 2001; Boon et al., 2002). Therefore, biosensors that measure these compounds could be of critical importance in the field of food nutrition and safety.

Antibiotics that can be translocated to various foods by way of feed and injection into livestock are important residual materials in food hygiene due to their conspicuous side effect. For example, β-lactam antibiotics, which consist of penicillin compounds, and cephalosporin, which hold a broad-action spectrum against Gram-positive and -negative bacteria, have also been known for their toxicity in humans and livestock (Park et al., 2004b). There are concerns associated with chloramphenicol (CAP), which is an effective and widely used veterinary antibiotic for the treatment of infectious diseases in cattle, because of its residual persistence in the human food chain. Due to the potential hazard and innate toxicity of CAP, the European Union (EU), Switzerland, and the United States no longer allow it to be used in animals in the human food chain, and a threshold value of 1 ppb has been set for foods imported from other countries (Kolosova et al., 2000; Jemmi, 2002). Sulfamethazine (SMZ), a representative sulfonamide, is a synthetic antibiotic with a broad-action spectrum and a suspected carcinogen. It has been found in meat, fish, milk, and cheese, and has been attributed to the major cause of violations involving sulfonamides in tissues (Ram et al., 1991; Clark et al., 2005). Based on its toxicity, a maximum residue level (MRL) of 25 ppb has been set in milk by the Codex Committee of FAO/WHO, and an MRL of 100 ppb for the sum of sulfonamides in the EU (Sternesjö et al., 1995; Situ et al., 2002). Our group has developed biosensors that can detect important residual antibiotics in food using potentiometric, piezoelectric, and optoelectronic transducers.

Penicillinase (β-lactamase, EC 3.5.2.6, from *B. cereus*) was immobilized onto an Immobilon cellulose nitrate membrane by soaking it in a penicillinase solution, and the resulting enzyme membrane was attached to the sensing component of a flat-bottomed pH electrode to prepare an enzyme electrode. The enzyme electrode was then connected to a pH/mV meter to develop a potentiometric biosensor system that can measure and detect β-lactam antibiotics. According to the substituent groups of the β-lactam ring, natural penicillins like penicillin G and V, and semisynthetic aminopenicillins such as ampicillin and amoxicillin are known (Pam et al., 1996). Upon enzyme action, the β-lactam ring is cleaved by hydrolysis, which converts the penicillin compound into penicilloic acid. The carboxyl group of penicilloic acid easily deprotonates H^+ ions, which are released to the surrounding medium and induce a pH decrease or potential increase that can be measured with the pH/mV meter. Therefore, the enzymes immobilized onto the Immobilon cellulose nitrate membrane were shown to significantly affect the biosensor response, and thus the sensor responses at two enzyme loadings of 50 and 100 units/ml were compared by measuring the increase in potential (Figure 10.22). When the enzyme was loaded at 50 units/ml, good correlations were found between analyte concentration and sensor response for all penicillin compounds. However, the detection limit was around 100 μM for the individual analyte. This was significantly higher than 0.4 to 1.3 ppm, which is regarded as a reference value for positive decision making on the presence of antibiotics in urine or serum of livestock in screening tests (Park et al., 2004b). When enzyme loading during the immobilization procedure was increased to 100 units/ml, linear equations with a good correlation (r values for penicillin G, amoxicillin, and ampicillin were 0.9985, 0.9999, and 0.9878, respectively) were acquired up to an analyte concentration of 1,000 μM, and the detection limit was around 1 μM. This met the established reference level described above and was quite good considering a measuring range of 0.01 to 3 mM was reported in previous studies that examined penicillin sensors (Cullen et al., 1974; Rusling et al., 1976). Compared to colorimetric and iodometric detection, the potentiometric biosensor seemed to have some advantages, like convenience, rapidity, and simplicity. The CV values after seven repetitive measurements were 2.28% for 1 mM penicillin G, 3.40% for 0.5 mM amoxicillin, and 6.20% for 4 mM ampicillin, which indicated good reproducibility.

To evaluate the performance of the biosensor on a real sample, a meat drip was prepared at 1,000 rpm on a tabletop centrifuge using a centrifuge tube installed with an acryl net, followed by filtration through a 0.45 μm syringe filter membrane to prepare the analytical sample. For comparison, the sample was analyzed by HPLC, the conventional method used to detect penicillin compounds, which is by

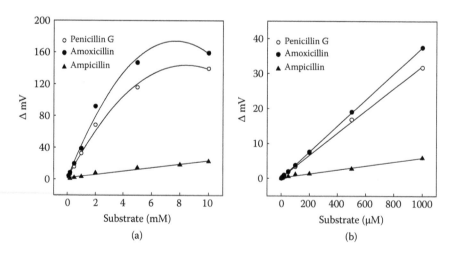

FIGURE 10.22   Calibration curves for the penicillin compounds at enzyme loadings of 50 (a) and 100 (b) units/ml. (Reprinted from I.-S. Park, D.-K. Kim, and N. Kim, *Journal of Microbiology and Biotechnology*, 14(4):698–706, 2004, with permission from the Korean Society for Microbiology and Biotechnology.)

far more complicated relative to our method (Park et al., 2004b). To analyze the samples by using the biosensor and HPLC, each penicillin compound at different concentrations was spiked into the model samples. A low ionic strength buffer, 2 mM sodium phosphate buffer (pH 7.2), was used as the reaction buffer for the biosensor analysis to improve the sensitivity of analysis. To preclude possible matrix effects, the sensor response was set to be the steady-state potential change obtained with the model sample minus that of the analytical sample. As shown in Table 10.8, the model samples spiked with each penicillin compound showed relative responses of 82.1 to 103.5%, which were comparable to the 79.5 to 106.1% obtained when HPLC was used. Based on this result, the potentiometric penicillin sensor holds promise for use as a highly sensitive and precise tool for rapid screening of penicillin compounds.

A direct-binding piezoelectric CAP immunosensor was prepared by coupling an anti-CAP antibody to SAMs of thiol or sulfide compounds on the gold electrode of QCM by 1-ethyl-3-(3-dimethylaminopropyl)carbodiimide hydrochloride and N-hydroxysulfosuccinimide-mediated immobilization (Vaughan et al., 1999). The thiol or sulfide compounds used were thioctic acid, thiosalicylic acid, thiodiglycolic

TABLE 10.8

Comparison of the Relative Responses for the Penicillin Compounds Obtained by the Biosensor Method and HPLC

	Biosensor method		HPLC	
Substrate	Spiked concentration (mM)	Relative response (%)[a]	Spiked concentration (mM)	Relative response (%)[a]
	0.2	98.6	2.5	92.4
	0.5	99.2	4.0	101.4
Penicillin G	1.0	103.5	5.0	84.0
	2.0	95.5		
	0.2	82.1	0.5	98.5
	0.5	96.2	0.8	106.1
Amoxicillin	0.75	87.4	1.0	86.1
	1.0	96.1		
	2.0	84.4	2.5	91.2
	3.0	100.7	4.0	99.1
Ampicillin	4.0	94.4	5.0	79.5
	5.0	97.1		

Source: Reprinted from I.-S. Park, D.-K. Kim, and N. Kim, *Journal of Microbiology and Biotechnology*, 14(4):698–706, 2004, with permission from the Korean Society for Microbiology and Biotechnology.

[a] Percentage of the responses of the model samples against those of the standard samples.

acid, 3,3-thiodipropionic acid, 3,3-dithiodipropionic acid, and 3-mercaptopropionic acid (MPA). A batch type well holder system having a nominal capacity of 500 µl was used for this system. Two hundred microliters of 0.01 M phosphate buffer (pH 7.4), the reaction buffer, was added into the reaction cell, followed by measurement of the steady-state resonant frequency of the 9 MHz quartz crystal (F_1). The standard sample (200 µl) was then injected into the reaction cell. The steady-state resonant frequency (F_2) was read again to calculate the frequency shift ($\Delta F = F_1 - F_2$) (Park et al., 2004c). Table 10.9 shows the responses of the CAP sensor during repetitive use according to the thiol or sulfide compounds used during antibody immobilization. Due to the diminution of the antibody layer after regeneration, the responses of the CAP sensor gradually decreased after each use in all cases. However, the degree of residual sensitivity after each measurement was quite different according to the thiol or sulfide compounds used for monolayer formation. The best result in reusability was obtained for MPA-based antibody immobilization, which resulted in residual sensitivities of 94.5 and 88.7% at the fourth and eighth measurement. This seems to be attributable to the fact that MPA forms the strongest cross-linker–antibody complex to the gold electrode of the quartz crystal due to the presence of the free–SH group with a straight carbon chain (Bain et al., 1989). This result demonstrates that the CAP sensor can be reused several times after regeneration for the purpose of screening CAP in real food samples.

Enhanced chemiluminescence reactions generally exploit co-oxidation of 5-amino-2,3-dihydro-1,4-phthalazinedione (luminol) and an enhancer by H_2O_2 in the presence of horseradish peroxidase (HRP). They are important in many sectors of analytical chemistry, including immunoassay, immunosensing based on flow injection analysis, DNA probe-based assay, and biochemical analysis, due to their high sensitivity and short analysis time (Hlavay et al., 1994; Osipov et al., 1996; Park and Kim, 2006). Therefore, they have a strong potential to be used for the rapid and sensitive detection of low molecular weight residual contaminants like antibiotics. A flow type direct-competitive chemiluminescent immunosensor system, which exploited the competition reaction between CAP as an analyte and CAP-HRP conjugate as a tracer for binding to an anti-CAP antibody, was devised (Figure 10.23). Binding between the tracer and antibody coated on the solid support elicits chemiluminescence upon the addition of the substrate mixture, including luminol. Antibody immobilization was conducted over the surface of positively charged Biodyne B membrane pieces that were 10 mm in diameter by immersing them into 125 µl of the antibody solution for 24 h at 4°C. After washing the membrane pieces with 20 mM Tris buffer (pH 8.6), the reaction buffer, they were blocked with 1% BSA for 30 min (Park and Kim, 2006).

The system was constructed as depicted in Figure 10.23. A homemade dark box (16 × 16 × 20 cm) blocking outside light and a flow-through cell having a nominal capacity of 20 µl were used. The reaction cell was fixed to a Styrofoam mold in the dark box and the photocathode of a photomultiplier (PMT) tube, having the maximum emission at 400 nm, was placed in front of it to receive the emitted light. A capillary tubing (0.4 mm i.d.) was connected to the reaction cell through two tiny holes in the upper part of the dark box, with the inlet part connected to an injector and a micro-dispensing peristaltic pump to allow for a flow of 0.155 ml/min. Before system

TABLE 10.9
Detection Sensitivity of the CAP Sensor for Repeated Use

Assay number	Frequency shift (Hz)[a]/residual sensitivity (%)					
	TCA	TSA	TDGA	TDPA	DTDPA	MPA
			Thiol or sulfide compound			
1	198 ± 0.9/100	390 ± 8.9/100	186 ± 1.9/100	193 ± 2.9/100	322 ± 9.8/100	399 ± 2.6/100
2	160 ± 3.3/80.8	334 ± 4.5/85.6	167 ± 2.2/89.8	181 ± 2.0/93.8	284 ± 4.6/88.2	388 ± 2.0/97.2
3	162 ± 3.2/81.8	303 ± 1.2/77.7	166 ± 3.5/89.3	172 ± 1.6/89.1	267 ± 5.0/82.9	382 ± 2.5/95.7
4	157 ± 2.0/79.3	290 ± 2.3/74.4	159 ± 1.7/85.5	150 ± 0.6/77.7	270 ± 3.7/83.9	377 ± 2.2/94.5
5	152 ± 2.6/76.8	298 ± 4.6/76.4	148 ± 1.8/79.6	148 ± 1.0/76.7	251 ± 6.7/78.0	379 ± 4.2/95.0
6	143 ± 2.1/72.2	226 ± 7.9/58.0	149 ± 2.3/80.1	148 ± 3.7/76.7	253 ± 3.0/78.6	369 ± 8.0/92.5
7	140 ± 3.3/70.7	207 ± 4.3/53.1	141 ± 1.7/75.8	143 ± 4.0/74.1	245 ± 1.3/76.1	359 ± 3.6/90.0
8	130 ± 1.7/65.7	197 ± 3.2/50.5	140 ± 1.2/75.3	137 ± 0.8/71.0	232 ± 1.7/72.1	354 ± 3.3/88.7

Source: Reprinted from I.-S. Park, D.-K. Kim, N. Adanyi, M. Varadi, and N. Kim, *Biosensors and Bioelectronics*, 19:667–74, 2004, with permission from Elsevier.

The frequency shifts were measured after 1 min regeneration with 0.1 M NaOH.

[a] Mean ± S.D. (n = 30), 30 values of frequency shift in 1 min intervals were averaged after the steady-state sensor response was attained.

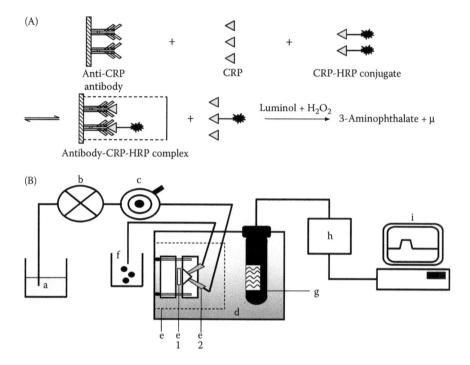

(A)

Anti-CRP antibody

CRP

CRP-HRP conjugate

Antibody-CRP-HRP complex

$$\text{Luminol} + H_2O_2 \longrightarrow \text{3-Aminophthalate} + \mu$$

(B)

FIGURE 10.23 Principle of direct competitive chemiluminescent immunosensing (A) and schematic diagram of a flow type immunosensor system for chemiluminescence measurements (B). Symbols: (a) buffer solution, (b) peristaltic pump, (c) injector, (d) black box, (e) flow-through cell, (e_1) antibody-immobilized membrane intercalated between two O-rings, (e_2) flow guide, (f) waste, (g) PMT tube, (h) PC Lab card, (i) computer. (Reprinted from I.-S. Park and N. Kim, *Analytica Chimica Acta*, 578:19–24, 2006, with permission from Elsevier.)

operation, the antibody-coated membrane was placed between two acryl holders of the reaction cell with two O-rings and was fixed by long joints. The following operating conditions of the biosensor were selected: substrate composition (0.25, 13.3, and 0.66 mM for luminol, H_2O_2, and *p*-iodophenol, respectively), injection volume of the substrate solution (200 μl), antibody concentration for immobilization (0.10 mg/ml), and tracer concentration (0.030 mg/ml). System operation was carried out by an initial flow of the reaction buffer, followed by the addition of the reaction mixture (tracer with or without analyte) through an injection loop. After washing with the reaction buffer, the substrate solution was added through the injection loop to elicit chemiluminescence. The emitted light was collected in the PMT tube, followed by conversion to an electrical current, which was interfaced with an IBM-compatible PC. The maximum light intensity of the peak was measured as a sensor signal. When measured using this system, the sensor responses at varying CAP concentrations were well fitted to a linear equation on the semilogarithmic scale, with a LOD value of 10^{-8} M for CAP (Park and Kim, 2006).

A label-free OGC immunosensor was prepared by immobilizing anti-SMZ antibodies over an amino-silane surface in stop-flow mode using the APTS protocol to detect and measure SMZ in the flow mode at a flow rate of 194 μl/min (Kim et al., 2007a). The reaction buffer for this biosensor system was 4 mM Tris-HCl (pH 7.2). The responses of the OWLS-based immunosensor at varying SMZ concentrations of 10^{-8} to 10^{-2} M were determined after regeneration with 10 mM HCl, and the corresponding calibration curve was plotted on a semilogarithmic scale (Figure 10.24). A typical sensorgram was characterized by an initial steep increase followed by a gradual decrease in surface coverage as reported by other flow type optoelectronic biosensors based on the affinity-binding principle (Hug et al., 2004; Leonard et al., 2004). The time required to obtain a sensor response was around 10 min, and the addition of the reaction buffer itself was not responsive, as shown by the arrow in Figure 10.24. A linear relationship was found between analyte concentration and sensor response, with an r value of 0.9620. This behavior had also been found in a previous grating coupler-based biosensor (Kuhlmeier et al., 2003). The LOD for SMZ detection was determined to be 10^{-8} M, by considering the normally accepted criterion of three times the standard deviation for baseline drift. The sensitivity obtained in this study was comparable to the levels obtained in a previous study that utilized in-tube solid-phase micro-extraction coupled to HPLC to monitor and detect sulfonamide residues (Wen et al., 2005), and was superior to SMZ detection by high-volume enzyme immunoassays, *B. stearothermophilus* tube test, and biosensor-based immunochemical screening assays (Ram et al., 1991; Bjurling et al., 2000; Shitandi et al., 2006). Based on the above results in terms of analysis time and sensitivity, the OWLS-based immunosensor holds great promise for use in the sensitive and rapid screening of residual SMZ, which may be present in food (Kim et al., 2008).

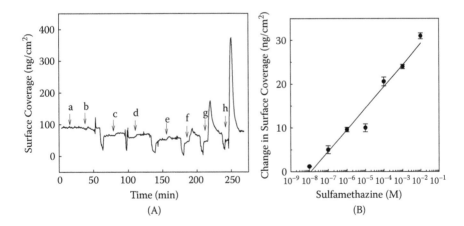

FIGURE 10.24 Concentration-dependent responses of an SMZ sensor (A) and calibration curve for SMZ (B). (A) Five hundred microliters of the reaction buffer (a) and SMZ solutions at concentrations of 10^{-8} (b), 10^{-7} (c), 10^{-6} (d), 10^{-5} (e), 10^{-4} (f), 10^{-3} (g), and 10^{-2} M (h) were individually injected into the immunosensor system after baseline stabilization following 10 mM HCl treatment. (B) Changes in surface coverage at the adjacent measuring time were determined and error bars were inserted. (Reprinted from N. Kim, D.-K. Kim, and W.-Y. Kim, *Food Chemistry*, 108:768–73, 2008, with permission from Elsevier.)

The majority of insecticides are cholinesterase (ChE) inhibitors, such as organo-phosphates and carbamates, which constitute 26% of total pesticides. Some of these pesticides show dose-related acute and chronic toxicity in human beings by inhib-iting ChE activity, which accumulates acetylcholine at cholinergic receptor sites, resulting in an excessive stimulation of the cholinergic receptors. Since this can lead to various adverse clinical effects, particularly for infants and children, very stringent allowance limits in food have been set for organophosphorus and carbam-ate pesticides. Thus, there is a growing need for faster and more sensitive screen-ing methods for these compounds. ChE-based biosensors measure the presence of organophosphates and carbamates as a sum parameter of enzyme inhibition (Skládal and Mascini, 1992; Bachmann et al., 2000). Compared to ChE inhibition tests, they are useful for on-site monitoring and are easily arrayed for real-time multisample analysis in a short time (Mulchandani et al., 2001).

A QCM-precipitation biosensor detecting acetylcholinesterase (AChE) inhibi-tion by organophosphates and carbamates was devised. The operating principle of this biosensor was based on measuring AChE activity. AChE was immobilized on one side of a 9 MHz QCM sensor by chemisorption of thiolated AChE through a heterobifunctional cross-linker, sulfo-LC-SPDP, which was simultaneously exposed to a substrate solution that was prepared by dissolving 3-indolyl acetate in dimethylformamide (DMF). Since the enzymatic reaction product is precipi-tated over the QCM surface after dimerization (Abad et al., 1998), the degree of the sensor response can be traced in real time by determining the decrease in frequency caused by mass deposition over the QCM surface. The precipitation procedure mediated by the immobilized AChE is depicted in Figure 10.25. By

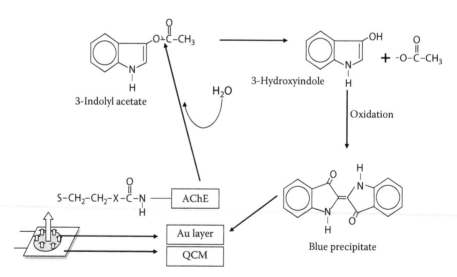

FIGURE 10.25   Schematic diagram of the action mechanism for the QCM-precipitation sensor prepared by chemisorption of the AChE thiolated with sulfo-LC-SPDP. X is the resi-due remaining after enzyme immobilization, which has the chemical structure of CO-NH-(CH$_2$)$_5$. (Reprinted from N. Kim, I.-S. Park, and D.-K. Kim, *Biosensors and Bioelectronics*, 22:1593–99, 2007, with permission from Elsevier.)

measuring the degree of inhibition in AChE activity caused by organophosphates, which generally exhibit noncompetitive inhibition, or by carbamates, which show competitive inhibition (Schulze et al., 2003), the presence of these pesticides or the concentration of one single compound can be determined with good sensitivity. The sensor system for this purpose was constructed by batch basis using a dip holder mounted with the AChE-coated sensor chip. The dip holder was inserted into a small beaker filled with 19.5 ml of 0.1 M potassium phosphate buffer (pH 8.0), the reaction buffer. The steady-state frequency shift obtained after the addition of 500 µl of the substrate solution (50 mg/ml of 3-indolyl acetate) dissolved in DMF was measured. To measure AChE inhibition, an organophosphate solution was added to the reaction cell simultaneously with the substrate solution. On the other hand, the sensor chip was preincubated with a carbamate solution, followed by the addition of the substrate solution. The QCM-precipitation sensor could detect a model organophosphorus pesticide ethyl-p-nitrophenyl thionobenzenephosphonate (EPN) and carbamate, a carbofuran, with high sensitivity and a detection limit of nanomolar concentrations (Kim et al., 2007b).

Heavy metals, which comprise about sixty-five metal elements in nature, are divided into essential heavy metals and xenobiotics. Zinc (Zn), iron (Fe), copper (Cu), and cobalt (Co) are considered essential heavy metals and are known to maintain the physiological function of organisms at trace concentrations. In contrast, mercury (Hg), lead (Pb), cadmium (Cd), and chromium (Cr) are environmental pollutants and very harmful to living organisms, and are members of the xenobiotics. They form organic complexes with endogenous materials from living organisms and accumulate in bone and internal organs such as the liver and kidney. The human body is exposed to xenobiotics through polluted water and air directly or through the intake of food materials like fish in the food chain. Since xenobiotics are responsible for fatal diseases after accumulation in the human body, xenobiotic contamination should be avoided. Because of the dangers of xenobiotics, the development of rapid and sensitive analytical methods for screening and quantification is needed. Heavy metal biosensors that employ immobilized enzymes like urease as biological components normally measure inhibition in enzyme activity in the presence of xenobiotics as a sum parameter. However, enzyme sensitivity and selectivity according to heavy metal species are quite different.

A urease-based ISFET biosensor for heavy metal detection was developed by the immobilization of urease onto a negatively charged Nafion polymer to allow for accumulation of xenobiotic cations in the biomatrix. That is, the enzyme membrane was prepared by depositing 0.1 µl of 10% urease in 5 mM Tris-HNO$_3$ (pH 7.0), the reaction buffer, onto the sensitive area of an ISFET, followed by drying at room temperature for 3 min. Nafion (0.5%) in ethanol was added over the enzyme membrane to cover the biomembrane. The biomembrane was then dried for 5 min. A homemade ISFET structure (Figure 10.10) was used to measure the pH change in the biomembrane due to enzymatic catalysis of urea, and the measurement was done in a glass cell (1.5 ml) filled with the reaction buffer. The enzyme membrane was inhibited by heavy metal ions in the order of Ag(I) > Hg(II) > Cu(II) like free urease. The sensitivity of the sensor, however, was strongly dependent on the thickness of the Nafion coating. As shown in Figure 10.26, the highest sensitivity for the Hg(II) ions

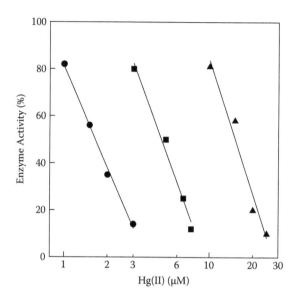

FIGURE 10.26 Dependence of biosensor response to Hg(II) on polymeric membrane thickness. Membrane was formed with 2 µl of 0.5% Nafion (•), 3 µl of 0.5% Nafion (■), and 4 µl of 0.5% Nafion (▲). (Reprinted from V. Volotovsky, Y. J. Nam, and N. Kim, *Sensors and Actuators B: Chemical*, 42:233–37, 1997, with permission from Elsevier.)

was obtained when the enzyme membrane was covered with 2 µl of 0.5% ethanolic Nafion. It has been reported that the inhibitor selectivity could be improved by selective rewashing (Tranh-Minh, 1985) or by properly chosen immobilization protocols (Gayet et al., 1993). To make the urease-based biosensor selective to mercury ions, the addition of small amounts of NaI up to 100 µM into the sample was proposed to suppress the sensitivity to silver ions and to rewash the biosensor in 100 mM of EDTA for 5 min to remove copper ions (Volotovsky et al., 1997).

The utilization of cyanide biosensors in the food industry has been connected with the sensitive determination of cyanogenic glycosides such as amygdalin in the agricultural products, including lima bean, bamboo shoots, and almond. Since some iron-containing enzymes such as cytochrome oxidase and peroxidase are inhibited by cyanide ions, cyanide biosensors based on these enzymes can be prepared for precise cyanide determination (Albery et al., 1990; Smit and Cass, 1990; Tatsuma and Oyama, 1996).

An ISFET biosensor based on enzyme inhibition with cyanide ions was fabricated by immobilizing HRP onto a positively charged PVPy polymer, which would allow for the accumulation of cyanide ions in the biomatrix and thus improve sensor sensitivity (Volotovsky and Kim, 1998d). For this purpose, the enzyme membranes containing peroxidase were prepared separately by glutaraldehyde cross-linking and PVPy deposition over the peroxidase in the sensitive area of the ISFET. The main reason for using the ISFET structure as a transducer was because a portable ISFET-based pH meter is already available as an inexpensive cyanide detector. The sensor response was first measured at room temperature in a 1.5 ml glass cell filled with a buffer solution containing 10 mM Tris, 15 mM L-ascorbic acid, and 100 mM NaCl. The solution was adjusted to pH 7.5 prior to detection and was intensively stirred during the measurement. After rinsing with the fresh buffer solution, the sensor chip was immersed into the above buffer solution containing a known concentration of inhibitor (KCN) for 1 min and the sensor response was then measured. The residual peroxidase activities of the glutaraldehyde-cross-linked and PVPy-deposited sensor chips were compared at a fixed H_2O_2 concentration of 5 mM and cyanide concentrations varying from 0.1 to 1,000 µM. As shown in Figure 10.27, a significant difference in the sensitivity to cyanide ions was observed between the two biomembranes. In the case of the glutaraldehyde-cross-linked membrane, the calibration curve covered a cyanide concentration range of 3.3 to 1,000 µM with a 50% enzyme inhibition at 80 µM KCN. On the other hand, the PVPy-deposited membrane was capable of measuring cyanide concentrations from 0.1 to 10 µM with a 50% enzyme inhibition at 0.6 µM KCN. This difference in sensitivity seemed to be due to the fact that a negative net charge of BSA (reported to be 18 at pH 7.4) can create a potential barrier against cyanide ions and suppress their diffusion into the biomembrane, and the charged PVPy film can increase the accumulation of cyanide ions into the biomatrix (Volotovsky and Kim, 1998d).

To improve the versatility of analysis, a multibiosensor that can measure several hazardous residual materials in food, including heavy metal, cyanide, and

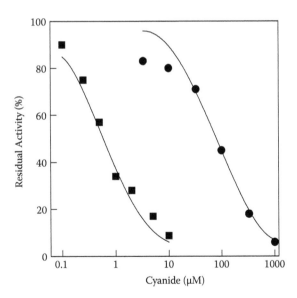

FIGURE 10.27 Response of the BSA-membrane (•) and PVPy-membrane (■) biosensor to cyanide. The electrolyte solution was 10 mM Tris buffer containing 15 mM ascorbic acid, 100 mM NaCl, and titrated with 0.1 M NaOH up to pH 7.5. The sensor was incubated in a cyanide solution for 1 min, and then the hydrogen peroxide sample was injected into a cell at a 5 mM concentration. Theoretical curves were calculated from the equation of $V/V_{max} = 1/(1 + I/K_{app})$ for $K_{app} = 80$ µM (•) and $K_{app} = 0.6$ µM (■). (Reprinted from V. Volotovsky and N. Kim, *Biosensors and Bioelectronics*, 13:1029–33, 1998, with permission from Elsevier.)

pesticide, was proposed by Cowell et al. (1995). In this study, an array of six enzymes was used to develop a computer model from which the resultant inhibition pattern could be interpreted using an artificial neural net. This model, however, was rather complex and inappropriate for use as a portable device for screening purposes. A simple alternative is the co-immobilization of several enzymes into the same biomatrix.

An ISFET-based multienzyme inhibition biosensor for the detection of mercury ions, cyanide, and pesticide was prepared by the co-immobilization of urease, HRP, and butyrylcholinesterase onto the sensitive area of the ISFET by glutaraldehyde cross-linking. Sample analysis and measurements were conducted at room temperature in a glass cell filled with 5 mM Tris buffer (pH 7.5) and 100 mM NaCl as the reaction buffer (1.5 ml) using the homemade ISFET structure shown in Figure 10.10. For measurements using the peroxidase-based probe, the buffer solution also contained 15 mM of ascorbic acid. To measure enzyme inhibition, the responses of the enzyme electrode to individually selected substrate solutions (urea, H_2O_2, and bytyrylcholine iodide) showing saturated sensor responses were first measured. After rinsing carefully, the enzyme electrode was immersed into the reaction buffer containing a known concentration of the individual toxicant for 1 to 10 min, followed by the measurement of the sensor response again after careful rinsing. The difference in the sensor signals before and after inhibition was proportional to the inhibitor concentration (Volotovsky and Kim, 2003). Inhibition of the three-enzyme biosensor at varying concentrations of mercury ions, KCN, and a carbamate pesticide carbofuran is depicted in Figure 10.28, together with the inhibition obtained using the one-enzyme counterpart. The loading of each enzyme in the three-enzyme biosensor was one-third of that in the one-enzyme electrode, which increased the sensitivity of the probe to the inhibitors (Scladal and Mascini, 1992). The three-enzyme electrode not only showed high responses to the corresponding inhibitors, but also produced good inhibitor specificities. That is, there was no sensitivity toward other inhibitors (e.g., cyanide and pesticide in the case of mercuric ion measurement) up to a toxicant concentration of 1 mM. As shown in Figure 10.28, a 1 mM inhibitor concentration conspicuously exceeded the dynamic range of the multienzyme bioprobe, which indicated that the three-enzyme electrode specifically responded to each inhibitor within the operational concentration range.

10.7 PERSPECTIVE

In the future, the importance of biosensing in the food industry is expected to grow steadily because it can produce a high added value to the bioindustry through the introduction of new trends in biotechnology, nanotechnology, and information technology. As a meaningful tool for rapid on-site detection for food quality and safety, the potential applications of food biosensors will expand worldwide, as shown by the accomplishments of the last decade. Perhaps new legislation like functional food laws and consumer interests might expedite this process and stimulate research on new biosensing methods. If the industrial needs and limits are provided, food biosensors will evolve into the development of new powerful tools that will close the gap not covered by traditional analytical methods. It is

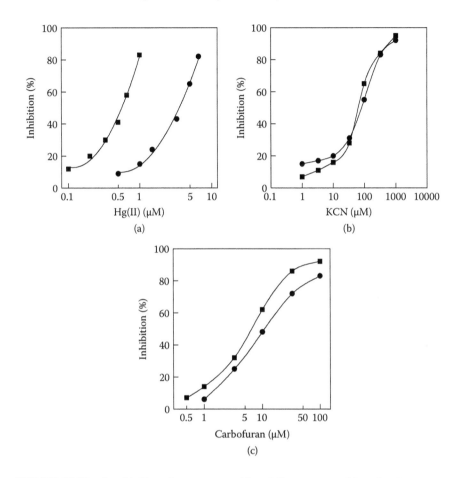

FIGURE 10.28 Sensitivities of one-enzyme (•) and three-enzyme (■) probes to mercury ions (a), KCN (b), and pesticide carbofuran (c). The percent inhibition was calculated as follows: $100\% \times (V - V_i)/V$, where V is the sensor's response to the substrate addition before inhibition and V_i is the sensor's response after inhibition. V and V_i were measured in duplicate and then averaged. (Reprinted from V. Volotovsky and N. Kim, *Journal of Microbiology and Biotechnology*, 13(3):373–77, 2003, with permission from the Korean Society for Microbiology and Biotechnology.)

expected that food functionality and safety will be the two major concerns to the consumers and manufacturers in the food industry. With respect to the former, rapid and sensitive detection of biomarkers and index materials in food matrices that are closely related with specific food functionalities like cardioprotective and antiaging effect will be of critical importance to the development of biosensors in the future. On the other hand, hazardous residual materials possibly present in food like endocrine disruptors, food poisoning bacteria like pathogenic *E. coli*, exotoxins such as staphylococcal toxin, GMOs, or life-modified organisms (LMOs) will be quite interesting targets for future biosensor studies. At the same time, novel biological components, including aptamers, abzymes, and molecularly imprinted polymers, will be developed to increase the stability and sensitivity of biosensors. In regards to the future development of food biosensors, nanomaterials including fluorescent silica nanoparticles, carbon nanotubes, and nanocolloids, sensor miniaturization comprising microarrays, biochips and cell chips, development of high-throughput detectors, and metabolite monitoring in single cells will be vital.

The most important factor to the expansion of commercialized biosensors is the development of market-oriented biosensors. Currently, biosensor production is often dependent on higher manual processing. In the future, automation in biosensor fabrication will be expanded according to progress made in transducer miniaturization and the development of high-throughput devices. Through the viewpoint of market-oriented food biosensors, sample preparation might be a crucial step during biosensor measurements, considering complex food matrices. Therefore, market-oriented food biosensors might not be commercialized on a massive scale; however, low-cost biosensors may be able to compete with other products without the need for extensive sample pretreatment.

REFERENCES

Abad, J. M., Pariente, F., Hernández, L., Abruña, H. D., and Lorenzo, E. 1998. Determination of organophosphorus and carbamate pesticides using a piezoelectric biosensor. *Anal. Chem.* 70:2848–55.

Akiyama, H., Toko, K., and Yamafuji, K. 1996. Detection of taste substances using impedance changes of phospholipid Langmuir Blodgett membrane. *Jpn. J. Appl. Phys.* 35:5516–5521.

Albery, W. J., Cass, A. E. G., Mangold, B. P., and Shu, Z. X. 1990. Inhibited enzyme electrodes. Part 3. A sensor for low levels of H_2S and HCN. *Biosens. Bioelectron.* 5:397–413.

Arukwe, A., Kullman, S. W., and Hinton, D. E. 2001. Differential biomarker gene and protein expressions in nonylphenol and estradiol-17βtreated juvenile rainbow trout (*Oncorhynchus mykiss*). *Comp. Biochem. Physiol. C* 129:1–10.

Atkins, P. W. 1996. *The elements of physical chemistry.* Oxford: Oxford University Press.

Bachmann, T. T., Leca, B., Villatte, F., Marty, J.-L., Fournier, D., and Schmid, R. D. 2000. Improved multianalyte detection of organophosphates and carbamates with disposable multielectrode biosensors using recombinant mutants of *Drosophila* acetylcholinesterase and artificial neural networks. *Biosens. Bioelectron.* 15:193–201.

Bae, Y. M. 2000. Development of electronic tongue using array of polymer membrane electrodes. PhD thesis, Seoul National University, Seoul, Korea.

Bain, C. D., Evall, J., and Whitesides, G. M. 1989. Formation of monolayers by the coadsorption of thiols on gold: Variation in the head group, tail group, and solvent. *J. Am. Chem. Soc.* 111:7157–66.

Benesch, J., Askendl, A., and Tengvall, P. 2000. Quantification of adsorbed human serum albumin at solid interfaces: A comparison between radioimmunoassay (RIA) and simple null ellipsometry. *Colloids Surf. B Biointerfaces* 18:71–81.

Bhatia, S. K., Shriver-Lake, L. C., Prior, K. J., et al. 1989. Use of thiol-terminal silanes and heterobifunctional crosslinkers for immobilization of antibodies on silica surfaces. *Anal. Biochem.* 178:408–13.

Birch, G. G., and Ogunmoyela, G. 1980. Taste properties of cocoa drinks with an added bitter/sweet sugar: Intensity/time effects. *J. Food Technol.* 15:549–55.

Bjurling, P., Baxter, G. A., Caselunghe, M., Jonson, C., O'Connor, M., and Persson, B. 2000. Biosensor assay of sulfadiazine and sulfamethazine residues in pork. *Analyst* 125:1771–74.

Boon, J. P., van Zanden, J. J., Lewis, W. E., Zegers, B. N., Goksøyr, A., and Arukwe, A. 2002. The expression of CYP1A, vitellogenin and zona radiata proteins in Atlantic salmon (*Salmo salar*) after oral dosing with two commercial PBDE flame retardant mixtures: Absence of short-term responses. *Mar. Environ. Res.* 54:719–24.

Choi, M. W., Kim, K. H., Kim, S. H., and Park, K. Y. 1997. Inhibitory effects of *Kimchi* extracts on carcinogen-induced cytotoxicity and transformation in C3H/10T1/2 cells. *J. Food Sci. Nutr.* 2:241–45.

Clark, S. B., Turnipseed, S. B., Madson, M. R., Hurlbut, J. A., Kuck, L. R., and Sofos, J. N. 2005. Confirmation of sulfamethazine, sulfathiazole, and sulfadimethoxine residues in condensed milk and soft-cheese products by liquid chromatography/tandem mass spectrometry. *J. AOAC Intl.* 88:736–43.

Cooke, J. R., and Moxon, R. E. D. 1982. Ascorbic acid in fruit juices and beverages. In *Vitamin C (ascorbic acid)*, ed. J. N. Counsell and D. H. Hornig, 123–37. London: Applied Science Publishers.

Cowell, D. C., Dowmann, A. A., and Ashcroft, T. 1995. The detection and identification of metal and organic pollutants in portable water using enzyme assays suitable for sensor development. *Biosens. Bioelectron.* 10:509–16.

Craggs, A., Moody, G. J., and Thomas, J. D. R. 1974. PVC matrix membrane ion-selective electrodes. *J. Chem. Educ.* 51:541–44.

Cullen, L. F., Rusling, J. F., Schleifer, A., and Papariello, G. J. 1974. Improved penicillin selective enzyme electrode. *Anal. Chem.* 46:1955–61.

Elwing, H. 1998. Protein absorption and ellipsometry in biomaterial research. *Biomaterials* 19:397–406.

Fung, S. H., and Wong, Y. Y. 2001. Self-assembled monolayers as the coating in a quartz crystal immunosensor to detect *Salmonella* in aqueous solution. *Anal. Chem.* 73:5302–9.

Gayet, J. C., Haouz, A., Geleso-Meyer, A., and Burstein, C. 1993. Detection of heavy metal salts with biosensors built with an oxygen electrode coupled to various immobilized oxidases and dehydrogenases. *Biosens. Bioelectron.* 8:177–83.

Hlavay, J., Haemmerli, S. D., and Guilbault, G. G. 1994. Fiber-optic biosensor for hypoxanthine and xanthine based on a chemiluminescent reaction. *Biosens. Bioelectron.* 9:189–95.

Homola, J., Yee, S. S., and Gauglitz, G. 1999. Surface plasmon resonance sensors: Review. *Sens. Actuator B Chem.* 54:3–15.

Huetz, P., Schaaf, P., Voegel, J. C., et al. 1995. Reactivities of antibodies on antigens adsorbed on solid surfaces. *Proteins Interfaces II* 602:334–49.

Hug, T. S., Prenosil, J. E., and Morbidelli, M. 2001. Optical waveguide lightmode spectroscopy as a new method to study adhesion of anchorage-dependent cells as an indicator of metabolic state. *Biosens. Bioelectron.* 16:865–74.

Ikezaki, H., Hayashi, K., Yamanaka, M., Tatsukawa, R., Toko, K., and Yamafuji, K. 1991. Multichannel taste sensor with artificial lipid membranes. *Trans. IEICE Jpn.* 174-C-II:443–42.

Imamura, T., Toko, K., Yanagisawa, S., and Kume, T. 1996. Monitoring of fermentation process of *Miso* (soybean paste) using multichannel taste sensor. *Sens. Actuator B Chem.* 37:179–85.

Jemmi, T. 2002. Analytical laboratories of the FVO. *FVO Magazine*, February, 18–19.

Kijk, P. J. 1994. Confirmation of chloramphenicol residues in bovine milk by gas chromatography/mass spectrometry. *J. AOAC Intl.* 77:34–40.

Kim, G. H., Rand, A. G., and Letcher, S. V. 2003. Impedance characterization of a piezoelectric immunosensor. Part II. *Salmonella typhimurium* detection using magnetic enrichment. *Biosens. Bioelectron.* 18:91–99.

Kim, J. P., Kim, J. B., and Park, I. W. 1988. K-value and nucleotide-degrading enzymes in fish muscles. *J. Kor. Soc. Food Nutr.* 17:33–41.

Kim, N. 1997. Optimization of enzyme membrane for a needle-type L-lactate sensor. *Foods Biotechnol.* 6:113–17.

Kim, N. 2005. Discriminant analysis of marketed liquor by a multi-channel taste evaluation system. 14:554–57.

Kim, N., Haginoya, R., and Karube, I. 1996. Characterization and food application of an amperometric needle-type L-lactate sensor. *J. Food Sci.* 61:286–90.

Kim, N., Kim, D.-K., and Kim, W.-Y. 2008. Sulfamethazine detection with direct-binding optical waveguide lightmode spectroscopy-based immunosensor. *Food Chem.* 108:768–73.

Kim, N., and Park, I.-S. 2003. Application of a flow-type antibody sensor to the detection of *Escherichia coli* in various foods. *Biosens. Bioelectron.* 18:1101–7.

Kim, N., Park, I.-S., and Kim, W.-Y. 2007a. *Salmonella* detection with a direct-binding optical grating coupler immunosensor. *Sens. Actuator B Chem.* 121:606–15.

Kim, N., Park, I.-S., and Kim, D.-K. 2007b. High-sensitivity detection for model organophosphorus and carbamate pesticide with quartz crystal microbalance-precipitation sensor. *Biosens. Bioelectron.* 22:1593–99.

Kim, N., Park, K.-R., Park, I.-S., Cho, Y.-J., and Bae, Y. M. 2005. Application of a taste evaluation system to the monitoring of *Kimchi* fermentation. *Biosens. Bioelectron.* 20:2283–91.

Kim, S.-D., Yoon, S.-H., Kang, M.-S., and Park, N.-S. 1986. Effect of subatmospheric pressure and polyethylene film package on the *Kakdugi* fermentation. *J. Kor. Soc. Food Nutr.* 15:39–44.

Kolosova, A. Y., Samsonova, J. V., and Egorov, A. M. 2000. Competitive ELISA of chloramphenicol: Influence of immunoreagents structure and application of the method for the inspection of food of animal origin. *Food Agric. Immunol.* 12:115–25.

Kuhlmeier, D., Rodda, E., Kolarik, L. O., Furlong, D. N., and Bilitewski, R. 2003. Application of atomic force microscopy and grating coupler for the characterization of biosensor surfaces. *Biosens. Bioelectron.* 18:925–36.

Lacorn, M., Lahrssen, A., Rotzoll, N., Simat, T. J., and Steinhart, H. 2001. Quantification of metallothionein isoforms in fish liver and its implications for biomonitoring. *Environ. Toxicol. Chem.* 20:140–45.

Leonard, P., Hearty, S., Quinn, J., and O'Kennedy, R. 2004. A generic approach for the detection of whole *Listeria monocytogenes* cells in contaminated samples using surface plasmon resonance. *Biosens. Bioelectron.* 19:1331–35.

Li, T. L., Chung-Wang, Y. J., and Shih, Y. C. 2002. Determination and confirmation of chloramphenicol residues in swine muscle and liver. *J. Food Sci.* 67:21–28.

Luong, J. H. T., Male, K. B., Masson, C., and Nguyen, A. L. 1992. Hypoxanthine ratio determination in fish extract using capillary electrophoresis and immobilized enzymes. *J. Food Sci.* 57:77–81.

Mellgren, C., Sternesjo, A., Hammer, P., Suhren, G., Bjorck, L., and Heeschen, W. 1996. Comparisons of biosensor, microbiological, immunochemical, and physical methods for detection of sulfamethazine residues in raw milk. *J. Food Prot.* 59:1223–26.

Miller, N. J., and Rice-Evans, C. A. 1997. The relative contributions of ascorbic acid and phenolic antioxidants to the total antioxidant activity of orange and apple fruit juices and blackcurrant drink. *Food Chem.* 60:331–37.

Minunni, M., Tombelli, S., Mariotti, E., and Mascini, M. 2001. Biosensors as new analytical tools for detection of genetically modified organisms (GMOs). *Fresenius J. Anal. Chem.* 369:589–93.

Mulchandani, A., Chen, W., Mulchandani, P., Wang, J., and Rogers, K. R. 2001. Biosensors for direct determination of organophosphate pesticides. *Biosens. Bioelectron.* 16:225–30.

Ng, C. S., Chin, Y. N., Lim, P. Y., et al. 1983. Changes in quality of white pomfret, Chinese pomfret and grouper during ice-storage. *Bull. Jpn. Soc. Sci. Fish.* 49:769–75.

Norma, P., Rey, G., Mario, N., et al. 2002. Liquid chromatographic determination of multiple sulfonamides, nitrofurans and chloramphenicol residues in pasteurized milk. *J. AOAC Intl.* 85:20–24.

Oh, B.-K., Kim, Y.-K., Park, K. W., Lee, W. H., and Choi, J.-W. 2004. Surface plasmon resonance immunosensor for the detection *of Salmonella typhimurium*. *Biosens. Bioelectron.* 19:1497–504.

Oh, K. C., Kim, K. A., Paeng, I. R., Baek, D., and Paeng, K. J. 1999. Anion-selective membrane electrodes based on polymer-supported metalloporphyrins. *J. Electroanal. Chem.* 468:98–103.

Oohira, K., and Toko, K. 1996. Theory of electric characteristics of the lipid/PVC/DOPP membrane and PVC/DOPP membrane in response to taste stimuli. *Biophys. Chem.* 61:29–35.

Osipov, A. P., Zaitseva, N. V., and Egorov, A. M. 1996. Chemiluminescent immunoenzyme biosensor with a thin-layer flow-through cell. Application for study of a real-time bimolecular antigen-antibody interaction. *Biosens. Bioelectron.* 11:881–87.

Pam, L., Rosa, L. S., and Mark, T. M. 1996. Electrochemiluminescence-based detection of β-lactam antibiotics and β-lactamases. *Anal. Chem.* 68:2426–31.

Park, I.-S., Cho, Y.-J., and Kim, N. 2000a. Characterization and meat freshness application of a serial three-enzyme reactor system measuring ATP-degradative compounds. *Anal. Chim. Acta* 404:75–81.

Park, I.-S., Kang, S.-J., Kim, J.-H., and Noh, B.-S. 1993. L-Lactate oxidase electrode and dissolved oxygen meter for specific determination of L(+)-lactic acid in *Kimchi* during fermentation. *Foods Biotechnol.* 2:39–43.

Park, I.-S., Kim, D.-K., Adanyi, N., Varadi, M., and Kim, N. 2004c. Development of a direct-binding chloramphenicol sensor based on thiol or sulfide mediated self-assembled antibody monolayers. *Biosens. Bioelectron.* 19:667–74.

Park, I.-S., Kim, D.-K., and Kim, N. 2004b. Characterization and food application of a potentiometric biosensor measuring β-lactam antibiotics. *J. Microbiol. Biotechnol.* 14:698–706.

Park, I.-S., and Kim, N. 1998. Thiolated *Salmonella* antibody immobilization onto the gold surface of piezoelectric quartz crystal. *Biosens. Bioelectron.* 13:1091–97.

Park, I.-S., and Kim, N. 1999. Simultaneous determination of hypoxanthine, inosine and inosine 5'-monophosphate with serially connected three enzyme reactors. *Anal. Chim. Acta* 394:201–10.

Park, I.-S., and Kim, N. 2006. Development of a chemiluminescent immunosensor for chloramphenicol. *Anal. Chim. Acta* 578:19–24.

Park, I.-S., Kim, W.-Y., and Kim, N. 2000b. Operational characteristics of an antibody-immobilized QCM system detecting *Salmonella* spp. *Biosens. Bioelectron.* 15:167–72.

Park, J. K., Shin, M. C., Lee, S. G., and Kim, H. S. 1995. Flow injection analysis of glucose, fructose, and sucrose using a biosensor constructed with permeabilized *Zymomonas mobilis* and invertase. *Biotechnol. Progr.* 11:58–63.

Park, K.-R., Bae, Y.-M., Park, I.-S., Cho, Y.-J., and Kim, N. 2004a. Discriminant analysis of marketed beverages using multi-channel taste evaluation system. *J. Kor. Soc. Appl. Biol. Chem.* 47:300–6.

Pfeiffer, D., Ralis, E. V., Makower, A., and Scheller, F. 1990. Amperometric bienzyme based biosensor for the detection of lactose—Characterization and application. *J. Chem. Tech. Biotechnol.* 49:255–65.

Polzius, R., Schneider, Th., Bier, F. F., Bilitewski, U., and Koschinski, W. 1996. Optimization of biosensing using grating couplers: Immobilization on tantalum oxide waveguides. *Biosens. Bioelectron.* 11:503–14.

Pungor, E. 1998. The theory of ion-selective electrode. *Anal. Sci.* 14:249–56.

Ram, B. P., Singh, P., Martins, L., Brock, T., Sharkov, N., and Allison, D. 1991. High-volume enzyme immunoassay test system for sulfamethazine in swine. *J. AOAC Intl.* 74:43–46.

Rusling, J. F., Luttrell, G. H., Cullen, L. F., and Papariello, G. J. 1976. Immobilized enzyme-based flowing-stream analyzer for measurement of penicillin in fermentation broths. *Anal. Chem.* 48:1211–15.

Scholze, J., Hampp, N., and Brauchle, C. 1991. Enzymatic hybrid biosensors. *Sens. Actuator B Chem.* 4:211–15.

Schulthess, P., Shijo, Y., Pham, H. V., Pretsch, E., Ammann, D., and Simon, W. 1981. A hydrogen ion-selective liquid-membrane electrode based on tri-*n*-dodecylamine as a neutral carrier. *Anal. Chim. Acta* 131:111–16.

Schulze, H., Vorlová, S., Villatte, F., Bachmann, T. T., and Schmid, R. D. 2003. Design of acetylcholinesterases for biosensor applications. *Biosens. Bioelectron.* 18:201–9.

Scladal, P., and Mascini, M. 1992. Sensitive detection of pesticides using amperometric sensors based on cobalt phthalocyanine-modified composite electrodes and immobilized cholinesterases. *Biosens. Bioelectron.* 7:335–43.

Sethi, R. S. 1994. Transducer aspects of biosensors. *Biosens. Bioelectron.* 9:243–64.

Shitandi, A., Oketch, A., and Mahungu, S. 2006. Evaluation of a *Bacillus stearothermophilus* tube test as a screening tool for anticoccidial residues in poultry. *J. Veter. Sci.* 7:177–80.

Situ, C., Crooks, S. R. H., Baxter, A. G., Ferguson, J., and Elliot, C. T. 2002. On-line detection of sulfamethazine and sulfadiazine in porcine bile using a multi-channel high-throughput SPR biosensor. *Anal. Chim. Acta* 473:143–49.

Skládal, P., and Mascini, P. 1992. Sensitive detection of pesticides using amperometric sensors based on cobalt phthalocyanine-modified composite electrodes and immobilized cholinesterases. *Biosens. Bioelectron.* 7:335–43.

Smit, M. H., and Cass, A. E. G. 1990. Cyanide detection using a substrate regenerating, peroxidase-based biosensor. *Anal. Chem.* 62:2429–36.

Sternesjö, Å., Mellgren, C., and Björck, L. 1995. Determination of sulfamethazine residues in milk by a surface plasmon resonance-based biosensor assay. *Anal. Biochem.* 226:175–81.

Tatsuma, T., and Oyama, N. 1996. H_2O_2-generating peroxidase electrodes as reagentless cyanide sensors. *Anal. Chem.* 68:1612–15.

Toko, K. 1998. Electronic tongue. *Biosens. Bioelectron.* 13:701–9.

Toko, K., and Fukusaka, T. 1997. Measurement of hydrophobicity of amino acids using a multichannel taste sensor. *Sens. Mater.* 9:171–76.

Tranh-Minh, C. 1985. Immobilized enzyme probes for determining inhibitors. *Ion-Selective Electrode Rev.* 7:41–75.

Turner, A. P. H. 1996. Biosensors: past, present and future. http://www.cranfield.ac.uk/health/researchareas/biosensorsdiagnostics/page18795.html

Uchiyama, S., Hasebe, Y., and Tanaka, M. 1997. L-Ascorbate sensor with polypyrrole-coated carbon felt membrane electropolymerized in a cucumber juice solution. *Electroanalysis* 9:176–78.

Vaughan, R. D., O'Sullivan, C. K., and Guilbault, G. G. 1999. Sulfur based self-assembled monolayers (SAMs) on piezoelectric crystals for immunosensor development. *Fresenius J. Anal. Chem.* 364:54–57.

Volotovsky, V., and Kim, N. 1998a. EDTA determination by urease-based inhibition biosensor. *Electroanalysis* 10:961–63.

Volotovsky, V., and Kim, N. 1998b. Ascorbic acid determination with an ion-sensitive field effect transistor-based peroxidase biosensor. *Anal. Chim. Acta* 359:143–48.

Volotovsky, V., and Kim, N. 1998c. Determination of glucose, ascorbic and citric acids by two-ISFET multienzyme sensor. *Sens. Actuator B Chem.* 49:253–57.

Volotovsky, V., and Kim, N. 1998d. Cyanide determination by an ISFET-based peroxidase biosensor. *Biosens. Bioelectron.* 13:1029–33.

Volotovsky, V., and Kim, N. 2003. Ion-sensitive field effect transistor-based multienzyme sensor for alternative detection of mercury ions, cyanide, and pesticide. *J. Microbiol. Biotechnol.* 13(3):373–77.

Volotovsky, V., Nam, Y. J., and Kim, N. 1997. Urease-based biosensor for mercuric ions determination. *Sens. Actuator B Chem.* 42:233–37.

Vörös, J., Ramsden, J. J., Csúcs, G., et al. 2002. Optical grating coupler biosensors. *Biomaterials* 23:3699–3710.

Warsinke, A. 1997. Biosensors for food analysis. In *Frontiers in biosensorics II, practical applications*, ed. F. W. Scheller, F. Schubert, and J. Fedrowitz, 121–39. Basel, Switzerland: Birkhäuser Verlag.

Water, C. V., and Haagsma, N. 1991. Analysis of chloramphenicol residues in swine tissues and milk: Comparative study using different screening and quantitative methods. *J. Chromatogr.* 566:173–85.

Wen, Y., Zhang, M., Zhao, Q., and Feng, Y. Q. 2005. Monitoring of five sulfonamides antibacterial residues in milk by in-tube solid-phase microextraction coupled to high-performance liquid chromatography. *J. Agric. Food Chem.* 53:8468–8473.

Zourob, M., Mohr, S., Treves Brown, B. J., Fielden, P. R., McDonnell, M. B., and Goddard, N. J. 2003. The development of a metal clad leaky waveguide sensor for the detection of particles. *Sens. Actuator B Chem.* 90:296–307.

11 Nanotechnology in Food Quality and Safety Evaluation Systems

Sanghoon Ko, Jae-Ho Kim, Bosoon Park, and Yong-Jin Cho

CONTENTS

11.1 INTRODUCTION

Improvement of food safety and quality in terms of prevention of food-borne illness and rapid detection of food shelf-life have been a critical issue globally for several decades. Although traditional techniques have been played an essential role in evaluating the quality and safety of foods so far, the sensitivity, specificity, and efficacy of detection are constrained by limitations in the sampling and laborious lab work. Additionally, current consumer food trends such as convenience, organic foods, fresh foods, and increased demand for raw and minimally processed foods make pathogen reduction or elimination difficult, and thereby necessitates the development of new concepts of pathogen detection. For food quality evaluation, the effectiveness of nutrients and bioactives needs to be measured rapidly and precisely at any point during production, processing, transport, retailing, meal preparation, intake, digestion, and absorption in the body. In order to evaluate the quality of various foods, well-designed detection technologies should be considered: the aroma in the food is related to sensory quality, the presence of vitamins is related to nutritious quality, and the effectiveness of the preservative agents is related to the food's shelf-life.

New trends in food consumption have spread throughout the world and people have become more interested in quick and precise evaluation of the quality and safety of foods. The new emerging science of nanotechnology could provide solutions to some of these concerns. Nanotechnology is concerned with the understanding and manipulation of materials at the atomic and molecular level, generally with structures less than 100 nm in size. The inherently small size and unusual optical, magnetic, catalytic, and mechanical properties of nanomaterials permit the development of novel devices for food applications. Recently, nanotechnology approaches to detect food hazards, and quality parameters have been applied in the food industry (Mandrell and Wachtelt, 1991; Batt, 1997). The use of nanotechnology will resolve various technological problems in evaluating food quality and safety. For example, nanosizing increases specific surface area, which improves detection time and sensitivity of sensing units (i.e., increase of reactivity).

One of the important applications of nanotechnology is the development of nanosensors for evaluating food quality and safety. Nanosensors are defined as analytical devices that combine physical, chemical, and biological specific recognition systems with a variety of quantitative signaling. Nanosensors have been designed for early detection of food contaminants even at low concentration. Nanosensors significantly increased sensitivity but reduced response time. Thus, nanosensors can detect food contaminants rapidly, accurately, and sensitively.

Nanotechnology is playing an increasingly important role in the development of biosensors. Various approaches to exploit the advances in nanoscience and nanotechnology for bioanalysis have been recently reported (Lai and Cheng, 2004; Jain, 2003; Haruyama, 2003; Vo-Dinh et al., 2001). Significant progress has been made so far in micro- and nanofabrication, optoelectronics, and electromechanics. Remarkable progress has also been made in the last two decades in the development of optical sensors and their utilization in food safety (Ligler et al., 2003) and quality evaluation (Wiskur and Anslyn, 2001). The concept of DNA biochips have been applied to determine the presence of pathogens in meat, fish, and dairy products. The biochip platform expanded to microarray sensors that can be used to identify pesticides on fruit and vegetables. Sensitivity and other attributes of nanosensors can be improved by using nanomaterials in their construction; they improve their spatial resolution down to molecular levels, reduce their detection volume to a few cubic micrometers, and speed up their signal response to milliseconds (Jinet al., 2003). Because of their submicron size, nanosensors enable rapid analysis of multiple substances.

Research on food nanotechnology has significantly grown in the recent past. It is assumed that food nanotechnology will attain significant importance in the near future in evaluating food quality and safety. Also food nanotechnology is assumed to provide innovation to protect consumers from risky and unfit food products as well as accelerate growth in the food industry.

In this review, we will focus on available emerging nanotechnology tools for evaluating food safety and quality. Nanotechnology-based quantitative analyses of food pathogens, mycotoxins, chemical contaminants, nutrients, bioactives, etc., are reviewed in this chapter. The nanotechnology approaches to evaluate shelf-life and sensory characteristics are also summarized. We have chosen examples that demonstrate the applications of both chemical and biological nanosensing technologies for both food quality and safety evaluations.

11.2 NANOTECHNOLOGY IN FOOD SAFETY EVALUATION

11.2.1 FOOD-BORNE PATHOGENS

The consequences of food-borne diseases can be manifested within a few minutes, days, or even years after the consumption of infected food (Altekruse et al., 1999; Kaittanis, 2010). Although the developed countries have advanced techniques to detect and control most infectious agents, resulting in a reduction in the rate of food-borne illnesses, developing countries have continued to suffer from infectious disease due to lack of proper detection mechanisms and poor hygienic control (Fleming, 1990). Food-borne infections are generally treated as a major public health problem in both developed and developing countries. In fact, the infectious diseases induced by food-borne pathogens such as *Escherichia coli*, *Bacillus cereus*, *Salmonella* sp., *Campylobacter jejuni*, and *Staphylococcus aureus* have not been decreased in spite of enhanced hygiene in developed countries (Swaminathan and Feng, 2003). Food-borne infectious diseases are caused by pathogens contaminating food during the manufacturing process, distribution, and storage period (Beuchat and Ryu, 1997; Heinitz and Johnson, 1998). Pathogens are categorized into groups, which are pathogenic bacteria, fungi, and viruses (Kaittanis et al., 2010). Thus, appropriate detecting and controlling techniques are important, depending on the specific strains.

Traditionally, various monitoring methods have been adopted for detecting the pathogens, which are associated with physical, chemical, and biological evaluations (Ivnitski et al., 1999). The conventional methods include visible cell counting, enzyme-linked immunosorbent assay (ELISA), and polymerase chain reaction (PCR) technique (Ko and Grant, 2006; Lazcka et al., 2007). The visible cell counting method is usually performed after culturing samples that are collected from the infected individual, and it require much time to achieve a reaction (Brooks et al., 2004). The fluorescent immunoassays are a very sensitive methods for detecting pathogens. However, these methods only detect molecules such as proteins and antigenic components released from pathogens, resulting in no direct information about the presence of pathogens. Moreover, they are time-consuming and require careful treatment and preservation to prevent the protein denaturation of antibody (Velusamy et al., 2010). The PCR technique has become a valuable tool for pathogen detection due to its high sensitivity and specificity, generous detection limit, direct detection, and low amount of pathogen requirement (Malorny, 2003; Olsen et al., 1995). But the PCR technique is affected by dead cells nucleic acid interference (Kaittanis et al., 2010; Allmann et al., 1995). Therefore, development of new monitoring methods that require a cost-effective and short reaction time as well as accuracy is desired for the detection of pathogens. Nanotechnology can provide an opportunity to develop new detecting and evaluating techniques. Furthermore, it can also lead to the improvement of sensitivity and accuracy in combination with other techniques, such as physical, biochemical, and electromechanical techniques (Jain, 2005, Rosi and Mirkin, 2005). Nanotechnological approaches used in pathogen detection include colorimetric immunoassay using metal nanocomposites, quantum dots (QDs), fluorescent polymeric nanoparticles, nanoarrays, and biosensors (Sotiropoulou and Chaniotakis, 2003; Lynch et al., 2004; Gerion et al., 2003; Chan and Nie, 1998; Kneipp et al., 2002; Hong et al., 2002; Betala et al., 2008). Nobel nanobiosensor technologies are listed in Table 11.1.

Biosensors are analytical devices that use biological components to react or bind with a target molecule and transduce this event into a detectable signal. When the biological component is an antibody or an antibody fragment, the biosensor is termed an immunosensor. Immunosensors may offer certain advantages over the more traditional antibody-based ELISA format. Immunoassay techniques have been successfully applied for detection of food pathogens since the property of highly specific molecular antibody recognition of antigens leads to highly selective assays. The extremely high affinity of antigen-antibody interactions also results in very high sensitivity. This allows sensitive and cost-effective analysis, rapid assays, reusable sensor elements, and the capacity for continuous monitoring. For these reasons, the number and types of immunosensors have expanded rapidly in recent years.

Colorimetric immunoassay, which is related to the biospecific molecular recognition interaction between antibodies and target molecules, is a simple and economical technique for detecting the target components, including proteins and pathogenic microorganisms. This technique has been developed based on metal-mediated

---

TABLE 11.1

Pathogens Detection Using Novel Nanotechnologies

Technology	Measurement	Pathogens	Sensitivity (CFU/ml)	Reference
Gold nanoparticle	Colorimetric detection	Ovalbumin *Escherichia coli*	—	(33)
Quantum dots	Fluorescence	*E. coli* O157:H7	10^4	(37, 138)
		S. typhimurium		
		E. coli	10^2–10^7	(38)
		S. aureus		
Carbon nanotube	FET	*S. infantis*	10^2	(43)
Nano material	SPR	*L. monocytogenes*	10^6	(55)
	QCM	*L. monocytogenes*	10^7	(56)
	SERS	*E. coli*	10^3	(57)
Nano optical fiber	FRET	*S. typhimurium* in PBS solution	10^3	(18)
		S. typhimurium in homogenized pork	10^5	

Note: FET = field effect transistor, SPR = surface plasmon resonance, QCM = quartz crystal microbalance, SERS = surface-enhanced Raman scattering, FRET = fluorescence resonance energy transfer.

immunoassay, which uses metal to amplify the signals resulting in high sensitivity (Liao and Huang, 2005). Betala et al. (2008) used the colorimetric immunoassay with a streptavidin-colloidal gold nanoparticle, which acts as a reducing agent for silver ions, with silver reduction and precipitation in an attempt to detect *Escherichia coli* and ovalbumin. Ovalbumin was easily detected at a 50 µM/ml concentration, and *Escherichia coli* exhibited specific binding to anti- *Escherichia coli*, which was observed under fluorescent microscopy.

Fluorescent methods using as fluorescein isothiocyanate (FITC) and dye-filled nanoparticles have been commonly used in the detection of bacteria and other pathogens. However, these conventional dyes are destroyed easily by light and their wide emission spectrum causes spectral confusion between channels, resulting in limitation for their application (Gu et al., 2004). Quantum dots, a colloidal semiconductor in a nanoscale diameter, are new fluorescent labeling materials that represent a narrow emission spectrum resistant to photobleaching and high brightness, which enhance the sensitivity of detection (Larson et al., 2003). Yang and Li (2006) identified food-borne pathogens such as *Escherichia coli* O157:H7 and *Salmonella typhimurium* using fluorescent QDs. They use the QDs in combination with magnetic separation for simultaneous detection. The established detection limit was 10^4 CFU/ml in this technique. Additionally, Xue et al. (2009) demonstrated rapid detection of *Escherichia coli* and *Staphylococcus aureus* using water-soluble QDs. These pathogens were detected in the range of 10^2 to 10^7 CFU/ml total count of the bacteria within 1 to 2 hours, and the low detection limit was 10^2 CFU/ml.

A nanobiosensor is an analytical device for detecting chemical signals, which is based on the combination of the biochemical and electromechanical techniques. The device converts the chemical signals produced due to binding of target components to a probe in machine-readable signals (Hall, 2002). The specific signals often originate from the biological binding reaction, biospecific interaction, nucleic acid interaction, and antigen-antibody interaction (Vo-Dinh et al., 2001). There have been many attempts to rapidly detect food-borne pathogens such as *Escherichia coli* O157:H7, *Staphylococcus aureus*, and *Campylobacter jejuni* (Chang et al., 2001; DeMarco et al., 1999). Villamizer et al. (2008) demonstrated fast and highly sensitive detection of *Salmonella infantis*, which causes gastroenteritis, using biosensors with carbon nanotube as a conductor. The anti-*Salmonella* antibodies were attached on the carbon nanotube and the carbon nanotube was coated with Tween 20 to improve specificity. The device detected a minimum of 100 CFU/ml within 1 hour, and additionally detected count as high as 500 CFU/ml under hindrance by *Streptococcus* and *Shigella*. Moreover, Junxue et al. (2008) prepared a silicon-gold nanorod array for detecting *Salmonella typhimurium*. The device represented high fluorescence and detection of *Salmonella*. Ko and Grant (2006) developed a fluorescence resonance energy transfer (FRET) to detect *Salmonella typhimurium*. They prepared two different solutions to determine the detection sensitivity. One solution containing *S. typhimurium* diluted in phosphate-buffered saline (PBS), and another one was a homogenized pork sample containing *Salmonella typhimurium*. The results indicated that the lowest detection limit was 10^3 CFU/ml for *S. typhimurium*–doped PBS solution, and 10^5 CFU/ml for homogenized pork.

Optical sensing techniques based on the plasmon resonance absorption exhibited by metallic films and particles have been widely used and played a significant role in biochemistry, biomaterials, and surface science due to their high sensitivity and facile implementation (Katz and Willner, 2004; Daniel and Astruc, 2004; Lee and Perez-Luna, 2005). The surface plasmon resonance (SPR) is the collective oscillations of surface electrons induced by visible light, and is responsible for the intense colors exhibited by colloidal metals (Dahlin et al., 2005; Raschke et al., 2004; Yonzon et al., 2004; Ghosh et al., 2004; Raschke et al., 2003; Nath and Chilkoti, 2002, 2004). This technique is advantageous in quantitative detection of analytes in their native concentrations. It is also useful in the characterization of binding kinetics by providing real-time monitoring of binding events reflected by changes in the optical properties of the system. Leonard et al. (2004) investigated the sensitive detection of *Listeria monocytogenes* using SPR detection of a pathogen without labeling. They detected and quantified 10^5 CFU/ml of *Listeria monocytogenes* within 30 min. Additionally, Vaughan et al. (2001) detected *Listeria monocytogenes* using a quartz crystal microbalance (QCM) immunosensor. The study reported the detection of a minimum of 10^7 CFU/ml in real time. Sengupta et al. (2006) investigated the sensitive detection of *Escherichia coli* using surface-enhanced Raman spectroscopy with colloidal silver nanoparticles. They observed a low detection limit, 10^3 CFU/ml of *Escherichia coli*.

Early detection of pathogenic bacteria and their toxins in food is one of the potential applications of nanotechnology. Nanosensors use an array of thousands of nanoparticles as biomarkers which are designed to fluoresce in different colors, when combine with various pathogens. The advantage of such a system is that thousands of nanoparticles can be placed on a single nanobiosensor for the rapid, accurate, and affordable detection of any number of different pathogens. Recently, highly sensitive sensors based on nanostructures having great potential for pathogen detection have emerged. Some of these methods include detections by fluorescent QDs, localized surface plasmon resonance (LSPR) of metallic nanoparticles, enhanced florescence, immobilized nanoparticles, and Raman reporter molecules with metallic nanoparticles. In single-component nanostructures, it can be difficult to immobilize the recognition molecules and signaling molecules simultaneously; hetero-nanostructure provides a promising platform to solve this problem. Using a hetero-nanostructured platform, different functional molecules can be immobilized to the different parts of the hetero-nanostructure to improve selectivity and specificity of detection. Since conventional detection methods such as cell isolation, PCR, antigen detection, and immunoassay are time-consuming, cumbersome, or lack sensitivity, new methods using nanotechnology with aligned Ag nanorod arrays prepared by oblique angle vapor deposition (OAD) as a substrate for surface-enhanced Raman scattering (SERS) are now emerging as the potential tools to identify food pathogens. The OAD substrate preparing method facilitates the selection of nanorod size, shape, density, alignment, orientation, and composition, while the procedure is reproducible and relatively simple to implement. This nanotechnology method indicates the fundamental nanostructural design of metallic nanorod arrays and their influence on SERS enhancement, as well as the development of a spectroscopic biosensor assay for pathogen detection.

11.2.2 PATHOGENIC VIRUS

Pathogenic viruses such as (human immunodeficiency virus (HIV), hepatitis C virus (HCV), influenza virus, adenovirus, etc.) are known to cause gastroenteritis and hepatitis. Pathogenic viruses invade the host via various routes, such as skin, respiratory tract, gastrointestinal tract, genitourinary tract, and conjunctiva, etc. These viruses are characterized by their very small size (15 to 400 nm); therefore cannot be observed using an optical microscope. Accordingly, they can be observed using an electron microscope (EM) (Kaittanis et al., 2010). Unlike other detection methods such as QDs, colorimetry, fluorescence imaging, etc., have recently been used (Hauck et al., 2010; Tallury et al., 2010) for sensitive detection of viruses.

With the recent changes in food habits and food processing, there are increasing risks for outbreaks due to food-borne viruses. Food-borne or waterborne pathogenic viruses causing gastroenteritis include Norwalk virus or Norwalk-like virus (also known as norovirus), sapporo-like calicivirus, rotavirus, adenovirus, astrovirus, hepatitis A virus (HAV), and hepatitis E virus (HEV) (Koopmans et al., 2002). Norovirus is one of the four main causative agents for food-borne gastroenteritis together with *Campylobacter, Salmonella,* and *Escherichia coli* O157:H7, and as a viral pathogen, it accounts for 50% of the total recent occurrences of gastroenteritis (Velusamy et al., 2010) worldwide.

EM and other visual confirmation methods, which have traditionally been used for the diagnosis of food-borne and waterborne viruses, can be used for diagnosis if there are 10^5 to 10^6 CFU/ml of stool samples. Using these techniques, there is a limitation in that certain detailed structures such as *specific amino acid sequences* cannot be observed. Therefore, An accurate identification of food-borne virus therefore cannot be made. Of the methods that have recently been used for virus identification, the antigen-directed ELISA and reverse transcription PCR (RT-PCR) techniques are more efficient and sensitive. ELISA is solely used in detecting rotavirus and adenovirus in stool samples. ELISA-based assays are also used to detect astrovirus and norovirus. In norovirus, however, the technique lacks broadness. Therefore, ELISA assays often have a limited sensitivity due to these limiting factors. Furthermore, diagnosis of food-borne viral gastroenteritis is not performed routinely in all cases. RT-PCR assays can produce false negative results due to the PCR inhibitors that may be present in the samples (Koopmans et al., 2002; Driskell et al., 2008).

Most of the food-borne viruses that have been described up to the present are identified from clinical samples. That is, these methods are useful in the diagnosis following the occurrence of food poisoning rather than at the prevention stage before foods are ingested. These methods are therefore ineffective. Of the methods that have recently been developed based on nanotechnological approaches for detection of food-borne virus, active studies on mass spectrometry (MS) that have been conducted to identify specific markers for food-borne viruses are worth noting. These methods usually detect proteins that are abundantly present in microorganisms as a target. Particularly in the detection of norovirus they use capsid protein as a target. Colquhoun et al. (2006) used three MS-based methods (scanning of intact protein analysis, peptide mass fingerprinting, and peptide sequencing) for the identification of norovirus capsid protein. Of these, only the peptide sequencing method using nanospray in tandem with mass spectrometry was effective for identification of the 250 femtomoles of the capsid protein.

Another method for the surface-enhanced Raman spectroscopy (SERS) is a technique with sensitivity and specificity for the qualitative detection and identification of bacteria. But this method has the disadvantage that it requires expensive equipment and the types of enhancing substrates are limited. To resolve these disadvantages, Driskell et al. (2008) smeared a silver nanorod array with an oblique angle on the surface. Using this technique, these authors detected rotavirus. This method is highly reproducible and inexpensive in comparison to conventional assay (Shanmukh et al., 2006).

11.2.3 MYCOTOXINS

Mycotoxins are toxic secondary metabolites produced by filamentous fungi (molds). Mycotoxins can enter the food chain either directly from the use of contaminated food ingredients or indirectly from the growth of toxigenic fungi on foods. Consumption of mycotoxin-contaminated food causes acute or chronic mycotoxicosis in humans and animals. Some mycotoxins are highly toxic, mutagenic, teratogenic, and carcinogenic, and have been implicated as causative agents in human hepatic and extrahepatic carcinogenesis. In the worst-case scenario, they may even cause death in humans and other mammals. Therefore, there is a public health concern over the potential risks with consumption of foods containing residues of mycotoxins and their metabolites.

Many methods have been used to detect mycotoxins in foods, of which chromatography and ELISA have been applied most extensively (Turner et al., 2009). Chromatography can detect various types of target materials from a single sample. Besides, it can also be used for quantitative purpose. In the ELISA method for the detection of mycotoxins, a competitive assay format of the antigens against antibodies is basically used.

Ochratoxin A (OTA) is one of the mycotoxins that are the most abundantly found in human blood, milk, animal tissues, and organs. For toxin detection, on a glass slide coated with indium–tin oxide (ITO), the nanostructured cerium oxide (NanoCeO$_2$) film was prepared. Then, the rabbit-immunoglobulin antibodies (r-IgGs) and bovine serum albumin (BSA) were fixed. Thus, immunos by which such mycotoxins as ochratoxin A (OTA) can be detected were prepared. In BSA/r-IgGs/NanoCeO$_2$/ITO immunoelectrode, the linear range (0.5 to 6 ng/dl), low detection limit (0.25 ng/dl), fast response time (30 s), and high sensitivity (1.27 μA/ng $\cdot$ dl $\cdot$ cm^2) were improved (Kaushik et al., 2009a). For the detection of ochratoxin A, the nanobiocomposite film of a chitosan (CH)–iron oxide (Fe$_3$O$_4$) nanoparticle was attached to the indium–tin oxide (ITO) electrode. Then, the rabbit immunoglobulin antibodies (IgG) were fixed to it. The IgGs/CH-Fe$_3$O$_4$ nanobiocomposite/ITO immunoelectrode is characterized by the presence of a low detection limit (0.5 ng/dl), fast response time (18 s), and high sensitivity (36 μA/ng $\cdot$ dl $\cdot$ cm^2) (Kaushik et al., 2008). A nanobiocomposite film consisting of fumed silica nanoparticles (NanoSiO$_2$) and chitosan (CH) was used to detect OTA by fixing rabbit-immunoglobulin antibodies (r-IgGs) and BSA. In this BSA/r-IgG/CH NanoSiO$_2$/ITO immunoelectrode, linearity (0.5 to 6 ng/dl), detection limit (0.3 ng/dl), response time (25 s), and sensitivity (18 $\mu A \cdot ng/dl \cdot cm^2$) were improved (Kaushik et al., 2009b).

Aflatoxin is a very detrimental carcinogenic mycotoxin that is commonly generated by many species of fungus in various types of agricultural products such as corn and peanuts (Radoi et al., 2008). By attaching aflatoxin B_1 antibody (anti-AFB$_1$) to the nanosize gold hollow balls (NGBs) with the dendritic surface, the aflatoxin B_1 can be detected. Using the quartz crystal microbalance (QCM) technique, it was shown that aflatoxin B_1 was sensitive at a volume of 0.05 ng/ml (Liao, 2007). In this method, the monoclonal anti-AFB$_2$ antibodies were attached to magnetic nanogold microspheres consisting of nano-Fe_2O_3 particles as the core and gold nanoparticles as the shell. Then, this was used to rapidly detect aflatoxin B_2. Using a 0.9 ng/ml AFB$_2$, this method was three times as effective in obtaining the sensitive and quantitative results as the previous methods within 15 min (Tang et al., 2009). To quantitatively detect aflatoxin B_1 from milk, using gold nanoparticles, which were strengthened with a competitive immunoreaction technique, a piezoelectric immunosensor was used. The detection limit reached as low as 0.01 ng/ml (Jin et al., 2009).

Sterigmatocystin is identified from food products and crops such as corn, peanuts, soybean, ham, and cheese. This is a mycotoxin that is associated with the occurrence of cancers such as stomach cancer and hepatoma cancer (Ma et al., 2003, Reiß, 1975). To examine the characteristics of reactions with a multiwalled carbon nanotubes–aflatoxin-detoxifizyme (MWNTs-0ADTZ) electrode, which was covalently bonded to sterigmatocystin, a biosensor with the three-electrode system was used. As a result, the oxidation peak of sterigmatocystin reached a value of +400 mV. Compared with the conventional assays, the sensitivity was approximately twice as great (Yao et al., 2006).

11.2.4 Chemical Contaminants

Chemical contaminants that are found in foods can be classified into agrochemical and environmental contaminants. Agrochemical contaminants include such pesticides as insecticides, herbicides, and rodenticides; veterinary drugs such as nitrofuran, fluoroquinolones, malachite green, and chloramphenicol; and plant growth regulators. Environmental contaminants include radionuclides (^{137}caesium and ^{90}strontium) and polycyclic aromatic hydrocarbons (PAHs), which are present in the air, and arsenic and mercury, which are present in water. In the soil, there are cadmium, nitrates, and perchlorates. Such substances as polychlorinated biphenyl (PCB), dioxins, and polybrominated diphenyl ethers (PBDEs) are present in not only air, water, and soil, but also the entire biosphere. Packaging materials (e.g., antimony, tin, lead, perfluorooctanoic acid (PFOA), semicarbazide, benzophenone, isopropylthioxanthone (ITX), bisphenol A, and processing/cooking equipment such as copper, or other metal chips, lubricants, and cleaning and sanitizing agents) also contribute to the chemical contamination of within foods.

Chemical contaminants due to these agrochemical and environmental contaminants are also of major concern from a worldwide perspective. Of these, pesticides, heavy metals (e.g., mercury, cadmium), and phenols are also important. Intoxication from residues of pesticides in agricultural products may cause problems such as eye pain, abdominal pain, convulsions, respiratory failure, paralysis, and even death

(Tran-Minh et al., 1990; Cremisini et al., 1995; Ray, 1998). Ingestion of mercury has been proven harmful to the brain, heart, kidneys, lungs, and immune system of humans of all ages (Darbha et al., 2007).

The detection and assessment of chemical contaminants in foods is therefore mandatory. As for the conventional methods for detecting chemical contaminants, there are liquid/gas chromatography (Leoni et al., 1991; Lacorte and Barcelo, 1994), high-performance liquid chromatography, and mass spectroscopy (Hall et al., 1997). These methods are very efficient and also allow for the differenciation of organophosphorus pesticide compounds from other types. However, the pretreatment process of samples is relatively longer, and a higher level of technical expertise and ingenious machinery is required. A cost-effective, sensitive screening analysis would therefore be essential. In most cases, it can be expected that a nanotechnology can be used to develop the excellent multianalyte detection systems in a cost-effective manner (Liu et al., 2008).

As the representative agrochemical contaminants, most pesticides have a higher degree of water solubility. They are commonly used for pest management in agricultural systems. Therefore, analytical methods for determination residual pesticides in foods should be very sensitive and selective (Valdés et al., 2009). He et al. (2008) characterized and quantified melamine, cyanuric acid, and melanine cyanurate using surface-enhanced Raman spectroscopy (SERS) combined with gold nanosubstrates. SERS coupled with gold nanoparticles provided vibrational spectroscopic properties, and they can measure even trace amount of melamine and its analogs. Using SERS, a detection limit for melamine was lowered to 2.6×10^{-7} mol/L (~33 ppb). Fluorescent nanoparticles for detection of organophosphate (OP) appreciated for their sensitivity, lower limit, and a wider range of detection. The basic requirements of fluorescent nanoparticles for contaminants detection include a stable, high, long-lived fluorescence intensity and a broad range of wavelengths. Vamvakaki and Chanintakis (2007) monitored the organophosphate pesticides dichlorvos and paraoxon with liposome-based nanobiosensors. This sensor is based on the acetylcholinesterase (AChE) inhibition mechanism. It has been also shown that AChE encapsulated by liposome can be stable even at a temperature of 4° (Vamvakaki and Chaniotakis, 2007). Simonian et al. (2005) developed a specific method for detecting OP neurotoxin based on the enhancement or quenching of fluorescence intensity and is a function of the distances between gold nanoparticle and fluorophore, as described in Figure 11.1. When decoy and organophosphorus hydrolase (OPH)–gold nanoparticle conjugates were present at a near equimolar level, the degree of paraoxon sensitivity was the highest. Some groups reported a biosensor based on OPH-combined (CdSe)ZnS core-shell QDs for the detection of paraoxon. A detection limit concentration of paraoxon using OPH-QD bioconjugates was measured to be 10^{-8} M. In the bioconjugates, when the OPH molar ratio was increased, the sensitivity of the biosensor was slightly elevated (Ji et al., 2005). Zhou et al. (2004) prepared nanocolloidal gold particles labeled with anti-carbofuran monoclonal antibody (Mab). The carbofuran was detected on the basis of colored band appeared on the test line. The sensitivity of detection of carbofuran exceeded 0.25 mg/L. Besides the advantage of being a one-step strip test for which the reagents are provided in the test device and the detection time is shorter than 10 min. A carbon nanotube (CNT) is a molecular-scale, new substance with high chemical stability, high mechanical strength, modulus, and high electrical

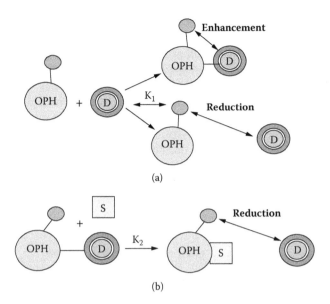

FIGURE 11.1 Detection of OP neurotoxin based on function between gold nanoparticle and fluorophore. (a) Schematic of decoy-enzyme interaction for enhancement in the absence of substrate. Decoy (D) binds to enzyme-nanogold conjugate (OPH), leading to a surface-enhanced fluorescence of the decoy. (b) Schematic of analyte (S) displacement of decoy (D) from OPH-gold complex (OPH), leading to a decreased fluorescence signal from the decoy (86). (From A. L. Simonian et al. 2005. Nanoparticle-based optical biosensors for the direct detection of organophosphate chemical warfare agents and pesticides. *Analytica Chemica Acta*, 534:69–77. With permission from Elsevier.)

conductivity. Pargaonkar et al. (2005) dispersed multiwall CNTs (MWCNTs) into the Nafion solution with a glassy carbon electrode (GCE), in order to prepare an MWCNTs/Nafion film. This can act as an ampermetric sensor and a chromatography detector that can be used to detect parathion. It was shown to have an efficient electrocatalytic effect on the electrochemical response of parathion. The detection limit was 1.0×10^{-9} mol/L, and this represents the lowest value obtained so far.

Many efforts have been made to isolate heavy metals that are prevailing in the wastewater, drinking water, and agricultural water for instance, the mechanism of isolation include ion exchange (Dabrowski et al., 2004), adsorption (Huang and Blankenship, 1984), precipitation (Naylor and Richard, 1975), and membrane separation (Huang and Koseoglu, 1993). Xu et al. (2009) examined the simple and sensitive methods for Hg^{2+} ions using DNA oligonucleotides and unmodified gold nanoparticles (DNA/AuNPs). The authors obtained the results around a sensitive linear range (0 to 5 μM) and a detection limit (0.5 μM) of Hg^{2+} ions. At a smaller number of T-T mismatches, the detection range became narrowed and the degree of sensitivity was found to be very high. Darbha et al. (2007) proposed gold nanoparticles based nanometal surface energy transfer probe by which a trace amount of Hg^{2+} can be detected from the environmental samples (1,100-fold). This has an excellent sensitivity (2×10^{-9} g/ml), for detection of mercury contaminated soil, water, and fish, and it can also be used as a selective, miniaturized, inexpensive, and battery-operated ultra-sensitive screening tool. Choi et al. (2006) synthesized ETS-10 and ETAS-10 with different Al_2O_3/TiO_2 ratios the nanosized, large-pored titanium silicate zedite materials, and successfully used for the removal of Pb^{2+} and Cd^{2+}. Sugunan et al. (2005) reported gold nanoparticles capped with chitosan for sensing heavy metal ions (Zn^{2+} and Cu^{2+}). The chelating properties of chitosan and the sensitivity of the optical properties of gold nanoparticles to aggromeration have been employed for the detection of trace amounts of heavy metal ions in water.

Arecchi et al. (2010) proposed a tyrosinase-modified electrode, which can be used as an amperometric biosensor detecting the phenolic compounds in food products. In this method has been the enzyme fixed on the glassy carbon electrode covered by a polyamidic nanofibrous membrane by drop coating deposition. A biosensor showed a reaction time of 16 s, a detection limit of 0.05 μM, and a linearity extended up to 100 μM.

11.2.5 Genetically Modified (GM) and Irradiated Foods

Genetically modified foods (GM foods) are the food materials that are developed by altering the genetic makeup of organisms such as animals, plants, or bacteria to enhance food quality, health, nutrition, and agricultural practice. Due to many advantages over traditional foods, the proportion of countries cultivating GM foods have been continuously increasing. GM foods that are mainly cultivated include corn, tomato, soybean, and sugar beet. GM technology has a great number of advantages. Due to suspected possible potential risks of new technology, however, scientists are interested in the detection of GM foods for the assessment of safety.

In the detection of GM foods, nanotechnology has recently been used for screening of modified DNA. DNA microarrays and biosensors are used as nanotechnology-based methods for detecting GM foods (Deisingh and Badrie, 2005). DNA microarrays may also be termed as DNA chips, which can determine GMOs or GM foods by comparing sequences that were analyzed to combine covalently bound sequences corresponding to DNA sequences of the sample with the solid phase (Götz, 2009).

DNA chips require a small amount of sample and are more convenient and sensitive than other detectors, and have a higher degree of screening throughput. To enhance the cost-effectiveness and sensitivity, Leimanis et al. (2006) developed a low-density DNA chip for simultaneous identification of GMOs with the use of six GMO specific capture probes. (Bt11, Bt176, GA21, MON810, Roundup-Ready Soybean, and StarLink) and five plants (corn, soybean, rapeseed, tomato, and sugar beet). Additionally, Xu et al. (2006) conducted studies on the detection of GMO from soybean, maize, and cotton, where GMOs were detected with a synchronous use of polymerase chain reaction (PCR) and microarray. The microarray technology that has been used in this study can be classified into three types according to the purpose: The first type may be referred as a chip consisting of probes, including promoter, reporter, and terminator genes, which are used to differentiate GMO from genetically unmodified plants. The second type is a species-specific chip, which can be used for samples whose target sequences are known. In soybean, microarray consisting of probes for lectin, 35S promoter, nos terminator, and EPSPS is used. In maize, a microarray consisting of probes for zein, 35S promoter, nos terminator, 35S terminator, Bar, PAT,

CryIA(b), and Cry9C is used. In cotton, a microarray consisting of probes for Rbcl, 35S promoter, nos terminator, NptII terminator, and CryIA(c) is used. The third type is an integrated chip. Because this is an integration of the chips, which are specific to various species, it is appropriate for detecting multiple targets. With a synchronous use of these three types of microarray, the detection can be done in a more accurate, prompt manner. Using these methods, 95% of commercial GMOs can be detected. Limits of detection were found to be 1% in maize and 0.5% in soybean.

Detection of GMO GM foods based on DNA chips can be easily performed in the seeds and grains. It remains problematic, however, in that the detection is still difficult to perform in foods; further studies are warranted to develop the technology by which the detection can be done both conveniently and promptly using nanotechnology from any kind of foods. Additionally, more inclusive methods of detection should also be developed for GMOs.

Foods are irradiated in order to suppress them from sprouting and eradicating pathogens and parasites. Gamma rays & x-rays are mainly used to irradiate foods, where a lower degree of environmental harmfulness are commonly preferred. In many countries, food irradiation is permitted, but the type of x-rays used to irradiate foods are restricted. According to the types of foods, a maximum dose has been established (Ziegelmann et al., 1998). Irradiated foods have shown no differences in their taste, smell, or sight, in order to confirm this, studies on detection technology would be essential (Gautam et al., 1998).

Electron spin resonance (ESR) measurement is commonly used to detect irradiated food. It remains problematic, however, since the detection is subject to the environmental impacts. Rogers et al. (2001) conducted a study to analyze radiation-induced DNA damage using a fiber optic biosensor. Figure 11.2 represents a schematic diagram of a fiber optic biosensor, whose mechanisms are based on an analysis of DNA

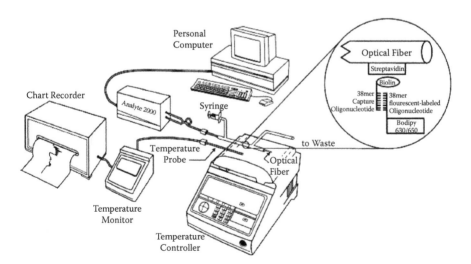

FIGURE 11.2 Schematic of the fiber optic biosensor using PCR thermal cycler for temperature control. (From K. R. Rogers et al. 2005. Fiber optic biosensor for detection of DNA damage. *Analytica Chimica Acta*; 444:51–60, with permission from Elsevier.)

damage through an evaluation of nonspecific oligonucleotides using an optical fiber. The result, compared with the previous detection methods, is prompt, cost-effective. Food irradiation may affect DNA or proteins contained in foods. But x-rays are not left in the foods. As for the detection of GMO, as well as irradiated food, there can be performed through an analysis of DNA or protein.

11.3 QUALITY

11.3.1 FOOD COMPONENTS AND ADDITIVES

The quality assurance of foods is essential because consumers demand wholesome foods that are beneficial for their safety and health (Neethirajan and Jayas, 2010). Using equipment and tools based on nanotechnology, the bioactives and nutrients can be detected on a nanotechnology scale. Besides, a detailed evaluation of the phenomena can also be made on a nanotechnology scale (Khosravi-Darani et al., 2007).

Sensor systems for food quality control, such as rapid detection of spoilage of food components, are possible through nanotechnology. To monitor the food quality through nanotechnology approaches nanosensors and surface plasmon resonance (SPR) are used. Nanosensors can provide quality assurance by tracking microorganisms, toxins, and contaminants throughout the food processing chain (Neethirajan and Jayas, 2010). Bodor et al. (2001) combined a nanosensor with the microfluidic device to analyze various food additives such as benzoate, sorbate, p-hydroxybenzoic, acid esters, and glutamate. Within a chip setup, an assessment was performed to evaluate the performance of different types of electrophoresis methods. In each type of food additive, different types of optimal detection methods have been identified.

To guarantee the quality until the food has been consumed, and to monitor the quality of foods during various stages of the logistic process, the implementation of low-cost nanosensors for food packaging has become possible (Neethirajan and Jayas, 2010). Neethirajan et al. (2009) monitored the quality of grains using polymer nanosensors at facilities storing the grains. These sensors respond to the analytes and volatiles in the food storage environment and thereby can detect the sources and types of grain spoilage. An advantage with this kind of sensor systems is that thousands of nanoparticles can be placed on a single sensor to accurately detect the presence of insects or fungus inside stored grain bulk in bins.

Ruengruglikit et al. (2004) developed an electronic tongue which has an array of nanosensors for inclusion in food packaging. According to these authors, nanosensors are extremely sensitive to gases released during food spoilage, therefore, food freshness can be detected based on the changes in sensor color. Legin et al. (2003) measured the flavor and taste of red wine produced in Italy using an electronic tongue and then compared them with the results of sensory evaluation. Basing on the comparison of the results of sensory function, Barbera d'Asti wine was found to have errors of prediction 6 to 27% in magnitude. Gutturnio wine had errors of prediction 4 to 13% in magnitude. This posed the possibility that the quality of wine could be predicted using an electrical nose system.

A surface plasmon resonance (SPR) sensor uses the resonance phenomenon of the surface plasma wave, which is generated when the photon energy is absorbed on the surface of a thin metal film. A resonance phenomenon is used as the standard measurement principle based on the measured degree of adsorption of a nanoparticle sample on the surface metals (gold (Au) or silver (Ag) in particular). This principle is commonly used for a biosensor, by which the amount of sample is detected based on the color change (Homola et al., 1999). Choi et al. (2008) examined the enhancement of sensitivity of the SPR sensor by coupling Au nanoparticles having specific size and surface density for the detection of methanol content. Au nanoparticles with a diameter of 10, 30, and 60 nm were conjugated by amine groups of cystamine-modified chips. The surface density of Au particles was controlled by the reaction time and the concentration of solution containing the particles. To investigate the enhancement of sensitivity of the SPR sensor, a comparison was made for shifts of resonance angle with or without. Au particles in methanol aqueous solution. In association with lower surface density, due to the coupling of localized surface plasmon effect (which was derived by the density and size of the surface plasmon wave as well as specific Au nanoparticles) and adsorption of 30 nm diameter particles, the sensitivity was increased by 57% (Choi et al., 2008). Cho et al. (2010) investigated the application of SPR in the detection of alcohol contained in wine and liquor. This study claims the SPR sensor had a higher degree of potential for the assessment of alcohol content. Using Langmuir-Blodgett (LB) methods, gold film with a nanostructured surface was devised. For the preparation of new SPR chips, a single layer of nanoscaled silica particles are coated on to a plain surface of gold film. Thereafter, the gold will be deposited in the template using an e-beam evaporator. Finally, after silica particles removal the nanostructured surface with a basin-like shape was obtained through the sonication methods. In this study two types of silica particles (130 and 300 nm) were used as template beads. The sensitivity of the new SPR chips was tested using ethanol solution. When tested in alcoholic beverages as model food systems, the sensitivity was improved compared with conventional types of SPR sensor.

11.3.2 SHELF-LIFE

Shelf-life is affected by food type, composition, formulation, packaging, and storage conditions, and it is decreased due to spoilage or other conditions. A shelf-life inclusively represents the period during which foods can be circulated in the market. In more detail, it also represents the period during which foods taste best, or they can be ingested without being decayed, or until the food is consumed.

In foods, a negative change in the quality gradually occur during the processing, transportation, and storage. This eventually leads to the deterioration of food quality. By determining the color, changes in the physicochemical property of various substances or pH of foods, or measuring the amount of microorganisms, the degree of the change of food quality can be confirmed. The methods for measuring shelf-life are called the accelerated shelf-life testing (ASLT) methods (Corradini and Peleg, 2007). ASLT helps predict shelf-life of processed foods such as dairy or meat products.

ASLT cannot be applied to all food products, but can be used for monitoring the conditions where foods decay occured due to a temperature elevation. There is also disadvantage that many erroneous results might occur in the prediction of a shelf-life (Hough, 2006).

Nanotechnology has the potential to resolve these disadvantages in measuring a shelf-life and SA 2009. A biosensor based on nanotechnology detects the alterations in environmental factors surrounding foods, such as temperature or water content, and thereby diagnoses the changes in physicochemical properties and contaminations due to microorganisms. A nanosensor can detect bacteria, viruses, toxins, or other contaminants both promptly and accurately with the use of a nanoscale chip that specifically binds to the target substances (Chen et al. 2004). Kilian et al. (1999) detected *Salmonella* in poultry using a silicon chip-based biosensor, where this biosensor detects *Salmonella* based on change in pH due to the reaction of urease-conjugated anti-*Salmonella* antibodies within sample membranes. Using this method, the detection could be done at even low concentration of *Salmonella* (119 CFU/ml). Waswa et al. (2007) detected *Escherichia coli* at a concentration of 10^2 to 10^3 CFU/ml using an SPR-based immunosensor in samples including milk, apple juice, and beef. This immunosensor specifically can detect *Escherichia coli* 0157:H7 even at lower concentrations of 10^2 CFU/ml without interference from other contaminated pathogenic organisms. Of particular note, an SPR-based biosensor can directly detect the pathogens from liquid products both promptly and accurately, unlike the previous methods.

In the case of microorganisms, the environmental conditions such as temperature, pH, and water content are essential for the growth. The amount of water contained in foods affects the growth of microorganisms and the chemical interactions between food constituents. This eventually alters a food's shelf-life. It is therefore crucial to consistently balance the amount of water contained in foods. In foods, changes in pH occur due to hydrolysis of food constituents because of metabolic activity of microorganisms, enzymes, or acidity during the processing, or in storage and distribution of the food products. Measurement of pH changes is the simplest and fastest method for monitoring food spoilage (Mello and Kubota, 2002). Recently, several studies have been conducted to examine the effectiveness of methods for evaluating the moisture content and pH changes in foods using nanotechnology on a real-time basis (Doussineau et al., 2009; Nohria et al., 2006; Talley et al., 2004).

In the processing of foods, for the purposes of enhancing the food preservation and improving the quality and reinforcement of nutrition, various chemical additives are added. But there are some cases in which undesirable chemical compounds are imported or contaminated with microorganisms (Valdés et al., 2009; Shapiro and Mercier, 1994). These undesirable chemical substances or food constituents that lower the nutritional value of foods or promoting spoilage should promptly be removed. Shao et al. (2008) detected dichlorvos, one of the types of organophosphate pesticides, using a biosensor based on a gold/silicon nanowire. A silicon nanowire, which was coated with gold nanoparticles, has a higher degree of electrical conductivity. Accordingly, it shows a high degree of sensitivity for detection. These authors promptly detected dichlorvos at a concentration of 8 ng/L. Besides, Zhu et al. (2009) proposed a hybrid Au-Ag triangular for detecting enterotoxin B formed by *Staphylococcus aureus* nanoparticles array-based nano-biosensor. To determine the specific binding detection, these authors prepared enterotoxin B solutions at concentrations of 1, 10, and 100 nm and then measured the refractive index using LSPR. Because the intensity of the spectrum was relatively lower at a concentration of 0.1 nm/ml, enterotoxin B was not detected. At a concentration of 1 ng/ml, however, it showed a very high sensitivity for detection.

11.3.3 Assessment of the Quality of Foods during Circulation

Consumers want safe, environmentally friendly packaging that is easy to handle, portable, and can maintain freshness during long-term storage. To accomplish this task, nanotechnology approaches can be used for product packaging and thereby enhanced shelf-life and the safety of packaged foods can be assured. Packaging can also be used to keep track of the distribution process and for maintaining safe food conditions. Furthermore, it is pro-environmental because it can be recycled and can efficiently maintain resistance to physical impact and heat. Even with the use of a thin-film packaging, the functions of previous packaging (including mechanical resistance, the prevention of oxidation, flavor protection, physical impact, and fluid control) could be upgraded to a higher level. Meanwhile, when a thin-film packaging was pierced or punctured, self-healing could be performed. Additionally, film packaging materials can lead to the preparation of an intelligent active packaging system that responds to environmental changes.

Of the methods for packaging food products, antimicrobial packaging is the most important; this is because the antimicrobial system prolongs the shelf-life of food products, reduces the proliferation of microorganism in the foods, and thereby enhances the safety of foods. In this kind of packaging, packaging material is coated with antimicrobial nanoparticles. Consequently, this reduces microbial access to the foods as well as prevents the growth of bacteria. Antimicrobial nanoparticle coatings in the matrix of the packaging material can prevent microorganisms from growing in nonsterilized and pasteurized foods by preventing post-contamination (Neethirajan and Jayas, 2010). With the direct addition of antimicrobial compounds to foods, due to a lack of homogeneous mixture, there would be no overall antimicrobial effects. But, with the use of a food packaging whose surface is coated with antimicrobial macromolecules or a nanomolecule film, however, there would be overall antimicrobial effects (COMA, 2008). Silver nanoparticles suppress the growth of gram-negative bacteria such as *E. coli*. therefore, these particles can be used as an antimicrobial agents. As shown in Figure 11.3, zinc oxide nanoparticles can also be used as an antimicrobial substance (Sondi and Salopek-Sondi, 2004; Rodriguez et al., 2008). In addition to these antimicrobial substances, wrapping sheets with antifungal activity have also been developed. It has been shown that the wrapping sheets whose surface was coated with cinnamon oil and wax paraffin suppressed the growth of *Rhizopusstolonifer*, one of the bacterial strains that are well proliferated in bread (Rodríguez et al., 2008).

Also a recently highlighted aspect is smart packaging, which displays the history of the product and information on the presence of pathogenic bacteria on a wrapping sheet. This technology is also based on nanotechnology approaches. With the

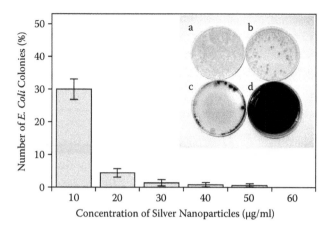

FIGURE 11.3　Number of *E. coli* colonies as a function of the concentration of silver nano-particles in LB agar plates expressed as a percentage of the number of colonies grown on silver-free control plates. The photograph inserted in the upper right corner shows LB plates containing different concentrations of silver nanoparticles: (a) 0 μm, (b) 10 μm, (c) 20 μm, (d) 50 μm,/cm³ (From I. Sondi et al. 2004. Silver nanoparticles as antimicrobial agent: a case study on *E. Coli* as a model for gram-negative bacteria. *Journal of Colloid and Interface Science*: 275: 177–82, with permission from Elsevier.)

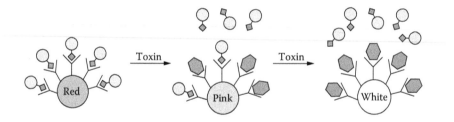

FIGURE 11.4　Indicator-analyte displacement between analog-conjugated gold nanopar-ticles and toxin molecules (■, latex microsphere; Y, toxin antibody; ○, gold nanoparticle; ◆, analog; ●, toxin; ●○, latex microsphere). (From S. Ko et al., 2010. Self-indicating nano-biosensor for detection of 2, 4-dinitrophenol. *Food Control*: 21: 155–161, with permission from Elsevier.)

use of a nanobarcode for a wrapping sheet, fraudulent actions such as the counter-feiting of products during the distribution of foods can also be prevented. Besides, methods where DNA sequences consistings of specific proteins are analyzed using gold nanoparticles and a DNA barcode is also based on nanotechnology (Nam et al., 2003).

There is a nanosensor that has been developed whose color is altered when latex microspheres bind to gold nanoparticles. When gold nanoparticles bind to the toxin analog, to 2,4-dinitrophenol–bovine serum albumin (DNP-BSA), the color becomes pinkish red. When the gold nanoparticles are displaced from host latex microspheres in the presence of toxin molecules, the color is changed to white (Figure 11.4) (Ko et al., 2008).

In this study, a model toxin, DNP-glycine has been detected and quantified via a competition that occurs between the analog-conjugated-gold nanoparticles and toxin molecules for the binding pocket in the anti-DNP antibody. Even in the foods which are sterilized and stored under good conditions, contamination can occur and their quality may be altered. Technique and equipment that can be used to assess the quality of foods during the distribution process is essentially needed. These equipment should be audited and assessed at regular intervals. First, it should be easy to carry. Portability requires a miniaturized instrumentation, and it is possible when the equipment assessing the quality of foods is developed based on the nanotechnology. A biosensor that is safe for foods can be used more frequently. In system that convert binding reaction with target substances to the readable signals, nanotechnology is frequently used (Hall, 2002).

In pork quality evaluation, a method for detecting *Salmonella typhimurium* uses a fiber optic portable biosensor, which functions based on a fluorescence resonance energy transfer (FRET) mechanism. This biosensor which functions based on which is used to measure the binding activity between *S. typhimurium* and fluorescent materials, is easy to carry and is convenient so that it can be used for analysis on spot. It can also persistently check the status of pork during circulation. Therefore, it can enhance safety as well as cost-effectiveness (Ko and Grant, 2006).

11.3.4 SENSORY EVALUATION

Sensory evaluation of foods is a scientific branch that applies principles of both experimental plans and statistical analysis using human senses (smell, sight, touch, taste, and hearing) for evaluating food products quality. By applying statistical techniques to the results, it is possible to make deductions and insights about the products under investigation. The branch requires a panel of assessors who test the products and record their answers. Nowadays, remarkable advances of nanotechnology support this process, which offers material, time, financial, and technical advantages. Nanosensors will lead this new trend for more effective sensory analysis.

Advanced nanobiosensors offering an enhanced electronic nose for food fragrance analysis will allow better assessment of food quality and food palatability while preserving common needs, which can be made possible by examining foods more easily than before, which is made possible as a result of the combination of sensorial physiology, chemical analysis, and nanotechnology. Akimov et al. (2008) made nanoscale sensing elements using olfactory receptors that are anchored onto functionalized gold nanoelectrodes used as odorant detectors as a new concept of a nanobioelectronic nose sensitive impedance metric measurement of a single receptor structural change upon odorant ligand binding, with a better specificity and lower limits of detection than usual physical sensors.

Nanosensors can offer quality assurance by tracing chemical and biological qualities through the whole of food processing via obtaining data for documents. Nanosensors also enable food packaging to examine food quality during different phases of the process to guarantee food safety until intake. Such nanosensors could be set into the packaging materials, where they would detect chemical nature of substances released during food spoilage. Ruengruglikit et al. (2004) have developed an electronic tongue for inclusion in food packaging that is made of arrays of nanosensors that are very sensitive to gases released by food as it spoils. In some cases, the sensors are designed in such a way that if the food in a package spoils, the particular sensor strip will change color gradually, as a result, and give a visible indication whether the food is fresh or not. Li et al. (2005) have created a nanobarcode detection system. This fluoresces under ultraviolet light in a combination of colors that can be read by a scanner. Food and biological samples including various combinations of some viruses have been investigated using this system, and several pathogens were obviously distinguished simultaneously by different color codes. This nanobarcode detection system can effectively monitor food quality by identifying changes in color and texture caused by pathogens in products.

Neethirajan et al. (2009) proposed nanosensors for grain quality monitoring, using conducting polymer sensors that consist of nanoparticles that respond to analytes and volatiles in the environment of food storage, and thereby identification of spoilages in sources already contaminated had been established.

Other types of nanosensors, called nanowire field effect sensors, are based on microfluidic devices. And these nanosensors can also be used to detect hydrogen ion concentration or pH sensing. These nanosensors function as receptors for hydrogen ions, which offer protonation and deprotonation reactions, thereby changing the net nanosensor's surface charge. Using this feature, nanowire field effect sensors can also be placed into the packaging material so that it would be used as an indicator of acidification, and can also work as an amperometric nano-electronic tongue in all kinds of foods affected by spoilage, ripening, and so on (Patolsky and Lieber, 2005).

Now these nanotechnology electronic tongues and noses are being developed all over the world. And new methods of estimating the quality of foods with nanotechnology are also being developed. Therefore the next generation of sensory analysis cannot improve further without combining the knowledge with the huge impact of nanotechnology.

11.4   CONCLUSIONS

Food nanotechnology has the potential for rapid and sensitive evaluation of food safety and quality-related attributes such as pathogens, mycotoxins, chemical contaminants, nutrients, bioactives, shelf-life, sensory characteristics, etc. In order to evaluate food hazards and quality parameters effectively, sciences and technologies in three categories should be improved: the advances in the detection unit, the consideration of detecting environments including the food matrix, and the in-line monitored and controlled food processing systems.

First, one of the important improvements in food nanotechnology is the advances of nanosensing units for detecting parameters directly or indirectly related to food quality and safety. Nanosensors recognize physical, chemical, and biological interactions for detection of food hazards or quality parameters, even at low concentration, rapidly, accurately, and sensitively. For the practical use of these nanosensors in the food industry, the specificity (nonspecific interactions), sensitivity, and reduction of false positive and negative values of nanosensors should be enhanced. For example, in a nanobiosensor, the type of antibody, concentration of signaling unit, time of incubation, and temperature are critical variables that control the adsorption of the target antigen in the binding sites of the antibodies. For achieving the desired results, novel nanotechnologies in food quality and safety evaluation systems may be extended for detecting a wide variety of parameters related to food safety and quality.

Second, for the practical application of the quality and safety evaluation systems, it is imperative that a sensing unit not only for detrimental factors of foods and quality factors, but also that should produce the maximal functions in the detection environment need to be extensively developed. Foods are morphologically classified into solid, semisolid, and liquid forms. Even if attempts are made to detect the detrimental factors and quality ones with the identical sensing unit, the reactivity can vary depending on the variability of the food matrix. As a result, this affects the efficacy, specificity, and sensitivity of a sensor. Besides, the effectiveness of the detection of detrimental factors and quality ones is also needed to be measured at any point during production, processing, transport, retailing, meal preparation, intake, digestion, or absorption in the body. In other words, it is essential to develop and optimize the sensing system with a total consideration of the morphology of the food matrix and the relevant detection environment.

Third, food nanotechnology will be applied to an automatic control of food processing in the future. A nanosensor will be installed via an in-line during the food processing, and thereby will provide the quantitative and qualitative data about detrimental and quality factors of foods. Based on these data, it can also be applied to different food processing states such as the management of raw materials, washing, sterilization, packaging, and storage on a real-time basis. Therefore, the food nanosensor technology can greatly contribute to the production of high-quality food products as well as their safety. Besides, nanosensors provide qualitative and quantitative data about contamination and impaired quality due to the additional detrimental factors occurring during the distribution of food products on a real-time basis, and thereby help customers to safely consume the food products.

REFERENCES

Akimov V, Alfinito E, Bausells J, Benilova I, Paramo I, Errachid A, Ferrari G, Fumagalli L, Gomila G, Grosclaude J, Hou Y, Jaffrezic-Renault N, Martelet C, Pajot-Augy E, Pennetta C, Persuy M-A, Pla-Roca M, Reggiani L, Rodriguez-Segui S, Ruiz O, Salesse R, Samitier J, Sampietro M, Soldatkin A, Vidic J, Villanueva G. 2008. Nanobiosensors based on individual olfactory receptors. *Analog Integrated Circuits and Signal Processing* 57:197–203.

Allmann M, Höfelein C, Köppel E, Lüthy J, Meyer R, Niederhauser C, Wegmüller B, Candrian U. 1995. Polymerase chain reaction (PCR) for detection of pathogenic microorganisms in bacteriological monitoring of dairy products. *Research in Microbiology* 146:85–97.

Altekruse SF, Stern NJ, Fields PI, Swerdlow DL. 1999. Campylobacter jejuni—An emerging foodborne pathogen. *Emerging Infectious Diseases* 5:28–35.

Arecchi A, Scampicchio M, Drusch S, Mannino S. 2010. Nanofibrous membrane based tyrosinase-biosensor for the detection of phenolic compounds. *Analytica Chimica Acta* 659:133–36.

Batt CA. 1997. Molecular diagnostics for dairy-borne pathogens. *Journal of Dairy Science* 80:220–29.

Betala P, Appugounder S, Chakraborty S, Songprawat P, Buttner W, Perez-Luna V. 2008. Rapid colorimetric detection of proteins and bacteria using silver reduction/precipitation catalyzed by gold nanoparticles. *Sensing and Instrumentation for Food Quality and Safety* 2:34–42.

Beuchat LR, Ryu JH. 1997. Produce handling and processing practices. *Emerging Infectious Diseases* 3:459–65.

Bodor R, Zúborová M, Ölvecká E, Madajová V, Masár M, Kaniansky D, Stanislawski B. 2001. Isotachophoresis and isotachophoresis-zone electrophoresis of food additives on a chip with column-coupling separation channels. *Journal of Separation Science* 24:802–9.

Brooks BW, Devenish J, Lutze-Wallace CL, Milnes D, Robertson RH, Berlie-Surujballi G. 2004. Evaluation of a monoclonal antibody-based enzyme-linked immunosorbent assay for detection of *Campylobacter* fetus in bovine preputial washing and vaginal mucus samples. *Veterinary Microbiology* 103:77–84.

Burdo OG. 2005. Nanoscale effects in food-production technologies. *Journal of Engineering Physics and Thermophysics* 78:90–96.

Chan WCW, Nie S. 1998. Quantum dot bioconjugates for ultrasensitive nonisotopic detection. *Science* 281:2016–18.

Chang YH, Chang TC, Kao EF, Chou C. 1996. Detection of protein A produced by *Staphylococcus aureus* with a fiber-optic-based biosensor. *Bioscience Biotechnology and Biochemistry* 60:1571–74.

Chen Jianrong MY, He Nongyue, Wu Xiaohua, Li Sijiao. 2004. Nanotechnology and biosensors. *Biotechnology Advances* 22:505–18.

Cho Y, Kim C, Kim N, Kim C, Kim T, Kim H, Kim J. 2010. Effect of SPR chip with nano-structured surface on sensitivity in SPR sensor. *Food Engineering Progress* 14:49–53.

Choi JH, Kim SD, Noh SH, Oh SJ, Kim WJ. 2006. Adsorption behaviors of nano-sized ETS-10 and Al-substituted-ETAS-10 in removing heavy metal ions, Pb^{2+} and Cd^{2+}. *Microporous and Mesoporous Materials* 87:163–69.

Choi S-W, Kim H-S, Kang W-S, Kim J-H, Cho Y-J, Kim J-H. 2008. Sensitivity enhancement by Au nanoparticles in surface plasmon resonance chemical sensors. *Journal of Nanoscience and Nanotechnology* 8:4569–73.

Colquhoun DR, Schwab KJ, Cole RN, Halden RU. 2006. Detection of norovirus capsid protein in authentic standards and in stool extracts by matrix-assisted laser desorption ionization and nanospray mass spectrometry. *Applied and Environmental Microbiology* 72:2749–55.

COMA Vr. 2008. Bioactive packaging technologies for extended shelf life of meat-based products. *Meat Science* 78:90–103.

Corradini MG, Peleg M. 2007. Shelf-life estimation from accelerated storage data. *Trends in Food Science and Technology* 18:37–47.

Cremisini C, Di Sario S, Mela J, Pilloton R, Palleschi G. 1995. Evaluation of the use of free and immobilised acetylcholinesterase for paraoxon detection with an amperometric choline oxidase based biosensor. *Analytica Chimica Acta* 311:273–80.

Dabrowski A, Hubicki Z, Podkoscielny P, Robens E. 2004. Selective removal of the heavy metal ions from waters and industrial wastewaters by ion-exchange method. *Chemosphere* 56:91–106.

Dahlin A, Zach M, Rindzevicius T, Kall M, Sutherland DS, Hook F. 2005. Localized surface plasmon resonance sensing of lipid-membrane-mediated biorecognition events. *Journal of the American Chemical Society* 127:5043–48.

Daniel MC, Astruc D. 2004. Gold nanoparticles: Assembly, supramolecular chemistry, quantum-size-related properties, and applications toward biology, catalysis, and nanotechnology. *Chemical Reviews* 104:293–346.

Darbha GK, Ray A, Ray PC. 2007. Gold nanoparticle-based miniaturized nanomaterial surface energy transfer probe for rapid and ultrasensitive detection of mercury in soil, water, and fish. *ACS Nanotechnology* 1:208–14.

Deisingh AK, Badrie N. 2005. Detection approaches for genetically modified organisms in foods. *Food Research International* 38:639–49.

DeMarco DR, Saaski EW, McCrae DA, Lim DV. 1999. Rapid detection of *Escherichia coli* O157:H7 in ground beef using a fiber-optic biosensor. *Journal of Food Protection* 62:711–16.

Doussineau T, Trupp S, Mohr GJ. 2009. Ratiometric pH-nanosensors based on rhodamine-doped silica nanoparticles functionalized with a naphthalimide derivative. *Journal of Colloid and Interface Science* 339:266–70.

Driskell JD, Shanmukh S, Yong-Jun L, Hennigan S, Jones L, Yi-Ping Z, Dluhy RA, Krause DC, Tripp RA. 2008. Infectious agent detection with SERS-active silver nanorod arrays prepared by oblique angle deposition. *IEEE Sensors Journal* 8:863–70.

Fleming AF. 1990. Opportunistic infections in AIDS in developed and developing countries. *Transactions of the Royal Society of Tropical Medicine and Hygiene* 84:1–6.

Gautam S, Sharma A, Thomas P. 1998. Improved bacterial turbidimetric method for detection of irradiated spices. *Journal of Agricultural and Food Chemistry* 46:5110–12.

Gerion D, Chen F, Kannan B, Fu A, Parak WJ, Chen DJ, Majumdar A, Alivisatos AP. 2003. Room-temperature single-nucleotide polymorphism and multiallele DNA detection using fluorescent nanocrystals and microarrays. *Analytical Chemistry* 75:4766–72.

Ghosh SK, Nath S, Kundu S, Esumi K, Pal T. 2004. Solvent and ligand effects on the localized surface plasmon resonance (LSPR) of gold colloids. *Journal of Physical Chemistry B* 108:13963–71.

Götz Fv. 2009. See what you eat—Broad GMO screening with microarrays. *Analytical and Bioanalytical Chemistry* 396:1961–67.

Gu YS, Decker EA, McClements DJ. 2004. Influence of pH and ι-carrageenan concentration on physicochemical properties and stability of β-lactoglobulin-stabilized oil-in-water emulsions. *Journal of Agricultural and Food Chemistry* 52:3626–32.

Hall GL, Mourer CR, Shibamoto T, Fitzell D. 1997. Development and validation of an analytical method for naled and dichlorvos in air. *Journal of Agricultural and Food Chemistry* 45:145–48.

Hall RH. 2002. Biosensor technologies for detecting microbiological foodborne hazards. *Microbes and Infection* 4:425–32.

Haruyama T. 2003. Micro- and nanobiotechnology for biosensing cellular responses. *Advanced Drug Delivery Reviews* 55:393–401.

Hauck TS, Giri S, Gao Y, Chan WCW. 2010. Nanotechnology diagnostics for infectious diseases prevalent in developing countries. *Advanced Drug Delivery Reviews* 62:438–48.

He L, Liu Y, Lin M, Awika J, Ledoux D, Li H, Mustapha A. 2008. A new approach to measure melamine, cyanuric acid, and melamine cyanurate using surface enhanced Raman spectroscopy coupled with gold nanosubstrates. *Sensing and Instrumentation for Food Quality and Safety* 2:66–71.

Heinitz ML, Johnson JM. 1998. The incidence of *Listeria* spp., *Salmonella* spp., and *Clostridium botulinum* in smoked fish and shellfish. *Journal of Food Protection* 61:318–23.

Homola J, Yee SS, Gauglitz G. 1999. Surface plasmon resonance sensors: Review. *Sensors and Actuators B: Chemical* 54:3–15.

Hong C, Ying X, Ningning Z, Pingang H, Yuzhi F. 2002. An electrochemical DNA hybridization assay based on a silver nanoparticle label. *The Royal Society of Chemistry* 127:803–8.

Hough G. 2006. Workshop summary: Sensory shelf-life testing. *Food Quality and Preference* 17:640–45.

Huang CP, Blankenship DW. 1984. The removal of mercury(II) from dilute aqueous solution by activated carbon. *Water Research* 18:37–46.

Huang Y-C, Koseoglu SS. 1993. Separation of heavy metals from industrial waste streams by membrane separation technology. *Waste Management* 13:481–501.

Ivnitski D, Abdel-Hamid I, Atanasov P, Wilkins E. 1999. Biosensors for detection of pathogenic bacteria. *Biosensors and Bioelectronics* 14:599–624.

Jain KK. 2003. Nanodiagnostics: Application of nanotechnology in molecular diagnostics. *Expert Review of Molecular Diagnostics* 3:153–61.

Jain KK. 2005. Nanotechnology in clinical laboratory diagnostics. *Clinica Chimica Acta* 358:37–54.

Ji X, Zheng J, Xu J, Rastogi VK, Cheng T-C, DeFrank JJ, Leblanc RM. 2005. (CdSe)ZnS quantum dots and organophosphorus hydrolase bioconjugate as biosensors for detection of paraoxon. *The Journal of Physical Chemistry B* 109:3793–99.

Jin RC, Wu GS, Li Z, Mirkin CA, Schatz GC. 2003. What controls the melting properties of DNA-linked gold nanoparticle assemblies? *Journal of the American Chemical Society* 125:1643–54.

Jin X, Jin X, Chen L, Jiang J, Shen G, Yu R. 2009. Piezoelectric immunosensor with gold nanoparticles enhanced competitive immunoreaction technique for quantification of aflatoxin B1. *Biosensors and Bioelectronics* 24:2580–85.

Junxue F, Park B, Siragusa G, Jones L, Tripp R, Zhao Y, Cho Y-J. 2008. An Au/Si heteronanorod-based biosensor for *Salmonella* detection. *Nanotechnology* 19:155502.

Kaittanis C, Santra S, Perez JM. 2010. Emerging nanotechnology-based strategies for the identification of microbial pathogenesis. *Advanced Drug Delivery Reviews* 62:408–23.

Katz E, Willner I. 2004. Integrated nanoparticle-biomolecule hybrid systems: Synthesis, properties, and applications. *Angewandte Chemie-International Edition* 43:6042–108.

Kaushik A, Solanki PR, Ansari AA, Ahmad S, Malhotra BD. 2008. Chitosan-iron oxide nanobiocomposite based immunosensor for ochratoxin-A. *Electrochemistry Communications* 10:1364–68.

Kaushik A, Solanki PR, Ansari AA, Ahmad S, Malhotra BD. 2009a. A nanostructured cerium oxide film-based immunosensor for mycotoxin detection. *Nanotechnology* 20:1–8.

Kaushik A, Solanki PR, Sood KN, Ahmad S, Malhotra BD. 2009b. Fumed silica nanoparticles-chitosan nanobiocomposite for ochratoxin-A detection. *Electrochemistry Communications* 11:1919–23.

Khosravi-Darani K, Pardakhty A, Honarpisheh H, Rao VSNM, Mozafari MR. 2007. The role of high-resolution imaging in the evaluation of nanosystems for bioactive encapsulation and targeted nanotherapy. *Micron* 38:804–18.

Kilian Dill LHS, Colin R. Young. 1999. Detection of salmonella in poultry using a silicon chip-based biosensor. *Biochemical and Biophysical Methods* 41:211–16.

Kneipp K, Haka AS, Kneipp H, Badizadegan K, Yoshizawa N, Boone C, Shafer-Peltier KE, Motz JT, Dasari RR, Feld MS. 2002. Surface-enhanced Raman spectroscopy in single living cells using gold nanoparticles. *Applied Spectroscopy* 56:150–54.

Ko S, Grant SA. 2006. A novel FRET-based optical fiber biosensor for rapid detection of *Salmonella typhimurium*. *Biosensors and Bioelectronics* 21:1283–90.

Ko S, Gunasekaran S, Yu J. 2008. Self-indicating nanobiosensor for detection of 2,4-dinitrophenol. *Food Control* 21:155–61.

Kokini N, Sa JL. 2009. Nanotechnology and its applications in the food sector. *Trends in Biotechnology* 27:82–89.

Koopmans M, Bonsdorff C-H, Vinjé J, Medici D, Monroe S. 2002. Foodborne viruses. *FEMS Microbiology Reviews* 26:187–205.

Lacorte S, Barcelo D. 1994. Rapid degradation of fenitrothion in estuarine waters. *Environmental Science and Technology* 28:1159–63.

Lai HM, Cheng HH. 2004. Properties of pregelatinized rice flour made by hot air or gum puffing. *International Journal of Food Science and Technology* 39:201–12.

Larson DR, Zipfel WR, Williams RM, Clark SW, Bruchez MP, Wise FW, Webb WW. 2003. Water-soluble quantum dots for multiphoton fluorescence imaging *in vivo*. *Science* 300:1434–36.

Lazcka O, Campo FJD, Muñoz FX. 2007. Pathogen detection: A perspective of traditional methods and biosensors. *Biosensors and Bioelectronics* 22:1205–17.

Lee S, Perez-Luna VH. 2005. Dextran-gold nanoparticle hybrid material for biomolecule immobilization and detection. *Analytical Chemistry* 77:7204–11.

Legin A, Rudnitskaya A, Lvova L, Vlasov Y, Di Natale C, D'Amico A. 2003. Evaluation of Italian wine by the electronic tongue: Recognition, quantitative analysis and correlation with human sensory perception. *Analytica Chimica Acta* 484:33–44.

Leimanis S, Hernández M, Fernández S, Boyer F, Burns M, Bruderer S, Glouden T, Harris N, Kaeppeli O, Philipp P, Pla M, Puigdomènech P, Vaitilingom M, Bertheau Y, Remacle J. 2006. A microarray-based detection system for genetically modified (GM) food ingredients. *Plant Molecular Biology* 61:123–39.

Leonard P, Hearty S, Quinn J, O'Kennedy R. 2004. A generic approach for the detection of whole *Listeria* monocytogenes cells in contaminated samples using surface plasmon resonance. *Biosensors and Bioelectronics* 19:1331–35.

Leoni O, Iori R, Palmieri S. 1991. Immobilization of myrosinase on membrane for determining the glucosinolate content of cruciferous material. *Journal of Agricultural and Food Chemistry* 39: 2322–26.

Liao K-T, Huang H-J. 2005. Femtomolar immunoassay based on coupling gold nanoparticle enlargement with square wave stripping voltammetry. *Analytica Chimica Acta* 538:159–64.

Liao JY. 2007. Construction of nanogold hollow balls with dendritic surface as immobilized affinity support for protein adsorption. *Colloids and Surfaces B: Biointerfaces* 57:75–80.

Ligler FS, Taitt CR, Shriver-Lake LC, Sapsford KE, Shubin Y, Golden JP. 2003. Array biosensor for detection of toxins. *Analytical and Bioanalytical Chemistry* 377:469–77.

Liu S, Yuan L, Yue X, Zheng Z, Tang Z. 2008. Recent advances in nanosensors for organophosphate pesticide detection. *Advanced Powder Technology* 19:419–41.

Lynch M, Mosher C, Huff J, Nettikadan S, Johnson J, Henderson E. 2004. Functional protein nanoarrays for biomarker profiling. *Proteomics* 4:1695–702.

Ma F, Misumi J, Zhao W, Aoki K, Kudo M. 2003. Long-term treatment with sterigmatocystin, a fungus toxin, enhances the development of intestinal metaplasia of gastric mucosa in helicobacter pylori-infected mongolian gerbils. *Scandinavian Journal of Gastroenterology* 38:360–69.

Malorny B, Tassios PT, Peter R, Cook N, Wagner M, Hoorfar J. 2003. Standardization of diagnostic PCR for the detection of foodborne pathogens. *International Journal of Food Microbiology* 83:39–48.

Mandrell RE, Wachtelt MR. 1999. Novel detection techniques for human pathogens that contaminate poultry. *Current Opinions in Biotechnology* 10:273–78.

Mello LD, Kubota LT. 2002. Review of the use of biosensors as analytical tools in the food and drink industries. *Food Chemistry* 77:237–56.

Nam J-M, Thaxton CS, Mirkin CA. 2003. Nanoparticle-based bio-bar codes for the ultrasensitive detection of proteins. *Science* 301:1884–86.

Nath N, Chilkoti A. 2002. A colorimetric gold nanoparticle sensor to interrogate biomolecular interactions in real time on a surface. *Analytical Chemistry* 74:504–9.

Nath N, Chilkoti A. 2004. Label free colorimetric biosensing using nanoparticles. *Journal of Fluorescence* 14:377–89.

Naylor LMD, Richard R. 1975. Simulation of lead removal by chemical treatment. *Journal of American Water Work Association* 67:560–65.

Neethirajan S, Gordon R, Wang L. 2009. Potential of silica bodies (phytoliths) for nanotechnology. *Trends in Biotechnology* 27:461–67.

Neethirajan S, Jayas DS. 2010. Nanotechnology for the food and bioprocessing industries. *Food and Bioprocess Technology*, online.

Nohria R, Khillan RK, Su Y, Dikshit R, Lvov Y, Varahramyan K. 2006. Humidity sensor based on ultrathin polyaniline film deposited using layer-by-layer nano-assembly. *Sensors and Actuators B: Chemical* 114:218–22.

Olsen JE, Aabo S, Hill W, Notermans S, Wernars K, Granum PE, Popovic T, Rasmussen HN, Olsvik. 1995. Probes and polymerase chain reaction for detection of food-borne bacterial pathogens. *International Journal of Food Microbiology* 28:1–78.

Pargaonkar N, Lvov YM, Li N, Steenekamp JH, de Villiers MM. 2005. Controlled release of dexamethasone from microcapsules produced by polyelectrolyte layer-by-layer nanoassembly. *Pharmaceutical Research* 22:826–35.

Patolsky F, Lieber CM. 2005. Nanowire nanosensors. *Materials Today* 8:20–28.

Radoi A, Targa M, Prieto-Simon B, Marty JL. 2008. Enzyme-linked immunosorbent assay (ELISA) based on superparamagnetic nanoparticles for aflatoxin M1 detection. *Talanta* 77:138–43.

Raschke G, Brogl S, Susha AS, Rogach AL, Klar TA, Feldmann J, Fieres B, Petkov N, Bein T, Nichtl A, Kurzinger K. 2004. Gold nanoshells improve single nanoparticle molecular sensors. *Nano Letters* 4:1853–57.

Raschke G, Kowarik S, Franzl T, Sonnichsen C, Klar TA, Feldmann J, Nichtl A, Kurzinger K. 2003. Biomolecular recognition based on single gold nanoparticle light scattering. *Nano Letters* 3:935–38.

Ray DE. 1998. Chronic effects of low level exposure to anticholinesterases—A mechanistic review. *Toxicology Letters* 102–103:527–33.

Reiß J. 1975. Semiquantitative determination of the mycotoxin sterigmatocystin on thin-layer chromatograms with a gray scale. *Fresenius' Journal of Analytical Chemistry* 275:30.

Rodríguez A, Nern C, Batlle R. 2008. New cinnamon-based active paper packaging against Rhizopusstolonifer food spoilage. *Journal of Engineering Physics and Thermophysics* 56:6364–69.

Rogers KR, Apostol A, J MS, Spencer CW. 2001. Fiber optic biosensor for detection of DNA damage. *Analytica Chimica Acta* 444:51–60.

Rosi NL, Mirkin CA. 2005. Nanostructures in biodiagnostics. *Chemical Reviews* 105:1547–62.

Ruengruglikit C, Kim H, Miller RD, Huang Q. 2004. Fabrication of nanoporous oligonucleotide microarrays for pathogen detection and identification. *Polymer Preprints* 45.

Sengupta A, Mujacic M, Davis EJ. 2006. Detection of bacteria by surface-enhanced Raman spectroscopy. *Analytical and Bioanalytical Chemistry* 386:1379–86.

Shanmukh S, Jones L, Driskell J, Zhao Y, Dluhy R, Tripp RA. 2006. Rapid and sensitive detection of respiratory virus molecular signatures using a silver nanorod array SERS substrate. *Nano Letters* 6:2630–36.

Shapiro A, Mercier C. 1994. Safe food manufacturing. *Science of the Total Environment* 143:75–92.

Sharma VK, Carlson SA. 2000. Simultaneous detection of *Salmonella* strains and *Escherichia coli* O157:H7 with fluorogenic PCR and single-enrichment-broth culture. *Applied and Environmental Microbiology* 66:5472–76.

Simonian AL, Good TA, Wang SS, Wild JR. 2005. Nanoparticle-based optical biosensors for the direct detection of organophosphate chemical warfare agents and pesticides. *Analytica Chimica Acta* 534:69–77.

Sondi I, Salopek-Sondi B. 2004. Silver nanoparticles as antimicrobial agent: A case study on *E. coli* as a model for gram-negative bacteria. *Journal of Colloid and Interface Science* 275:177–82.

Sotiropoulou S, Chaniotakis N. 2003. Carbon nanotube array-based biosensor. *Analytical and Bioanalytical Chemistry* 375:103–5.

Su S, He Y, Zhang M, Yang K, Song S, Zhang X, Fan C, Lee ST. 2008. High-sensitivity pesticide detection via silicon nanowires-supported acetylcholinesterase-based electrochemical sensors. *Applied Physics Letters* 93:023113.

Sugunan A, Thanachayanont C, Dutta J, Hilborn JG. 2005. Heavy-metal ion sensors using chitosan-capped gold nanoparticles. *Science and Technology of Advanced Materials* 6:335–40.

Swaminathan B, Feng P. 2003. Rapid detection of food-borne pathogenic bacteria. *Annual Review of Microbiology* 48:401–26.

Talley CE, Jusinski L, Hollars CW, Lane SM, Huser T. 2004. Intracellular pH sensors based on surface-enhanced Raman scattering. *Analytical Chemistry* 76:7064–68.

Tallury P, Malhotra A, Byrne LM, Santra S. 2010. Nanobioimaging and sensing of infectious diseases. *Advanced Drug Delivery Reviews* 62:424–37.

Tang D, Sauceda JC, Lin Z, Ott S, Basova E, Goryacheva I, Biselli S, Lin J, Niessner R, Knopp D. 2009. Magnetic nanogold microspheres-based lateral-flow immunodipstick for rapid detection of aflatoxin B2 in food. *Biosensors and Bioelectronics* 25:514–18.

Tran-Minh C, Pandey PC, Kumaran S. 1990. Studies on acetylcholine sensor and its analytical application based on the inhibition of cholinesterase. *Biosensors and Bioelectronics* 5:461–71.

Turner NW, Subrahmanyam S, Piletsky SA. 2009. Analytical methods for determination of mycotoxins: A review. *Analytica Chimica Acta* 632:168–80.

Valdés M, Valdés González A, García Calzón J, Díaz-García M. 2009. Analytical nanotechnology for food analysis. *Microchimica Acta* 166:1–19.

Vamvakaki V, Chaniotakis NA. 2007. Pesticide detection with a liposome-based nano-biosensor. *Biosensors and Bioelectronics* 22:2848–53.

Vaughan RD, O'Sullivan CK, Guilbault GG. 2001. Development of a quartz crystal microbalance (QCM) immunosensor for the detection of *Listeria* monocytogenes. *Enzyme and Microbial Technology* 29:635–38.

Velusamy V, Arshak K, Korostynska O, Oliwa K, Adley C. 2010. An overview of food-borne pathogen detection: In the perspective of biosensors. *Biotechnology Advances* 28:232–54.

Villamizar RA, Maroto A, Rius FX, Inza I, Figueras MJ. 2008. Fast detection of *Salmonella infantis* with carbon nanotube field effect transistors. *Biosensors and Bioelectronics* 24:279–83.

Vo-Dinh T, Cullum BM, Stokes DL. 2001. Nanosensors and biochips: Frontiers in biomolecular diagnostics. *Sensors and Actuators B: Chemical* 74:2–11.

Waswa J, Irudayaraj J, DebRoy C. 2007. Direct detection of *E. coli* O157:H7 in selected food systems by a surface plasmon resonance biosensor. *LWT-Food Science and Technology* 40:187–92.

Wiskur SL, Anslyn EV. 2001. Using a synthetic receptor to create an optical-sensing ensemble for a class of analytes: A colorimetric assay for the aging of scotch. *Journal of the American Chemical Society* 123:10109–10.

Xu J, Miao H, Wu H, Huang W, Tang R, Qiu M, Wen J, Zhu S, Li Y. 2006. Screening genetically modified organisms using multiplex-PCR coupled with oligonucleotide microarray. *Biosensors and Bioelectronics* 22:71–77.

Xu X, Wang J, Jiao K, Yang X. 2009. Colorimetric detection of mercury ion (Hg^{2+}) based on DNA oligonucleotides and unmodified gold nanoparticles sensing system with a tunable detection range. *Biosensors and Bioelectronics* 24:3153–58.

Xue X, Pan J, Xie H, Wang J, Zhang S. 2009. Fluorescence detection of total count of *Escherichia coli* and *Staphylococcus aureus* on water-soluble CdSe quantum dots coupled with bacteria. *Talanta* 77:1808–13.

Yang L, Li Y. 2006. Simultaneous detection of *Escherichia coli* O157H7 and *Salmonella* typhimurium using quantum dots as fluorescence labels. *Analyst* 131:394–401.

Yao D-S, Cao H, Wen S, Liu D-L, Bai Y, Zheng W-J. 2006. A novel biosensor for sterigmatocystin constructed by multi-walled carbon nanotubes (MWNT) modified with aflatoxin-detoxifizyme (ADTZ). *Bioelectrochemistry* 68:126–33.

Yonzon CR, Jeoungf E, Zou SL, Schatz GC, Mrksich M, Van Duyne RP. 2004. A comparative analysis of localized and propagating surface plasmon resonance sensors: The binding of concanavalin a to a monosaccharide functionalized self-assembled monolayer. *Journal of the American Chemical Society* 126:12669–76.

Zhou P, Lu Y, Zhu J, Hong J, Li B, Zhou J, Gong D, Montoya A. 2004. Nanocolloidal gold-based immunoassay for the detection of the N-methylcarbamate pesticide carbofuran. *Journal of Agricultural and Food Chemistry* 52:4355–59.

Zhu S, Du C, Fu Y. 2009. Localized surface plasmon resonance-based hybrid Au-Ag nanoparticles for detection of *Staphylococcus aureus* enterotoxin B. *Optical Materials* 31:1608–13.

Ziegelmann B, Bogl KW, Schreiber GA. 1998. Thermoluminescence and electron spin resonance investigations of minerals for the detection of irradiated foods. *Journal of Agricultural and Food Chemistry* 46:4604–9.

Index

Milton Keynes UK
Ingram Content Group UK Ltd.
UKHW031141141024
449569UK00024B/1160